Stem Cells in Regenerative Medicine

METHODS IN MOLECULAR BIOLOGY™

John M. Walker, SERIES EDITOR

502. Bacteriophages: *Methods and Protocols, Volume 2: Molecular and Applied Aspects,* edited by *Martha R. J. Clokie and Andrew M. Kropinski* 2009

501. Bacteriophages: *Methods and Protocols, Volume 1: Isolation,* Characterization, and Interactions, edited by *Martha R. J. Clokie and Andrew M. Kropinski* 2009

499. Candida Albicans: *Methods and Protocols,* edited by *Dr. Ronald L. Cihlar and Richard A. Calderone,* 2009

496. DNA and RNA Profiling in Human Blood: *Methods and Protocols,* edited by *Peter Bugert,* 2009

493. Auditory and Vestibular Research: *Methods and Protocols,* edited by *Bernd Sokolowski,* 2009

490. Protein Structures, Stability, and Interactions, edited by *John W. Schriver,* 2009

489. Dynamic Brain Imaging: *Methods and Protocols,* edited by *Fahmeed Hyder,* 2009

485. HIV Protocols: *Methods and Protocols,* edited by *Vinayaka R. Prasad and Ganjam V. Kalpana,* 2009

484. Functional Proteomics: *Methods and Protocols,* edited by *Julie D. Thompson, Christine Schaeffer-Reiss, and Marius Ueffing,* 2008

483. Recombinant Proteins From Plants: *Methods and Protocols,* edited by *Lóic Faye and Veronique Gomord,* 2008

482. Stem Cells in Regenerative Medicine: *Methods and Protocols,* edited by *Julie Audet and William L. Stanford,* 2009

481. Hepatocyte Transplantation: *Methods and Protocols,* edited by *Anil Dhawan and Robin D. Hughes,* 2008

480. Macromolecular Drug Delivery: *Methods and Protocols,* edited by *Mattias Belting,* 2008

479. Plant Signal Transduction: *Methods and Protocols,* edited by *Thomas Pfannschmidt,* 2008

478. Transgenic Wheat, Barley and Oats: *Production and Characterization Protocols,* edited by *Huw D. Jones and Peter R. Shewry,* 2008

477. Advanced Protocols in Oxidative Stress I, edited by *Donald Armstrong,* 2008

476. Redox-Mediated Signal Transduction: *Methods and Protocols,* edited by *John T. Hancock,* 2008

475. Cell Fusion: *Overviews and Methods,* edited by *Elizabeth H. Chen,* 2008

474. Nanostructure Design: *Methods and Protocols,* edited by *Ehud Gazit and Ruth Nussinov,* 2008

473. Clinical Epidemiology: *Practice and Methods,* edited by *Patrick Parfrey and Brendon Barrett,* 2008

472. Cancer Epidemiology, Volume 2: *Modifiable Factors,* edited by *Mukesh Verma,* 2008

471. Cancer Epidemiology, Volume 1: *Host Susceptibility Factors,* edited by *Mukesh Verma,* 2008

470. Host-Pathogen Interactions: *Methods and Protocols,* edited by *Steffen Rupp and Kai Sohn,* 2008

469. Wnt Signaling, Volume 2: *Pathway Models,* edited by *Elizabeth Vincan,* 2008

468. Wnt Signaling, Volume 1: *Pathway Methods and Mammalian Models,* edited by *Elizabeth Vincan,* 2008

467. Angiogenesis Protocols: *Second Edition,* edited by *Stewart Martin and Cliff Murray,* 2008

466. Kidney Research: *Experimental Protocols,* edited by *Tim D. Hewitson and Gavin J. Becker,* 2008

465. Mycobacteria, Second Edition, edited by *Tanya Parish and Amanda Claire Brown,* 2008

464. The Nucleus, Volume 2: *Physical Properties and Imaging Methods,* edited by *Ronald Hancock,* 2008

463. The Nucleus, Volume 1: *Nuclei and Subnuclear Components,* edited by *Ronald Hancock,* 2008

462. Lipid Signaling Protocols, edited by *Banafshe Larijani, Rudiger Woscholski, and Colin A. Rosser,* 2008

461. Molecular Embryology: *Methods and Protocols, Second Edition,* edited by *Paul Sharpe and Ivor Mason,* 2008

460. Essential Concepts in Toxicogenomics, edited by *Donna L. Mendrick and William B. Mattes,* 2008

459. Prion Protein Protocols, edited by *Andrew F. Hill,* 2008

458. Artificial Neural Networks: *Methods and Applications,* edited by *David S. Livingstone,* 2008

457. Membrane Trafficking, edited by *Ales Vancura,* 2008

456. Adipose Tissue Protocols, Second Edition, edited by *Kaiping Yang,* 2008

455. Osteoporosis, edited by *Jennifer J. Westendorf,* 2008

454. SARS- and Other Coronaviruses: *Laboratory Protocols,* edited by *Dave Cavanagh,* 2008

453. Bioinformatics, Volume 2: *Structure, Function, and Applications,* edited by *Jonathan M. Keith,* 2008

452. Bioinformatics, Volume 1: *Data, Sequence Analysis, and Evolution,* edited by *Jonathan M. Keith,* 2008

451. Plant Virology Protocols: *From Viral Sequence to Protein Function,* edited by *Gary Foster, Elisabeth Johansen, Yiguo Hong, and Peter Nagy,* 2008

450. Germline Stem Cells, edited by *Steven X. Hou and Shree Ram Singh,* 2008

449. Mesenchymal Stem Cells: *Methods and Protocols,* edited by *Darwin J. Prockop, Douglas G. Phinney, and Bruce A. Brunnell,* 2008

448. Pharmacogenomics in Drug Discovery and Development, edited by *Qing Yan,* 2008

447. Alcohol: *Methods and Protocols,* edited by *Laura E. Nagy,* 2008

446. Post-translational Modifications of Proteins: *Tools for Functional Proteomics, Second Edition,* edited by *Christoph Kannicht,* 2008

445. Autophagosome and Phagosome, edited by *Vojo Deretic,* 2008

444. Prenatal Diagnosis, edited by *Sinhue Hahn and Laird G. Jackson,* 2008

443. Molecular Modeling of Proteins, edited by *Andreas Kukol,* 2008

METHODS IN MOLECULAR BIOLOGY™

Stem Cells in Regenerative Medicine

Edited by

Julie Audet and William L. Stanford

University of Toronto, Toronto, ON, Canada

Editors
Julie Audet
University of Toronto, Toronto, ON,
Canada
julie.audet@utoronto.ca

William L. Stanford
University of Toronto, Toronto, ON,
Canada
william.stanford@utoronto.ca

Series Editor
John M. Walker
University of Hertfordshire
Hatfield, Herts.
UK

ISSN: 1064-3745
ISBN: 978-1-58829-797-6
DOI 10.1007/978-1-59745-060-7

e-ISSN: 1940-6029
e-ISBN: 978-1-59745-060-7

Library of Congress Control Number: 2008939410

© Humana Press, a part of Springer Science+Business Media, LLC 2009
All rights reserved. This work may not be translated or copied in whole or in part without the written permission of the publisher (Humana Press, c/o Springer Science + Business Media, LLC, 233 Spring Street, New York, NY 10013, USA), except for brief excerpts in connection with reviews or scholarly analysis. Use in connection with any form of information storage and retrieval, electronic adaptation, computer software, or by similar or dissimilar methodology now known or hereafter developed is forbidden.
The use in this publication of trade names, trademarks, service marks, and similar terms, even if they are not identified as such, is not to be taken as an expression of opinion as to whether or not they are subject to proprietary rights.

Printed on acid-free paper

springer.com

Preface

Regenerative Medicine is devoted to replacing diseased cells, tissues, or organs, or repairing tissues in vivo by augmenting natural or inducing latent regenerative processes. Underlying these goals is the manipulation – both expansion and directed differentiation – of stem cells, which are the primary source of de novo tissue regeneration and maintenance of organ homeostasis. In this book, *Stem Cells in Regenerative Medicine*, we aim to provide biomedical researchers, clinicians and biomedical engineers an updated representation of the landscape of stem cell-based therapies in a wide spectrum of tissue systems and ontogenic stages, starting from the isolation and culture of stem cells to their actual use in vivo. In this first edition, we have attempted to compile a foundation of protocols which can be refined in subsequent publications. We hope that this series of protocols will contribute to the definition of standardized procedures for the manipulation of somatic and embryonic stem cells in research and clinical applications.

William L. Stanford, Ph.D.
Julie Audet, Ph.D.

Contents

Preface .... *v*
Contributors .... *xi*
Color Plates .... *xv*

PART I: STEM CELL SOURCES AND MANIPULATION

1 Derivation and Manipulation of Murine Embryonic Stem Cells *3*
  Alexander Meissner, Sarah Eminli, and Rudolf Jaenisch

2 Patterning Mouse and Human Embryonic Stem Cells Using
  Micro-contact Printing *21*
  *Raheem Peerani, Celine Bauwens, Eugenia Kumacheva,
  and Peter W. Zandstra*

3 Efficient Gene Knockdowns in Human Embryonic Stem Cells
  Using Lentiviral-Based RNAi *35*
  Asmin Tulpule and George Q. Daley

4 Measurement of Cell-Penetrating Peptide-Mediated Transduction
  of Adult Hematopoietic Stem Cells *43*
  Aziza P. Manceur and Julie Audet

5 Stem Cell Sources for Regenerative Medicine *55*
  Ali M. Riazi, Sarah Y. Kwon, and William L. Stanford

PART II: REGENERATION OF THE BLOOD SYSTEM

6 Investigating the Interactions Between Haemopoietic Stem Cells
  and Their Niche: Methods for the Analysis of Stem Cell Homing
  and Distribution Within the Marrow Following Transplantation *93*
  Brenda Williams and Susan K. Nilsson

7 Ex Vivo Megakaryocyte Expansion and Platelet Production
  from Human Cord Blood Stem Cells *109*
  Valérie Cortin, Nicolas Pineault, and Alain Garnier

8 Ex Vivo Generation of Human Red Blood Cells: A New Advance
  in Stem Cell Engineering *127*
  Luc Douay and Marie-Catherine Giarratana

PART III: REGENERATION OF THE NERVOUS SYSTEM

9 Isolation and Manipulation of Mammalian Neural Stem Cells In Vitro *143*
  Claudio Giachino, Onur Basak, and Verdon Taylor

10 Isolation, Expansion, and Differentiation of Mouse
   Skin-Derived Precursors *159*
   Karl J.L. Fernandes and Freda D. Miller

viii   Contents

11 Xenotransplantation of Embryonic Stem Cell-Derived Motor Neurons
    into the Developing Chick Spinal Cord *171*
    Hynek Wichterle, Mirza Peljto, and Stephane Nedelec
12 Transplantation of Neural Stem/Progenitor Cells into Developing
    and Adult CNS *185*
    Tong Zheng, Gregory P. Marshall II, K. Amy Chen, and Eric D. Laywell
13 Transplantation of Embryonic Stem Cell-Derived Dopaminergic
    Neurons in MPTP-Treated Monkeys *199*
    Jun Takahashi, Yasushi Takagi, and Hidemoto Saiki

PART IV: REGENERATION OF THE EPIDERMIS
14 Isolation and Culture of Epithelial Stem Cells *215*
    Jonathan A. Nowak and Elaine Fuchs
15 Regeneration of Skin and Cornea by Tissue Engineering *233*
    *Danielle Larouche, Claudie Paquet, Julie Fradette,
    Patrick Carrier, François A. Auger, and Lucie Germain*

PART V: REGENERATION OF THE MUSCULOSKELETAL SYSTEM
16 Prospective Isolation of Mesenchymal Stem Cells from Mouse
    Compact Bone *259*
    Brenton J. Short, Nathalie Brouard, and Paul J. Simmons
17 Isolation, Propagation, and Characterization of Human Umbilical Cord
    Perivascular Cells (HUCPVCs) *269*
    *Rahul Sarugaser, Jane Ennis, William L. Stanford,
    and John E. Davies*
18 Bone Marrow-Derived Mesenchymal Stem Cells: Isolation, Expansion,
    Characterization, Viral Transduction, and Production
    of Conditioned Medium *281*
    Massimiliano Gnecchi and Luis G. Melo
19 Template DNA-Strand Co-Segregation and Asymmetric Cell Division
    in Skeletal Muscle Stem Cells *295*
    Vasily Shinin, Barbara Gayraud-Morel, and Shahragim Tajbakhsh
20 Isolation and Grafting of Single Muscle Fibres *319*
    Charlotte A. Collins and Peter S. Zammit

PART VI: REGENERATION OF THE VASCULAR SYSTEM
21 Differentiation and Dynamic Analysis of Primitive Vessels
    from Embryonic Stem Cells *333*
    Gefei Zeng and Victoria L. Bautch
22 Derivation of Contractile Smooth Muscle Cells from Embryonic
    Stem Cells *345*
    Sanjay Sinha, Mark H. Hoofnagle, and Gary K. Owens

PART VII: REGENERATION OF THE PANCREAS AND LIVER
23 Islet-Derived Progenitors as a Source of In Vitro Islet Regeneration *371*
    Stephen Hanley and Lawrence Rosenberg

24 Isolation and Characterization of Hepatic Stem Cells, or "Oval Cells,"
    from Rat Livers *387*
    *Thomas D. Shupe, Anna C. Piscaglia, Seh-Hoon Oh, Antonio Gasbarrini,
    and Bryon E. Petersen*

25 Reprogramming of Liver to Pancreas *407*
    *Shifaan Thowfeequ, Wan-Chun Li, Jonathan M.W. Slack,
    and David Tosh*

Index *419*

Contributors

JULIE AUDET, PhD • *Terrence Donnelly Centre for Cellular and Biomolecular Research, Institute of Biomaterials and Biomedical Engineering, University of Toronto, Toronto, Ontario, Canada*

FRANÇOIS A. AUGER, MD • *Experimental Organogenesis Laboratory/LOEX and Department of Surgery and Ophthalmology, Laval University, Sainte-Foy, Quebec, Canada*

ONUR BASAK • *Department of Molecular Embryology, Max Planck Institute of Immunobiology, Freiburg, Germany*

VICTORIA L. BAUTCH, PhD • *Department of Biology, Carolina Cardiovascular Biology Center, University of North Carolina at Chapel Hill, Chapel Hill, NC, USA*

CELINE BAUWENS, MSc • *Terrence Donnelly Centre for Cellular and Biomolecular Research, Institute of Biomaterials and Biomedical Engineering, University of Toronto, Toronto, Ontario, Canada*

NATHALIE BROUARD, PhD • *Peter MacCallum Cancer Centre, Victoria, Australia*

PATRICK CARRIER, MSc • *Experimental Organogenesis Laboratory/LOEX and Department of Surgery and Ophthalmology, Laval University, Sainte-Foy, Quebec, Canada*

K. AMY CHEN • *Department of Neuroscience, McKnight Brain Institute, University of Florida, Gainesville, FL, USA*

CHARLOTTE A. COLLINS, PhD • *Wellcome Trust Centre for Stem Cell Research, University of Cambridge, Cambridge, United Kingdom*

VALÉRIE CORTIN, PhD • *Département de Recherche et Développement, Héma-Québec, Québec City, Québec, Canada*

GEORGE Q. DALEY, MD, PhD • *Harvard Medical School, Karp Family Research Building, Division of Hematology/Oncology, Children's Hospital Boston, Boston, MA, USA*

JOHN E. DAVIES, PhD, D Sc • *Institute of Biomaterials and Biomedical Engineering, University of Toronto, Toronto, Ontario, Canada*

LUC DOUAY, MD, PhD • *Unité de recherché, Faculté de Médecine, Université Pierre et Marie Curie, Paris, France, and Service d'Hématologie Biologique, Assistance Publique Hôpitaux de Paris, Hôpital Armand Trousseau, Paris, France*

SARAH EMINLI, PhD • *Center for Regenerative Medicine and Cancer Center, Massachusetts General Hospital, Harvard Medical School and Harvard Stem Cell Institute, Boston, MA, USA*

JANE ENNIS, MASc • *Institute of Biomaterials and Biomedical Engineering, University of Toronto, Toronto, Ontario, Canada*

KARL J.L. FERNANDES, PhD • *Département de pathologie et biologie cellulaire, Université de Montréal, Montréal, Québec, Canada*

JULIE FRADETTE, PHD • *Experimental Organogenesis Laboratory/LOEX and Department of Surgery and Ophthalmology, Laval University, Sainte-Foy, Quebec, Canada*

ELAINE FUCHS, PHD • *Howard Hughes Medical Institute, The Rockefeller University, New York, NY, USA*

ALAIN GARNIER, PHD • *Department of Chemical Engineering, and Centre de Recherche sur la Fonction, la Structure et l'Ingénierie des Protéines, Université Laval, Québec City, Québec, Canada*

ANTONIO GASBARRINI, MD • *Department of Internal Medicine and Gastroenterology, Catholic University of Rome, Rome, Italy*

BARBARA GAYRAUD-MOREL, PHD • *Stem Cells & Development, Department of Developmental Biology, Pasteur Institute, Paris, France*

LUCIE GERMAIN, PHD • *Experimental Organogenesis Laboratory/LOEX and Department of Surgery and Ophthalmology, Laval University, Sainte-Foy, Quebec, Canada*

CLAUDIO GIACHINO, PHD • *Department of Molecular Embryology, Max Planck Institute of Immunobiology, Freiburg, Germany*

MARIE-CATHERINE GIARRATANA, PHD • *Service d'Hématologie Biologique, Assistance Publique Hôpitaux de Paris, Hôpital Armand Trousseau, Paris, France*

MASSIMILIANO GNECCHI, MD • *Department of Cardiology, Fondazione IRCCS Policlinico San Matteo and University of Pavia, Pavia, Italy*

STEPHEN HANLEY • *Department of Surgery, and Centre for Pancreatic Diseases, McGill University Health Centre, Montreal, Quebec, Canada*

MARK H. HOOFNAGLE, BA • *Department of Molecular Physiology and Biological Physics, Cardiovascular Research Center, The University of Virginia, Charlottesville, VA, USA*

RUDOLF JAENISCH, PHD • *Whitehead Institute for Biomedical Research, Massachusetts Institute of Technology, Cambridge, MA, USA*

EUGENIA KUMACHEVA, PHD • *Department of Chemistry, University of Toronto, Toronto, Ontario, Canada*

SARAH Y. KWON, B ENG • *Department of Chemical Engineering and Applied Chemistry, University of Toronto, Toronto, Ontario, Canada*

DANIELLE LAROUCHE, PHD • *Experimental Organogenesis Laboratory/LOEX and Department of Surgery and Ophthalmology, Laval University, Sainte-Foy, Quebec, Canada*

ERIC D. LAYWELL, PHD • *Department of Anatomy & Cell Biology, College of Medicine, University of Florida, Gainesville, FL, USA*

WAN-CHUN LI, PHD • *Centre for Regenerative Medicine, Department of Biology and Biochemistry, University of Bath, Bath, United Kingdom*

AZIZA P. MANCEUR, MSC • *Terrence Donnelly Centre for Cellular and Biomolecular Research, Institute of Biomaterials and Biomedical Engineering, University of Toronto, Toronto, Ontario, Canada*

GREGORY P. MARSHALL II, PHD • *Department of Anatomy and Cell Biology, McKnight Brain Institute, University of Florida, Gainesville, FL, USA*

ALEXANDER MEISSNER, PHD • *Whitehead Institute for Biomedical Research, Massachusetts Institute of Technology, Cambridge, MA, USA*

LUIS G. MELO, PHD • *Department of Physiology, Queen's University, Kingston, Ontario, Canada*

FREDA D. MILLER, PHD • *Programs in Developmental Biology and Brain and Behaviour, Toronto Hospital for Sick Children, and Departments of Molecular & Medical Genetics and Physiology, University of Toronto, Toronto, Ontario, Canada*

STEPHANE NEDELEC, PHD • *Departments of Pathology, Neurology, and Neurobiology and Behavior, Columbia University Medical Center, New York, NY, USA*

SUSAN K. NILSSON, PHD • *Australian Stem Cell Centre, Monash University, Clayton Victoria, Australia*

JONATHAN A. NOWAK, PHD • *Howard Hughes Medical Institute, The Rockefeller University, New York, NY, USA*

SEH-HOON, OH • *Department of Pathology, Immunology and Laboratory Medicine, College of Medicine, University of Florida, Gainesville, FL, USA*

GARY K. OWENS, PHD • *Department of Molecular Physiology and Biological Physics, Cardiovascular Research Center, The University of Virginia, Charlottesville, VA, USA*

CLAUDIE PAQUET • *Experimental Organogenesis Laboratory/LOEX and Department of Surgery and Ophthalmology, Laval University, Sainte-Foy, Quebec, Canada*

RAHEEM PEERANI, PHD • *Terrence Donnelly Centre for Cellular and Biomolecular Research, Institute of Biomaterials and Biomedical Engineering, University of Toronto, Toronto, Ontario, Canada*

MIRZA PELJTO, MSC • *Departments of Pathology, Neurology, and Neurobiology and Behavior, Columbia University Medical Center, New York, NY, USA*

BRYON E. PETERSEN, PHD • *Department of Pathology, Immunology and Laboratory Medicine, College of Medicine, University of Florida, Gainesville, FL, and Program for Stem Cell Biology, University of Florida – Shands Cancer Center, Gainesville, FL, USA*

NICOLAS PINEAULT, PHD • *Département de Recherche et Développement, Héma-Québec, Québec City, Québec, Canada, and Department of Biochemistry and Microbiology, Université Laval, Québec, Canada*

ANNA C. PISCAGLIA, MD • *Department of Pathology, Immunology and Laboratory Medicine, College of Medicine, University of Florida, Gainesville, FL, and Department of Internal Medicine and Gastroenterology, Catholic University of Rome, Rome, Italy*

LAWRENCE ROSENBERG, MD, PHD • *Department of Surgery, McGill University and Centre for Pancreatic Diseases, McGill University Health Centre, Montreal, Quebec, Canada*

ALI M. RIAZI • *Institute of Biomaterials and Biomedical Engineering, University of Toronto, Toronto, Ontario, Canada*

HIDEMOTO SAIKI, MD • *Department of Neurology, Kitano Hospital, Osaka, Japan, and Department of Neurology, Kyoto University Graduate School of Medicine, Kyoto, Japan*

RAHUL SARUGASER, PHD • *Institute of Biomaterials and Biomedical Engineering, University of Toronto, Toronto, Ontario, Canada*

VASILY SHININ, PHD • *Stem Cells & Development, Deptartment of Developmental Biology, Pasteur Institute, Paris France*

BRENTON J. SHORT, PHD • *Peter MacCallum Cancer Centre, Victoria, Australia*

PAUL J. SIMMONS. PHD • *Peter MacCallum Cancer Centre, Victoria, Australia*

THOMAS D. SHUPE, PHD • *Department of Pathology, Immunology and Laboratory Medicine, College of Medicine, University of Florida, Gainesville, FL, USA*
SANJAY SINHA, MBBChir, PHD • *Division of Cardiovascular Medicine, University of Cambridge, Addenbrooke's Hospital, Cambridge, United Kingdom*
JONATHAN M.W. SLACK, PHD • *Centre for Regenerative Medicine, Department of Biology and Biochemistry, University of Bath, Bath, United Kingdom*
WILLIAM L. STANFORD, PHD • *Institute of Biomaterials and Biomedical Engineering, University of Toronto, Toronto, Ontario, Canada*
SHAHRAGIM TAJBAKHSH, PHD • *Stem Cells & Development, Department of Developmental Biology, Pasteur Institute, Paris, France*
YASUSHI TAKAGI, MD, PHD • *Department of Neurosurgery, Kyoto University Graduate School of Medicine, Kyoto, Japan*
JUN TAKAHASHI, MD, PHD • *Department of Biological Repair, Institute for Frontier Medical Sciences, Kyoto University, Kyoto, Japan*
VERDON TAYLOR, PHD • *Department of Molecular Embryology, Max Planck Institute of Immunobiology, Freiburg, Germany*
SHIFAAN THOWFEEQU • *Centre for Regenerative Medicine, Department of Biology and Biochemistry, University of Bath, Bath, United Kingdom*
DAVID TOSH, PHD • *Centre for Regenerative Medicine, Department of Biology and Biochemistry, University of Bath, Bath, United Kingdom*
ASMIN TULPULE • *Harvard Medical School, Karp Family Research Building, Division of Hematology/Oncology, Children's Hospital Boston, Boston, MA, USA*
HYNEK WICHTERLE, PHD • *Departments of Pathology, Neurology, and Neurobiology and Behavior, Columbia University Medical Center, New York, NY, USA*
BRENDA WILLIAMS, PHD • *Australian Stem Cell Centre, Monash University, Clayton Victoria, Australia*
PETER S. ZAMMIT, PHD • *Randall Division of Cell and Molecular Biophysics, King's College London, London, United Kingdom*
PETER W. ZANDSTRA, PHD • *Institute of Biomaterials and Biomedical Engineering, Terrence Donnelly Centre for Cellular and Biomolecular Research, University of Toronto, Toronto Ontario, Canada*
GEFEI ZENG, PHD • *Department of Biology, Carolina Cardiovascular Biology Center, University of North Carolina at Chapel Hill, Chapel Hill, NC, USA*
TONG ZHENG, PHD • *Department of Neuroscience, McKnight Brain Institute, University of Florida, Gainesville, FL, USA*

Color Plates

Color Plate 1 Tetraploid blastocyst injection. This procedure can be used in order to decrease the time required to generate mice from targeted ES cells. Two-cell embryos are fused as described in **Section 3.11** and injected with the diploid-targeted ES cells (here shown as GFP-positive ES cells). The tetraploid cells of the blastocysts will only give rise to extra-embryonic tissues such as the placenta, whereas injected diploid ES cells will give rise to the embryo *(17)*. This procedure takes only a few weeks to generate completely ES-derived mice, in contrast to first generating chimeras and then breeding to obtain germline transmission. (*See* discussion on p. 8)

Color Plate 2 Analysis of patterned H9 hESC colonies (D = 400 μm, P = 500 μm). (**A**) Immunohistochemistry to identifying Oct4+ cells (*green*), all nuclei by Hoechst 33342 (*blue*). Individual Oct4+ cells (*light green*) and Oct4- cells (*dark blue*) are identified by the Target Activation algorithm of the Cellomics^{TM} Arrayscan V^{TI} software by drawing overlay masks around the nuclei. (**B**) Representative flow cytometry plots of H9 hESCs before and after patterning by immunocytochemistry. Data demonstrating how the Morphology Explorer algorithm of the Cellomics^{TM} Arrayscan V^{TI} software can be used to identify cells and colonies, (**C**) calculate their area, (**D**) diameter, and (**E**) the number of Oct4+ cells in the colony. Error bars represent the standard deviation over 150 colonies. (*See* discussion on p. 31)

Color Plate 3 Infection of hESCs with GFP marked lentiviral RNAi vectors. (**A**) hESC colony infected with a GFP marked lentivirus displays a patchwork of infected cells. (**B**) hESC colony after enrichment of GFP+ cells by FACS. (*See* discussion on p. 41)

Color Plate 4 The iliac crest with a line through where to remove the flat triangular piece of cartilage (**A**) and where the acetabular notch is removed ready to be flushed (**B**). (*See* discussion on p. 97)

Color Plate 5 Transplanted LSK cells detected using a spatial distribution assay. Cells are designated as either central (**C**) or endosteal (**E**). (*See* discussion on p. 103)

Color Plate 6 Donor cells migrate normally following engraftment. Parasagittal fluorescence montage through the brain of an adult mouse host following transplantation of GFP+ neurogenic astrocytes into the lateral ventricle. Sagittal sections were immunostained with antibodies against GFP (*green*) and NeuN (*red*) 28 days following engraftment. A large number of GFP+ donor cells can be seen within the rostral migratory stream (RMS) and olfactory bulb (OB). A few donor cells are also present along the roof of the lateral ventricle (LV), and in the septum (S). CC: corpus callosum. (*See* discussion on p. 187)

Color Plate 7   Higher magnification of GFP+ donor cells within the RMS and olfactory bulb. (**A**) Parasagittal section of a host brain immunolabeled for GFP (*green*) and β-III tubulin (*red*). Donor cells are distributed with in the SEZ–RMS–OB system, as well as lining the wall of lateral ventricle (LV). (**B**) Parasagittal confocal image of a donor cell that migrated into the olfactory bulb and differentiated into a granule interneuron. (*See* discussion on p. 187)

Color Plate 8   Donor-derived cells are capable of differentiating into neurons and astrocytes. (**A**) and (**B**): Confocal Z-series of GFP+ donor cells (*green*) located in the granular cell layer of olfactory bulb that are double immunolabeled with the neuron markers (*red*) NeuN (A) and β-III tubulin (B). (**C**) A GFP+ (*green*) donor cell that has differentiated into a cortical astrocyte, and is immunolabeled with the astrocyte marker GFAP (*red*). (*See* discussion on p. 188)

Color Plate 9   Grafted cells survive, migrate, and differentiate upon transplantation into the cerebellum of both wild type and weaver transgenic mouse pups. (**A**) Donor GFP+ cells (*green*) are seen migrating and differentiating within the cerebellum 3 weeks post-engraftment in the wild-type mouse. (**B**) High magnification confocal z-series of a donor cell transplanted into the cerebellum of a weaver mutant mouse. Pεδ= β-III tubulin in A, and NeuN in B. (*See* discussion on p. 188)

Color Plate 10 Monkey ES cells. Undifferentiated ES cells on STO feeder (**A**), and neural progenitors induced from monkey ES cells. Detached ES cell colonies formed spheres similar to those of neural progenitor cells (**B**). These spheres give rise to Tuj1- (green, neuronal marker) and TH-positive (red, DA neuron marker) cells in vitro. Bar=100 μm (A, B), 100 μm (**C**). (B, C: Reproduced from ref. *(6)* with permission). (*See* discussion on p. 205)

Color Plate 11 Parkinson's disease model monkey induced by MPTP. Neurological scores of MPTP-treated monkeys (**A**, $n = 34$). Immunofluorescence study revealed that neuro-terminals of the nigro-striatal tract, which is composed of DA neuron fibers and immunoreactive for TH, are severely reduced (**B, C**). (*See* discussion on p. 210)

Color Plate 12 Function of ES cell-derived neurospheres in MPTP-treated monkeys. Neurological scores of ES cell-transplanted ($n=6$) and sham-operated animals ($n=4$) are plotted. All values are the mean ± SD. *$p<0.05$. (*See* discussion on p. 211)

Color Plate 13 HUCPVCs expressing α-actin (**a**), vimentin (**b**), and desmin (**c**). (*See* discussion on p. 275)

Color Plate 14   HUCPVCs reduce lymphocyte proliferation, even if added 3 and 5 days into a 6 day culture. Addition of HUCPVCs showed a significant decrease in lymphocyte cell number compared to control (no HUCPVCs) over 6 days in a two-way MLC. There is no significant difference among HUCPVCs added on day 0, 3, or 5 ($n = 6$). This figure shows the average cell numbers, + standard deviations. (*See* discussion on p. 276)

Color Plate 15   Transduction efficiency by fluorescent microscopy. (**A**) Phase contrast microphotograph of P5 bone marrow-derived MSCs. (**B**) The same microscopic field observed under fluorescent light shows that the majority of the cells express GFP, meaning that they were successfully transduced. In our hands, this protocol allows a transduction efficiency higher than 80%, as confirmed also by FACS analysis. (*See* discussion on p. 288)

Color Plate 16   Conditioned media production and different in vitro and in vivo assays. (**A**) Stem cells are expanded in normal conditions until they are 90% confluent. The growth medium is then exchanged with medium not containing serum and the cells are left for 24 h in a CO_2 incubator. The medium is then collected and tested either in vitro or in vivo. (**B**) For the in vitro experiments, the conditioned medium is transferred into culture dishes containing different kind of cells according to the goal of the specific experiment. Several different properties of the medium can be tested in vitro. For example, the cytoprotective effects exerted by the conditioned medium can be tested on murine cardiomyocytes. After exposing the cardiomyocytes to hypoxia in the presence of control medium or conditioned medium, apoptosis and necrosis assays are performed and the results compared. To verify if the conditioned medium contains chemotactic factors, a specific cell type (i.e., endothelial cell or cardiac stem cell) is seeded on the membrane of the upper chamber of a dual chamber dish. The number of cells migrating into the lower chamber, containing either conditioned medium or control medium, is then counted.

To test the pro-angiogenetic properties of conditioned medium, endothelial cells are seeded on a matrigel and the number of capillaries is quantified after exposure to conditioned medium or control medium. To verify if cell metabolism is influenced by factors present in the medium, the cell type of interest (i.e., murine cardiomyocytes) is exposed to the conditioned and to the control medium and then collected to perform metabolic assays. Another example: cardiomyocyte contractility may be assessed in the presence of control or conditioned medium; if inotropic factors are present, then cell contractility will be increased in the presence of conditioned medium. Proliferation assay may also be performed. Finally, proteomic analysis of conditioned medium may allow the discovery of new therapeutic molecules and targets. (**C**) Conditioned medium may be tested also in vivo using different experimental disease models. For example, the effects of conditioned medium on ischemic myocardium may be assessed in a murine model of myocardial infarction. Small volumes of concentrated conditioned medium obtained by ultrafiltration are injected at the infarct border zone after left coronary ligation. At established time points, heart function is analyzed by echo (an M-mode image of the left ventricle is depicted in the figure) or other methods. The heart may also be collected for histology to determine, for example, the infarct size (a cross-section of mouse heart stained with Masson Trichrome is depicted in the figure). Furthermore, immunohistochemistry staining may performed to determine different parameters, such as cardiac regeneration, neoangiogenesis, apoptosis. (*See* discussion on p. 289)

Color Plate 17 Lineage progression of an adult skeletal muscle satellite cell to a differentiated cell. (**A**) Isolated fibre from the Tibialis Anterior muscle of a $Myf5^{nlacZ/+}$ mouse showing nuclear β-galactosidase activity by X-gal staining. Hoechst staining reveals myonuclei inside the fibre as well as the nucleus of the satellite cell on the fibre. (**B**) Lineage progression of a quiescent to an activated satellite cell, hallmarked by MyoD expression, to a myoblast. Myoblasts fuse homotypically or to pre-existing differentiated fibres or myotubes after leaving the cell cycle. Differentiated fibres are characterised by the expression of myosin heavy chain (MyHC). The expression of different commonly used markers is indicated. (*See* discussion on p. 296)

Color Plate 18 Dissections of TA, EDL and Soleus muscles. The skin was removed from a 4-week-old mouse to reveal the underlying muscles. (**A–D**) Removal of TA and EDL muscles. (**E-F**) Removal of Soleus muscle. Lower tendon (1) is sectioned first, the muscles are lifted, then the upper Soleus tendon (2) is sectioned. Note in older mice, the soleus appears more red and distinguishable. G) Actual sizes of dissected muscles with associated tendons (white arrowheads). Muscle attachment proximal to the knee is at the top of the photo. See main text for details. Red arrowhead, cartilage bridge in foot under which tendons are transit. White arrowheads, tendons. (*See* discussion on p. 301)

Color Plate 19 Asymmetric segregation of Numb in non-differentiating satellite cell-derived myoblasts. (**A**) Two examples of asymmetric distribution of endogenous Numb protein in mitotic satellite cell-derived myoblasts after 4 days in culture; immunocytochemistry with anti-Numb and anti-Ki67 antibodies; (**B**) p66Numb-EGFP and H2B-mRFP fusion proteins were overexpressed in satellite cell-derived myoblasts using pCAG-p66Nb-EGFP and pCAG-H2B- mRFP plasmids (transfected at 96 h after plating of myofibres). At 4 days after plating, cells were harvested by mitotic shake-off, re-plated at low density and grown for several hours on poly-D-lysine-coated dishes. The cells were fixed and stained with anti-Numb antibody (green), and visualised using EGFP and mRFP epifluorescence. Phase contrast on left; reconstituted confocal stack on right was rendered with Imaris software. Note asymmetric distribution of the majority of the Numb-GFP protein to one pole of this cell. (*See* discussion on p. 306)

Color Plate 20 Asymmetric segregation of template DNA strands in label retaining cells. (**A**) Asymmetric segregation of BrdU label to only one daughter satellite cell after mitosis, on a freshly isolated EDL fibre. $Myf5^{nlacZ/+}$ mice were pulsed with BrdU from P3-P7; chase for 7 days. (**B**) Clonal analysis of LRCs after 4 weeks chase in vivo, 72 h in culture. Antibody stainings for detecting Myo protein and BrdU in the nuclei of satellite cell derived myoblasts. The LRC in this case has a lower expression of MyoD. Scale bars: A, 15 μm; B, 40 μm. (*See* discussion on p. 311)

Color Plate 21 Time-lapse imaging of ES cell differentiation cultures. An ES cell line expressing endothelial cell-specific H2B-GFP was differentiated to day 8. It was imaged on a Perkin Elmer spinning disk confocal microscope. Culture is fixed right after imaging and stained for PECAM-1. (A) Frames of a time-lapse movie. Time is in minutes at the top right corner of each panel. (B) Last frame of the movie in green (**a**), the same field with PECAM-1 stain in red (**b**), and the overlay of the two images (**c**). (*See* discussion on p. 341)

Color Plate 22 Immunofluorescence for SMαA in intact EBs and puromycin purified SMCs. Immunostaining was carried out as described in 3.5 on intact day 28 EBs (**A**) and on cells following enzyme dispersal and overnight selection (**B**). A "sheet" of SMCs may be seen in an intact day 28 EB with adjacent non-staining non-SMC at the bottom of the image (**A**). Such groups of SMC may be seen to contract spontaneously under the microscope. Following puromycin selection, relatively pure populations of cells with a typical SMC morphology were seen (**B**). Red staining: SMαA immunofluorescence, Blue: DAPI nuclear staining. (*See* discussion on p. 356)

Color Plate 23 Transformation of embedded islets into DLS, and subsequent regeneration of ILS. (**A**) Immediately after embedding, islets are characterized by a solid-spheroid shape. (**B**) DLS formation appears to initiate in specific foci, (**C**) until a DLS replaces the islet. (**D**) Treatment with INGAP induces the budding of regenerating ILS from DLS (bar = 100 μm). (*See* discussion on p. 377)

Color Plate 24 Immunocytochemistry. Analysis of islets, DLS, and ILS demonstrates that while islets and ILS express endocrine hormones (insulin, glucagon, somatostatin, and pancreatic polypeptide) in approximately the same proportions, these hormone[+] cells are absent in DLS. Conversely, staining for a ductal cell marker (pan-cytokeratin) is observed primarily in DLS. Furthermore, while few or no ductal cells are observed in islets, ILS can be observed to be "budding" from ductal structures (bar = 100 μm). (*See* discussion on p. 380)

Part I

Stem Cell Sources and Manipulation

Part I

Chapter 1

Derivation and Manipulation of Murine Embryonic Stem Cells

Alexander Meissner, Sarah Eminli, and Rudolf Jaenisch

Abstract

Pluripotent embryonic stem (ES) cell lines were first isolated over 25 years ago and remain an essential tool in molecular and developmental biology to this day. In particular, the use of homologous recombination and subsequent generation of ES-derived mice has greatly facilitated research across all fields. Moreover, ES cells represent an extremely attractive model to study events in early development. In this chapter, we will describe the derivation and propagation of murine ES cells. This is followed by a description of targeting ES cells and a protocol for the generation of mice by diploid and tetraploid blastocyst injections.

Key words: Embryonic stem cells, pluripotency, homologous recombination.

1. Introduction

In 1981, the first embryonic stem (ES) cells were derived from the inner cell mass (ICM) of murine embryos *(1, 2)*. Shortly after these initial reports, ES cells were shown to contribute to all tissues of chimeric mice following injection into blastocysts. In contrast to embryonic carcinoma (EC) cells, which had been derived earlier, ES cells could efficiently and reproducibly contribute to the germline and therefore to the offspring of chimeras *(3)* (reviewed in *(4)*).

Stem cells are defined as cells that have the ability to self-renew and to generate more restricted, differentiated cells. A functional definition for pluripotent ES cells should include: (i) maintenance of the pluripotent state over extended periods in culture and (ii) a demonstration of pluripotency by contribution to all three germ layers (chimera or teratoma formation).

In the past decades, many intrinsic factors, such as the transcription factors Oct-4 and Nanog *(5–7)* and extrinsic factors, such as leukaemia inhibitory factor (LIF) *(8)* that are essential for maintaining an ES cell phenotype have been identified (reviewed in *(9)*). However, many biological questions, in particular the complex regulatory networks that maintain them in a pluripotent state, remain to be addressed.

Mouse ES cells represent an important model system for studying gene function during development and disease. For example, gene targeting by homologous recombination in murine ES cells allows for the introduction of mutations into genes. Likewise, overexpression can be achieved by the insertion of transgenes into ES cells. Subsequent blastocyst injections of targeted ES cells allows for the analysis of their consequence in vivo. Moreover, conditional and inducible gene targeting technologies such as the Cre/lox and FLPe/frt systems in combination with other genetic tools (drug-inducible systems, shRNA-mediated gene silencing, fluorescent proteins) make them useful instruments to induce precise mutations in a spatio-temporal manner.

In this chapter, we describe the derivation of ES cells from blastocysts and provide details on how to culture and manipulate them. In particular, we describe how ES cells can be targeted via homologous recombination and the subsequent generation of mice by diploid or tetraploid blastocyst injections.

2. Materials

2.1. Equipment

1. Plastic ware: V-bottom 96-well plates (Corning Incorporated #3894), 4-well plates (Nunc #176740), 6-well plates (Corning Incorporated #3506), 12-well plates (Corning Incorporated #3512), 24-well plates (Corning Incorporated #3527), T25 cm^2 flasks (Corning Incorporated #430639), T75 cm^2 (Corning Incorporated #3376), 10 cm plates (Corning Incorporated, #430167), 15 cm plates (Corning Incorporated #430599) and Petri dish (Falcon, # 351029).
2. 15 ml tubes (Sarstedt #62.554.002).
3. Freezing tubes (Nunc Cryo Tube Vials #363401).
4. Disposable underpads (Kendall #7105).
5. Sterile glass or plastic pipettes (2 ml, 5 ml, 10 ml, 25 ml).
6. Pasteur pipettes.
7. Pipettes and sterile filter tips.
8. Multichannel pipettor.
9. Mouth pipettes: VWR international (# 53432-783).

10. Needles (BD): 18G 1 1/2 (#305196), 21G 1 1/2 (#305167), 23G 1 (#305145), 26G 3/8 (#305110) and 27G 1/2 (#305109).

11. Syringes (BD): 1 ml (#309628), 3 ml (#309585) and 5 ml (#309603).

12. Radiation source (e.g. Gammacell 40 Exactor MDS Nordion www.mdsnordion.com/products/blood-research-irradiators-r.htm) for generating feeders.

13. Biorad Gene Pulser Xcell electroporator (#617BR1 03869).

14. 4 mm cuvette Biorad (#165-2088).

15. Nikon Eclipse TE 300 inverted scope with micromanipulators for blastocyst injections.

16. LF101 Electro Cell Fusion unit (Protech International, Inc.).

17. Fusion chamber (CUY 5010, Protech International, Inc.).

18. Piezo (Primetech, Ibaraki, Japan).

2.2. Tissue Culture Reagents

1. MEF medium: 450 ml DMEM (Gibco, 11965-092), 50 ml fetal bovine serum (FBS, heat inactivated, 56°C for 30 min Hyclone #SH30071.03), 5 ml non-essential amino acids (100× stock from Gibco, 11140-050), 5 ml penicillin/streptomycin (100× stock from Gibco, 15140-122), 5 ml glutamine (200 mM from Gibco, 25030-081), 4 µl β-mercaptoethanol (Sigma #M7522), Filter sterilize (Fisher Scientific 0.22 µm #13-678-6b) and store at 4°C for up to 3–4 weeks.

2. ES cell medium: 425 ml DMEM (Gibco/Invitrogen, knockout #10829-018), 75 ml fetal bovine serum (FBS, heat inactivated, 56°C for 30 min, hyclone characterized #SH30071.03) (see **Note 1**), 5 ml non-essential amino acids, 5 ml penicillin/streptomycin, 5 ml glutamine, 4 µl β-mercaptoethanol, 50 µl LIF ESGRO (1×10^7 units ESGRO/ml; Chemicon #ESG1106), Filter sterilize and store at 4°C up to 3–4 weeks.

3. Serum Replacement medium = Knockout SR, Gibco #1028-028 (see **Note 1**).

4. Gelatin solution: 0.2% gelatin (Type A from porcine skin from Sigma #G2500) in water, autoclave, store at room temperature.

5. 0.25% trypsin–EDTA (Gibco #25200-114), store in 10 ml aliquots at −20°C.

6. HEPES (for 10.8 l HEPES, pH 7.3, OSMO 290): 0.1 g Phenol Red, 0.4 g KH_2PO_4, 0.8 g $Na_2HPO_4 7H_2O$, 4.0 g KCl, 10.9 g dextrose, 47.66 g HEPES salt, 81.14 g NaCl, 1 ml HCl, add ddH_2O; filter (0.2 µm) into sterile bottles.

7. Hepes/EDTA: take 5 l from the Hepes above and add 0.18 g of EDTA, filter into sterile bottles.
8. 100 ml 2× freezing solution (20% FBS): 60 ml DMEM (Gibco, #10829-018), 20 ml DMSO (Sigma #D8418), 20 ml FBS; Filter sterilize and store at 4°C.
9. DNA lysis buffer: 100 mM Tris HCl (pH 8.5), 5 mM EDTA, 0.2% SDS, 200 mM NaCl; add Proteinase K 100 μg/ml fresh each time before use.
10. Pregnant Mare Serum (PMS) (Calbiochem #367222, 1,000 IU) add 20 ml saline (0.9% NaCl) for 50 IU/ml aliquots and store at −20°C.
11. Human chorionic gonadotrophin (HCG) (Calbiochem #230734; 3,100 IU). Make HCG stock with 500 IU/ml in saline, store 100 µl aliquots at −20°C until use. Dilute 1/10 just before use (50 IU/ml final) in saline.
12. KSOM (Speciality Media, MR-020P-5F or MR-020P-D).
13. Hyaluronidase: M2 containing 0.1% bovine testicular hyaluronidase (Sigma, type IV-S-BTE #H4272).
14. PD98059 (Mek1 inhibitor, Cell Signaling Technology, #9900). 5 mg add 373.3 µl DMSO to make a 1,000× stock (50 mM). Use 50 µm final concentration, sterile filtrate after adding PD to ES medium (0.22 μm filter) and store at 4°C.
15. PBS with $MgCl_2$/ $CaCl_2$ (Gibco # 14040-117).
16. PBS w/o $MgCl_2$/ $CaCl_2$ (Gibco #14190-136).
17. Proteinase K (Promega #V3021).
18. Acid Tyrode's solution, AT (Sigma, Tyrode's Solution, Acidic #T1788).
19. G418 (Gibco #11811-031).
20. Hygromycin (Roche #10843555001).
21. Puromycin (Sigma #P8833).
22. M2 media (Sigma #M7167).
23. 4 N Fusion buffer: 0.3 M mannitol (Sigma #M-9546), 0.3% BSA (Sigma #A3311-100G).
24. Serum replacement medium.

3. Methods

ES cells are routinely and most efficiently derived from the ICM of the blastocyst stage embryo (**Fig. 1.1A**). It has been shown that human and mouse ES cells can be derived from morula stage

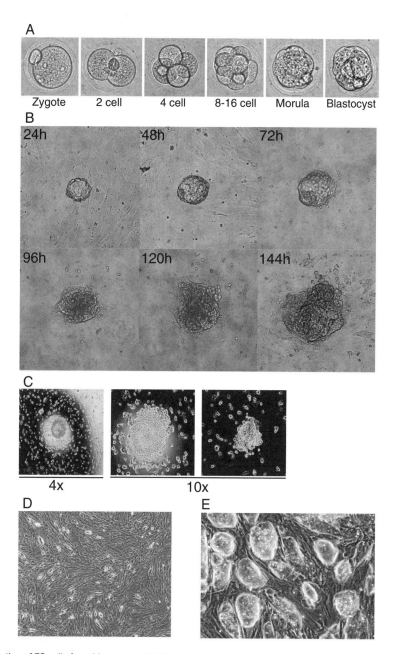

Fig. 1.1. Derivation of ES cells from blastocysts. (**A**) Murine pre-implantation embryos. (**B**) The initial growth of zona-free, explanted blastocysts. Typically after 3–5 days, an ICM outgrowth can be seen and dissociated. (**C**) Whole-well trypsinization of initial outgrowth. The outgrowth typically dissociates after the feeders are already lifted off. It is important to wait until it is almost completely dissociated and should be further treated by multiple cycles of pipetting up and down. (**D**) 2–3 days after dissociation of the outgrowth, colonies (4× magnification) should appear and (**E**) grow into an ES cell line (10× magnification).

embryos *(10, 11)* and, although very inefficiently, even from 8-cell stage blastomeres *(12)*. In this chapter, we will only focus on the derivation of ES cells from blastocysts.

Most established ES cell lines are male by genotype (XY), likely due to the increased instability of female (XX) ES cell lines *(13)*. ES cell lines have consistently been derived only from a few genetic backgrounds, such as the 129, C57BL/6, DBA strains or hybrids thereof. Some specialized protocols have been described to establish lines from other strains *(14)*. One major advancement for increasing the overall efficiency and allowing derivation of previously refractory strains such as CBA, was the addition of the Mek1 kinase inhibitor PD98059, which is currently used for all ES derivations in our laboratory *(15)*.

A classical knockout experiment involves the generation of a DNA construct in which a copy of the desired gene has been altered such as to cripple its function, e.g. by introducing a neomycin gene (neo) cassette which functions both as a mutagen and as a selection marker. The construct is then electroporated into ES cells, where the engineered copy of the gene replaces one copy of the wild-type gene through homologous recombination *(16)*. Gain-of-function experiments are similar, except that the construct is designed to increase the function of the gene, usually by providing extra copies of the gene or inducing synthesis of the protein more frequently. One way to gain information about the expression pattern of a gene is to fuse the transcription start site with a fluorescence marker gene such as green fluorescent protein (GFP) that will allow easy visualization and permits isolation of subsets of cells *in vitro* and *in vivo* (*see* **Fig. 1.2** and Color Plate 1).

Targeted ES cells can be used to generate chimeras through diploid blastocyst injections. Breeding of high-degree chimeras with wild-type animals will, in most cases, results in germline transmission and generally the derivation of homozygous animals after further breeding. This is the common route to generate mutant mice; however, it is very time consuming. The use of

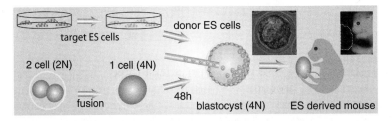

Fig. 1.2. Tetraploid blastocyst injection. This procedure can be used in order to decrease the time required to generate mice from targeted ES cells. Two-cell embryos are fused as described in **Section 3.11** and injected with the diploid-targeted ES cells (here shown as GFP-positive ES cells). The tetraploid cells of the blastocysts will only give rise to extra-embryonic tissues such as the placenta, whereas injected diploid ES cells will give rise to the embryo *(17)*. This procedure takes only a few weeks to generate completely ES-derived mice, in contrast to first generating chimeras and then breeding to obtain germline transmission. (*See* Color Plate 1)

tetraploid embryo complementation (**Fig. 1.2**) can significantly accelerate the generation of the desired mouse strain by omitting the chimera step *(17)*. Moreover, the time to generate homozygous animals can be further decreased by identifying XO sub-clones of the original XY line (frequency for loss of Y is about 2% of sub-clones) and generating male and female mice from the same targeted ES cell line *(18)*.

3.1. Isolation of Mouse Embryonic Fibroblasts (MEFs; see Note 2)

1. Euthanize pregnant females (E13.5), remove the complete uterus and transfer into HEPES buffer (*see* **Note 3**).
2. Carefully remove embryos from the uterus using forceps and scissors.
3. Wash several times in HEPES.
4. Dissect under microscope, remove all organs and the head.
5. Put into syringe barrel, add some MEF medium (*see* **Note 4**), and press through 18G needle (avoid too much force) into a 15 ml tube.
6. Repeat through a smaller needle (*see* **Note 5**).
7. Plate onto 15 cm plate (one embryo/plate) and add 25 ml of MEF medium.
8. MEFs will attach after a few hours.
9. Change MEF medium on the next day and let MEFs grow until they are confluent.
10. Wash twice with HEPES/EDTA and add 3–4 ml of trypsin.
11. After a few minutes, MEFs should lift off the plate and become single cells.
12. Add 5 ml of MEF medium (to stop trypsin) and pellet cells by centrifugation at 1,000 rpm (~800 g) for 4 min.
13. Discard supernatant.
14. Add equal amount of MEF medium and 2× freezing medium.
15. Freeze (see below) and store at –80°C.

3.2. Preparation of Irradiated Feeders

1. Thaw primary MEFs (vials containing MEFs from one embryo, can be plated onto at least three 15 cm plates) and expand for around —three to four passages (split 1:4 or 1:5) (*see* **Note 6**).
2. Trypsinize confluent MEFs, pellet cells by centrifugation and resuspend in an appropriate volume (choose the volume, so that after adding 2× freezing medium aliquots of the desired size can be made, e.g. one 15 cm plate gives roughly feeders for 4–6xT25 plates; resuspend 15 cm pellet in 2–3 ml medium (after irradiation add 2–3 ml of freezing medium and freeze as 1 ml aliquots for T25 plates).

3. Keep MEFs on ice before and after irradiation.
4. Irradiate MEFs. The irradiation time depends on the indicated dose rate on the machine (which depends on the dose decay chart). Dose rate unit is Gy/min. Divide the desired number of Grey (for MEFs 26 Grey) by the dose rate, e.g. if the actual dose rate is 1.029 Gy/min, the resulting radiation time is then 25 min and 16 sec.
5. Add freezing medium, distribute as described above, freeze and store at −80°C.

3.3. Derivation of Murine ES Cells From Blastocysts

1. Coat 4-well or 24-well plates with 0.2% gelatin for 5 min at RT (*see* **Note 7**).
2. Remove gelatin and plate MEFs (*see* **Note 8**).
3. Take a sterile non-coated Petri dish and add small drops (~50 µl) of ES medium and AT.
4. Under a dissecting scope (ideally within a tissue culture hood): Place individual blastocysts into AT using the mouth pipette and monitor until the zona pellucida disappears; this will take only seconds to a few minutes, depending on how much medium was carried over (*see* **Note 9**).
5. Carefully collect zona-free blastocysts and place into a drop of ES medium.
6. Transfer the zona-free blastocysts into 4-well or 24-well plates containing MEFs and 0.5 ml (for the 4-well plate) ES medium plus PD98059 inhibitor (*see* **Note 1**).
7. Wait until blastocysts attach to the feeder layer (typically 24–48 h).
8. Add a few drops of ES medium plus PD98059, if required.
9. Normally after 96–144 h (sometimes it might take more than a week), an ICM outgrowth (*see* **Fig. 1.1B**) should appear and continue to grow.
10. For whole-well trypsinization (*see* **Note 10**) wash the wells containing the ICM outgrowth twice with HEPES/EDTA.
11. Add 2–3 drops of trypsin.
12. Wait until the ES colonies are completely dissociated (*see* **Fig. 1.1C**).
13. Add 500 µl of ES medium plus PD98059 and pipette up and down with P-1000 (use sterile filter tips).
14. Transfer into 12-well plate with fresh feeders and add 1 ml of ES medium plus PD98059.
15. Next day, change medium (ES medium plus PD98059).
16. After 2–3 days ES colonies should appear (**Fig. 1.1D**).

17. Wait until ES colonies are grown up (**Fig. 1.1E**) and expand for —one to two passages as described in the next section (*see* **Section 3.4**), before freezing the new ES cell lines.

3.4. Culture and Passage of ES Cells

1. Plate MEFs (*see* **Note 6 and 7**).
2. Next day, thaw ES cells (see below for thawing and freezing cells) and plate onto feeder cells.
3. ES medium is changed every day and ES cells are usually split every 2–3 days (*see* **Note 11**).
4. Wash the ES cells twice with HEPES/EDTA or PBS (w/o $MgCl_2$/ $CaCl_2$).
5. Add 1 ml trypsin–EDTA and incubate at 37°C for 5 min.
6. Add 5 ml ES medium and generate single-cell suspension by pipetting up and down several times.
7. Transfer into 15 ml tube.
8. Pellet cells by centrifugation at 1,000 rpm (~800 g) for 4 min and split cells at a 1:3 ratio or higher, if ES cells are more dense (*see* **Note 11**).
9. Alternatively, ES cells can also be frozen at this point (see below).

3.5. Freezing ES Cells

1. Wash ES cells twice with HEPES/EDTA.
2. Add trypsin and wait 3–5 min at RT or 37°C until the cell layer lifts off.
3. Add ES medium and pipette up and down to generate a single cell suspension.
4. Pellet cells by centrifugation for 4 min at 1,000 rpm (~800 g).
5. Aspirate supernatant.
6. Resuspend in one volume ES medium and one volume 2× freezing medium.
7. Freeze ES cells without delay in –80°C freezer slowly at –1°C/minute through use of an appropriate container (polystyrene box).
8. Store temporary at –80°C or long term in liquid N_2.

3.6. Thawing ES Cells

1. Remove a frozen stock vial from liquid nitrogen.
2. Immerse vial in a 37°C water bath.
3. Once thawed, remove vial without delay from water bath.
4. Transfer cells into a 15 ml tube containing 5 ml ES medium.
5. Pellet cells by centrifugation 4 min at 1,000 rpm (~800 g) to remove cryoprotectant (DMSO) from cell suspension.

6. Aspirate supernatant.
7. Reconstitute the pellet in ES medium.
8. Transfer suspension onto feeder cells.

3.7. Targeting ES Cells

1. Seed two or more 10 cm plates with DR4 MEFs.
2. Linearize around 30–60 µg of the targeting vector with an enzyme that cuts once in plasmid backbone (the day before targeting).
3. Perform a phenol extraction and precipitate the linearized DNA with ethanol.
4. Dissolve the DNA in 50–100 µl sterile PBS in tissue culture hood.
5. Grow ES cells on a T25 plate with feeders until the flask is 70% confluent.
6. Trypsinize ES cells, add 5 ml ES medium and centrifuge to pellet cells.
7. Discard supernatant and wash twice with 5 ml PBS.
8. Centrifuge to pellet cells and resuspend the pellet in 500 µl PBS.
9. Electroporate 500 µl of ES cells with 50–100 µl (30–60 µg) of the linearized targeting construct in a 4 mm cuvette.
10. Electroporation settings: Time constant program, 800 V, 0.2 ms.
11. Plate the electroporated ES cells onto the prepared 10 cm plates in 10 ml of ES medium (*see* **Note 12**).
12. Start selection after 24 h by changing to ES cell medium containing the drug. Feed cells every day.
13. Depending on the selection marker used, non-targeted ES cells will die after a few days.

3.8. Picking and Expanding ES Clones

1. Seed 24-well plates with DR4 MEFs (one day prior to picking/sub-cloning).
2. In general, one can start picking the resistant clones 1–2 weeks after beginning drug selection (*see* **Table 1.1**).
3. Add 60 µl PBS without MgCl/CaCl to the wells of a V-bottom shape 96-well plate.
4. Wash the 10 cm plates containing the ES colonies to be picked twice with PBS with MgCl/CaCl and add 10 ml of PBS to the plate.
5. Pick the colonies with small 1–10 µl pipette tips (set the P-20 pipetman to 5 µl) by carefully scraping around individual colonies and transferring each of the colonies to a well in the 96-well dish.

Table 1.1
Conditions for antibiotic selection after targeting ES cells

Selection drug	Concentration (μg/ml)	Estimated time before picking clones (days)
Neomycin (G418)	350*#	7–10
Puromycin	2	5–7
Hygromycin	120–140	8–12

*active form
Neomycin can be used at a wide range of concentrations. Up to 6,000 μg/μl (active form) has been used to select for homozygous ES cell lines after targeting *(19)*.

6. After the desired number of resistant clones is picked, add one drop of trypsin (20 μl) to each well using a multichannel pipettor and incubate at 37°C for 5 min.
7. Resuspend the trypsinized clones in 100 μl ES media by pipetting up and down 5–10 times.
8. Transfer half of the suspension (six clones at a time) from the 96-well dish to a 24-well dish containing fresh selection medium using a tip at every other position on a 12-channel pipettor (*see* **Note 13**).
9. Proceed with the second half of the ES cells as described in the next section (*see* **Section 3.9**).
10. Let the colonies grow to confluence on the 24-well plate while feeding every day with ES medium containing the selection drug.
11. These cells will be harvested for DNA analysis to identify properly targeted clones for further expansion (*see* **Section 3.10**).

3.9. Freezing and Thawing ES Cells in the 96-Well Plates

1. Add 40 μl of 2× freezing medium using a multichannel pipettor to the remaining half of the ES cell suspension in each 96-well (from point 8 of **Section 3.8**). Wrap the plate in bench layer (disposable underpads) or put into a polystyrene box and freeze at −80°C.
2. After positive clones have been identified, they need to be expanded from the frozen 96-well plates.
3. Seed 24-well plates with DR4 MEFs (the day before thawing).
4. Add 50 μl of ES medium to each of the 96-wells that should be expanded and incubate at 37°C to accelerate thawing.
5. Transfer the clones immediately to a centrifugation tube, add 5 ml ES medium and centrifuge to pellet cells, 4 min at 1,000 rpm (∼800 g).

6. Discard supernatant.
7. Resuspend the pellets in ES medium and plate on 24-well plate.
8. *Optional:* Wrap the 96-well plate again and freeze the remaining clones at −80°C.
9. Let the cells grow until the well is 70% confluent and expand as described above.

3.10. DNA Analysis

1. Aspirate the ES medium from the 24-well plates (**Section 3.8**) and add 500 µl DNA lysis buffer with Proteinase K directly into each well and incubate overnight at 37°C.
2. Transfer the lysed cells into a centrifugation tube and perform an isopropanol precipitation.
3. Confirm homologous recombination by Southern blot analysis with an appropriate probe.

3.11. Generation of Mice by Diploid and Tetraploid Blastocyst Injections

1. BDF1 (C57BL/6xDBA/2 F1) females are superovulated by i.p. injection of 100 µl PMS.
2. Followed by a second injection 46–48 h later of 100 µl HCG.
3. After administration of HCG, females are mated with BDF1 males.
4. BDF2 zygotes (C57BL/6xDBA/2 F2) are collected 18–24 h post-HCG injection (0.5 days post coitum (dpc)).
5. Isolate oviducts in M2 medium, remove as much fat as possible and transfer to drops of M2 medium containing 0.1% hyaluronidase. Shred oviducts in the drop to release zygotes.
6. Collect cumulus-free zygotes using a mouth pipette.
7. Wash extensively by transferring through several microdrops of M2 medium.
8. After washing, transfer to a new culture dish containing microdrops of pre-equilibrated KSOM under mineral oil and place at 37°C, 5% CO_2.
9. For diploid blastocysts injection, embryos are continually cultured in KSOM until they become fully expanded blastocysts at 3.5 dpc (*see* **Note 14**).
10. For generating tetraploid embryos (*see* **Fig. 1.2**), zygotes are cultured overnight in KSOM to obtain two-cell embryos.
11. Blastomeres of two-cell embryos are electrofused as follows to produce one-cell tetraploid embryos.
12. Settings for electrofusion unit: Power ON, Timer = 5 s (shown as 05), Frequency = 10 × 0.1 MHz (shown as 10/×0.1 MHz), AC volts = 5 (shown as 05), DC volts = 75 (shown as 075), pulse number = 2 (shown as 02), pulse

width 35 μsec, Times = 02 (different from #6), Interval = 5 (shown as 05) × 0.1 s., post-fusion button ON.

13. Connect electrodes of fusion chamber (rinse with distilled H₂O, then apply a drop of fusion buffer and connect).
14. Equilibrate two-cell embryos in fusion buffer for 1 min.
15. Transfer embryos between electrodes.
16. Wait until embryos are aligned in AC current (~20 s), AC is ON when the electrodes are connected.
17. Push 'START' button.
18. After the two pulses wash embryos through drops of KSOM (pre-equilibrated at 37°C) and return to incubator.
19. Fused cells may appear 5 min after current application, but usually it takes 15–60 min to observe fusion of most embryos. Those that did not fuse must be separated after 1 h and discarded, since they will form regular diploid blastocysts. Expect more than 90% embryos to fuse (see **Note 15**).
20. Culture for 48 h until expanded 4 N blastocysts develop.

3.12. Blastocyst Injection

1. Diploid or tetraploid blastocysts (94–98 h post-HCG injection) are placed in microdrops of DMEM with 10% FBS (or into M2) under mineral oil.
2. A flat tip microinjection pipette with an internal diameter of 12–15 μm is used for ES cell injection.
3. Fifty or more ES cells can be picked up in the end of the injection pipette.
4. The blastocysts are held in the vicinity of the ICM with a holding pipette (inner diameter should be around 20 μm).
5. The injection pipette containing the ES cells is pressed against the zona opposite the ICM.
6. A brief pulse of the Piezo is applied, and the injection needle is simultaneously pushed through the zona and trophectoderm layer into the blastocoel cavity.
7. A controlled number of ES cells (typically 5–15) are injected into the blastocyst cavity.
8. After injection, blastocysts are returned to KSOM and placed at 37°C until transferred to recipient females. The transfer is generally done 0–3 h after injection.

3.13. Recipient Females and Caesarean Sections

1. Ten to twelve injected blastocysts are transferred to each uterine horn of 2.5 dpc pseudopregnant BDF1 females (see **Note 16**).
2. Recipient mice should give birth at 19.5 dpc (17 days after the embryo transfer).

3. If pups are not naturally delivered, c-section should be performed to recover ES-derived or chimeric pups.

4. If necessary, foster to lactating BALB/c mothers.

4. Notes

1. For general ES cell culture, FBS from different suppliers can be used (e.g. Hyclone, Invitrogen, Chemicon, Stem Cell Technologies). We have previously used Hyclone and Invitrogen for the successful derivation of pluripotent and germline competent ES cells (others have not been tested and may equally work). However, recently, the success rate and reproducibility for the derivation of ES cells was significantly improved by using Invitrogen's replacement serum (Knockout SR, Gibco #1028-028) for cultivating the initial outgrowth rather than FBS. The replacement serum is defined, and hence less subject to batch variability. We also noted that the occurrence of trophoblast-derived cells is significantly reduced and the initial outgrowth typically looks more like an ES colony (**Fig. 1.3**). The efficiency with normal serum can vary between 30 and 100% for fertilized embryos (129, C57Bl/6 and DBA strains), whereas with replacement serum it was constantly near 100%.

2. To maintain germline competent ES cells for extended periods in culture, they have to be grown on feeder cells (mitotically inactivated fibroblasts), which provide important cytokines and other essential factors. For the isolation of DNA or analysis that would otherwise be affected by the presence of feeder DNA, RNA or proteins, it is possible to

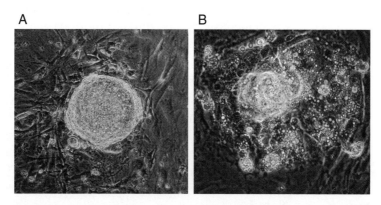

Fig. 1.3. Initial ICM outgrowth with ES medium containing replacement serum or FBS. (**A**) In most cases, when ES cells are derived in ES medium plus PD98059 containing Invitrogen's replacement serum, the appearance of trophectoderm cells is significantly reduced, (**B**) when compared to the outgrowth in ES medium plus PD98059 containing FBS.

grow ES cells on 0.2% gelatin for several passages. In this chapter, we only describe the generation of irradiated MEFs. If no radiation source is available, it is also possible to use mitomycin C.

3. Most common lab strains can be used for preparing feeder cells. However, manipulations such as targeting, viral infection and many more will require drug selection at some point. Therefore, the DR4 mouse strain (Tg(DR4)1Jae/J (Stock Number: 003208)), which has multiple drug resistance genes is highly recommended. Feeder cells generated from homozygous DR4 embryos tolerate drug concentration required for efficient ES cell selection (for details see http://jaxmice.jax.org/strain/003208.html).

4. All tissue culture reagents (medium, trypsin, etc.) should always be at least hand warm and ideally be pre-warmed to 37°C.

5. Alternatively, the embryos can be minced between two scalpels and a few drops of trypsin. Streak out the resulting mass onto 15 cm plate and wait for 1–2 min before adding MEF medium (this will ensure that the small pieces stick to the plate).

6. Do not passage the primary MEFs too extensively. MEFs should remain narrow shaped and doubling roughly every 24 h. When the cells start to flatten, it is a sign of senescence and they should no longer be used for the preparation of feeders. Culturing the primary MEFs under low oxygen conditions may delay senescence.

7. For ES cell culture and derivation, all plates should be coated with 0.2% gelatine for a few minutes prior to seeding the feeder cells. This is not required for expanding MEFs.

8. Feeders should ideally be plated 24–48 h before explanting blastocysts, thawing or splitting ES cells. However, we have not observed any detrimental effects when plating feeders simultaneously.

9. The AT treatment is not absolutely required, but will in most cases facilitate the initial attachment of the blastocyst. Zona-free blastocysts are very sticky. It is helpful to take up always some medium in the mouth pipette before collecting the zona-free embryos.

10. The initial ICM outgrowth can be picked similar to the subcloning described in the targeting section. However, in our hands, whole-well trypsinization is equally efficient and requires less time.

11. ES cells can be grown and split at wide-range of densities. **Figure 1.1E** shows ES cells at a good density and good size.

Bigger colonies will start to differentiate. If cells are plated too densely, trypsinize them the next day and replate at lower density. ES cells grow better at higher density, but established lines can be split at high ratios (1:100 or higher) to allow easier sub-cloning.

12. It is important for the sub-cloning that the ES cell colonies (post-selection) are not too densely grown. Therefore, it is recommended to either split the electroporated cells onto multiple plates or plate different dilutions (e.g. 1:2, 1:10, 1:100, 1:200).

13. 96 wells contain low cell numbers, therefore an alternative is to first expand the targeted cells by transferring the *entire* 96-well to 24-wells containing fresh selection medium using a tip at every other position on a 12-channel pipettor. Once confluent a small fraction of the 24-well can be frozen in 96-well plates as described in **Section 3.9** or individually as described in **Section 3.5**. The remaining cell can now be used for DNA isolation.

14. Isolating the embryos as zygotes has no negative effect on the efficiency of development after diploid blastocyst injection. However, if culture conditions for embryos are not established or pre-implantation media are not available, then it is also possible to isolate the blastocysts at day 3.5 by flushing the uteri with M2 medium.

15. Manipulation of a single group of embryos should take less than 5 min. When working with large number of embryos, do fusion in batches, rinse electrodes with distilled H_2O in between, since mannitol may precipitate.

16. The embryo transfer is a very complicated and complex procedure and its description goes beyond this chapter. It is necessary to consult more comprehensive literature and get specific training to perform the surgery.

Acknowledgements

We would like to thank Caroline Beard, Chris Lengner and Konrad Hochedlinger for critical reading of the manuscript. R.J. is supported by grants from the NIH.

References

1. Evans, M.J. and M.H. Kaufman, *Establishment in culture of pluripotential cells from mouse embryos.* Nature, 1981. **292**(5819): p. 154–6.

2. Martin, G.R., *Isolation of a pluripotent cell line from early mouse embryos cultured in medium conditioned by teratocarcinoma stem cells.* Proc Natl Acad Sci USA, 1981. **78**(12): p. 7634–8.

3. Bradley, A., et al., *Formation of germ-line chimaeras from embryo-derived teratocarcinoma cell lines.* Nature, 1984. **309**(5965): p. 255–6.
4. Solter, D., *From teratocarcinomas to embryonic stem cells and beyond: a history of embryonic stem cell research.* Nat Rev Genet, 2006. **7**(4): p. 319–27.
5. Nichols, J., et al., *Formation of pluripotent stem cells in the mammalian embryo depends on the POU transcription factor Oct4.* Cell, 1998. **95**(3): p. 379–91.
6. Mitsui, K., et al., *The homeoprotein Nanog is required for maintenance of pluripotency in mouse epiblast and ES cells.* Cell, 2003. **113**(5): p. 631–42.
7. Chambers, I., et al., *Functional expression cloning of Nanog, a pluripotency sustaining factor in embryonic stem cells.* Cell, 2003. **113**(5): p. 643–55.
8. Smith, A.G., et al., *Inhibition of pluripotential embryonic stem cell differentiation by purified polypeptides.* Nature, 1988. **336**(6200): p. 688–90.
9. Boiani, M. and H.R. Scholer, *Regulatory networks in embryo-derived pluripotent stem cells.* Nat Rev Mol Cell Biol, 2005. **6**(11): p. 872–84.
10. Strelchenko, N., et al., *Morula-derived human embryonic stem cells.* Reprod Biomed Online, 2004. **9**(6): p. 623–9.
11. Tesar, P.J., *Derivation of germ-line-competent embryonic stem cell lines from preblastocyst mouse embryos.* Proc Natl Acad Sci USA, 2005. **102**(23): p. 8239–44.
12. Chung, Y., et al., *Embryonic and extraembryonic stem cell lines derived from single mouse blastomeres.* Nature, 2006. **439**(7073): p. 216–9.
13. Zvetkova, I., et al., *Global hypomethylation of the genome in XX embryonic stem cells.* Nat Genet, 2005. **37**(11): p. 1274–9.
14. Brook, F.A. and R.L. Gardner, *The origin and efficient derivation of embryonic stem cells in the mouse.* Proc Natl Acad Sci U S A, 1997. **94**(11): p. 5709–12.
15. Buehr, M. and A. Smith, *Genesis of embryonic stem cells.* Philos Trans R Soc Lond B Biol Sci, 2003. **358**(1436): p. 1397–402; discussion 1402.
16. Thomas, K.R. and M.R. Capecchi, *Site-directed mutagenesis by gene targeting in mouse embryo-derived stem cells.* Cell, 1987. **51**(3): p. 503–12.
17. Eggan, K., et al., *Hybrid vigor, fetal overgrowth, and viability of mice derived by nuclear cloning and tetraploid embryo complementation.* Proc Natl Acad Sci U S A, 2001. **98**(11): p. 6209–14.
18. Eggan, K., et al., *Male and female mice derived from the same embryonic stem cell clone by tetraploid embryo complementation.* Nat Biotechnol, 2002. **20**(5): p. 455–9.
19. Okano, M., et al., *DNA methyltransferases Dnmt3a and Dnmt3b are essential for de novo methylation and mammalian development.* Cell, 1999. **99**(3): p. 247–57.

Chapter 2

Patterning Mouse and Human Embryonic Stem Cells Using Micro-contact Printing

Raheem Peerani, Celine Bauwens, Eugenia Kumacheva, and Peter W. Zandstra

Abstract

Local micro-environmental cues consisting of soluble cytokines, extra-cellular matrix (ECM), and cell–cell contacts are determining factors in stem cell fate. These extrinsic cues form a 'niche' that governs a stem cell's decision to either self-renew or differentiate into one or more cell types. Recently, it has been shown that micro-patterning stem cells in two- and three-dimensions can provide direct control over several parameters of the local micro-environment, including colony size, distance between colonies, ECM substrate, and homotypic or heterotypic cell–cell contact. The protocol described here uses micro-contact printing to pattern ECM onto tissue culture substrates. Cells are seeded onto the patterned substrates in serum-free media and are confined to the patterned features. After patterning, stem cell phenotype is analyzed using quantitative immunocytochemistry and immunohistochemistry.

Key words: Micro-contact printing, soft lithography, embryonic stem cell, flow cytometry, quantitative immunohistochemistry, ECM–cell interactions, high throughput screening.

1. Introduction

The controlled differentiation of embryonic stem cells (ESCs) for the purposes of regenerative medicine will require in depth knowledge of the cues required to differentiate ESCs into desired cell types as well as the development of technologies that deliver these cues in a robust, reproducible, and efficient fashion. Micro-patterning embryonic and adult stem cells has shown to provide inductive cues that can modulate cell fate (1, 2). Furthermore, it can be used as a platform for the high-throughput screening of various extra-cellular matrix (ECM) and soluble growth factors on stem cell differentiation (1, 3). The approaches used to pattern mammalian cells are numerous and include:

1. Spatially restricting cell attachment by seeding cells on patterned ECM substrates created through micro-contact printing (μCP) or micro-arraying robotic equipment *(4–6)*.
2. Seeding cells into micro-wells formed through capillary force lithography (CFL) or other methods *(7, 8)*.
3. Depositing cells onto a substrate through a micro-fluidic device *(9)*.
4. Positioning cells using electromagnetic, dielectrophoretic, optical forces using micro-electrodes and optical tweezers *(10)*.

A technique common to many micro-patterning approaches is 'soft lithography' (reviewed in *(11)*). Soft lithography refers to the casting of a elastomeric pre-polymer, such as poly(dimethylsiloxane) (PDMS) onto a master silicon wafer that has been patterned with a photoresist. These wafers are patterned with micro-scale features that can be either wells for the direct deposition of cells, posts for micro-contact printing (μCP) and capillary force lithography (CFL), or channels for a micro-fluidic device. PDMS replicas of these wafers are biocompatible and can be used in cell culture.

The protocol provided here is a simple method to pattern mouse and human ESC using micro-contact printing of ECM onto common tissue culture-treated slides or dishes. The process is inherently scalable allowing for features sizes ranging from 50 to 1,500 μm to be printed on dishes varying from 9 to 100 mm in diameter. Once fabricated in a cleanroom facility, template moulds can be repeatedly used to cast PDMS stamps. Furthermore, the PDMS stamps themselves can be routinely cleaned and reused several times. As described below, the protocols to micro-pattern mouse (mESC) and human ESCs (hESC) differ in some regards including cell culture media, ECM, and seeding density. After seeding, micro-patterned cultures of ESCs can be analyzed by immunocytochemistry through flow cytometry as well as quantitative immunohistochemistry using a high-throughput automated fluorescence microscope to detect single cell level protein expression. Quantitative immunohistochemistry can also be used to verify and measure the number of colonies, the area and diameter of individual colonies, and the number of cells per colony.

2. Materials

2.1. Cell Culture

1. For mESC patterning, serum-free seeding media (mESC-SFM) included Dulbecco's modified Eagle's medium (DMEM) (Gibco-BRL, Bethesda, MD) supplemented with 15% KNOCKOUT™ Serum Replacement (cat. #: 10828,

Gaithersburg MD) 50 μg penicillin–streptomycin (Gibco-BRL), 2 mM l-glutamine (Gibco-BRL), 0.1 mM 2-mercaptoethanol (Sigma, St. Louis, MO), and 500 pM leukemia inhibitory factor (LIF, Chemicon, Temecula, CA).

2. For hESC patterning, serum-free seeding media (hESC-SFM) included X-Vivo 10™ completely defined media (XV, cat. #: 04-380Q, Biowhittaker, Walkersville, MD) supplemented with 40 ng/mL basic fibroblast growth factor (bFGF, cat. #: 100-18B, Peprotech, Rocky Hill, NJ) and 0.1 ng/mL transforming growth factor-beta (TGF-β, cat. #: 240-B-010, R&D Systems, Minneapolis, MN), 4 mM Glutamax (cat. #: 35050, Gibco-BRL, Grand Island, NY), and 0.1 mM 2-mercaptoethanol.

3. Solution of 0.25% trypsin and 1 mM ethylenediamine tetraacetic acid (EDTA) (Trypsin–EDTA, cat. #: 25200-072, Gibco-BRL, Grand Island, NY).

4. Trypsin-inactivating media which is mESC-SFM with 15% fetal bovine serum (FBS, cat. #: SH30070.03, Hyclone, Logan, Utah).

5. 40 μm Nylon cell strainers (BD Falcon™ CA21008-949, BD Biosciences, Bedford, MA).

2.2. Soft Lithography

1. 3" <100> silicon wafers polished 1-side (Silicon Sense Inc., Nashua, NH).

2. Wafer process boats (3" boat, Model #: OX-13, ASQ Technology, Inc., San Clemente, CA).

3. SU-8 25 photoresist (Microchem Corp., Newton, MA).

4. SU-8 developer (Microchem Corp.).

5. 10 cm Petri dishes (cat. #: 633180, Greiner Bio-One, Germany).

6. Acetone, ACS grade.

7. Methanol, ACS grade.

8. Hexanes, 98.5% ACS grade.

9. Isopropyl alcohol, ACS grade.

10. Nitrogen gas (N_2, cat. #: W726, BOC gases, Mississauga, ON).

11. Poly(dimethyl siloxane) elastomer kit (Sylgard® 184 Silicone Elastomer Kit, Dow Corning, Midland, MI).

12. Computer aided design (CAD) software (Macromedia® FreeHand® MX, Adobe, San Jose, CA).

13. Ultrasonic cleaner (Model #: B-300, Branson, Danbury, CT) (*see* **Note 1**).

14. Spin coater (Model #: WS-400B-6NPP-LITE, Laurell Technologies, North Wales, PA).

15. Programmable contact hot plate (Model #: PMC DATA-PLATE 730-0024, Barnstead, Dubuque, Iowa).
16. Magnetic stir plate (Model #: S7725, Barnstead).
17. Mask Aligner (Model #: Suss MA6, SUSS MicroTec Inc., Waterbury Center, VT).

2.3. Preparation of ECM Solutions

1. Sterile double distilled water (ddH$_2$O, filtered using Millipore Q-Gard®1, cat. #: QGARD00R1, Millipore).
2. Dibasic potassium phosphate.
3. Sodium chloride.
4. Monobasic sodium phosphate.
5. Bottle top filters (0.22 μm, cat. #: SCGPT02RE, Millipore, Billerica, MA).
6. For **hESC** patterning, growth factor reduced Matrigel™ (GFR-Matrigelm, cat. #: 354230, BD Biosciences, Oakville, ON). Matrigel can be separated into aliquots of 1–2 mL before use and stored at −20°C.
7. For **mESC** patterning, 0.1% fibronectin from bovine plasma (cat. #: F1141, Sigma) to be stored at 4°C until used.
8. For **mESC** patterning, gelatin from bovine skin, Type B powder (cat. #: G9391, Sigma).

2.4. Micro-contact Printing

1. Prepare 5% (w/v) Pluronic F-127 (cat. #: P2443-iKG, Sigma) in autoclaved ddH$_2$O and filter sterilize using a bottle top filter.
2. Tissue-culture treated 3"×1" microscope slides (cat. #: 160004, Nalge Nunc International, Rochester, NY).
3. Tissue culture treated 60 mm dishes (cat. #: 353002, BD Falcon, Franklin Lakes, NJ).
4. 70% Ethanol – made by diluted 95% ethyl alcohol (Commerical Alcohols Inc., Brampton, ON).
5. Silicone gaskets: (JTR8R-A2.5 or JTR20-A-2.5, Grace Bio-Labs, Bend, OR) (*see* **Note 2**).
6. Nitrogen gas (N$_2$).
7. Plexiglass Humidity Chamber (VP473, V&P Scientific Inc., San Diego, CA).
8. Humidifier (Sunbeam Model 697, V&P Scientific Inc.).
9. Ultrasonic Cleaning Solution (Fisherbrand® 1533580, Fisher Scientific, Pittsburgh, PA).
10. Ultrasonic Cleaner (B-300, Branson).
11. Sterile (autoclaved) ddH$_2$O.

2.5. Immunocytochemistry

1. Intraprep 1 Fixing Reagent (cat. #: 2388, Immunotech, Marseilles, France).
2. Intraprep 2 Permeabilization Reagent (cat. #: 2389, Immunotech, Marseilles, France).
3. Blocking buffer: 5% bovine serum albumin, fraction V (cat. #: in phosphate buffered saline (PBS, cat. #: 14190, Gibco-BRL, Grand Island, NY)).
4. Wash buffer: Hanks buffered saline solution (HBSS, cat. #: 14175, Gibco-BRL) supplemented with 2% FBS (2%HF).
5. Primary mouse and human Oct3/-4 antibody (cat. #: 611203, BD Biosciences, Oakville, ON).
6. Goat anti-mouse IgG FITC secondary antibody (cat. #: F-2772, Sigma).

2.6. Immunohistochemistry

1. Prepare 3.7% (v/v) formaldehyde (cat. #: FX0410-5-, EMD Chemicals) in PBS.
2. 100% methanol, ACS grade (cat. #: MX0475, EMD Chemicals).
3. Blocking Buffer: 10% FBS in PBS.
4. Wash Buffer: 0.5% FBS in PBS.
5. Primary mouse and human Oct3/-4 antibody (cat. #: 611203, BD Biosciences, Oakville, ON).
6. Goat anti-mouse IgG Alexafluor 488 secondary antibody (cat. #: A11029, Invitrogen-Molecular Probes, Eugene, ON).
7. Hoechst 33342 (cat. #: H3570, Invitrogen-Molecular Probes).

3. Methods

The general procedure for micro-patterning stem cells involves two major steps as outlined in **Fig. 2.1**. First, a master template mould is fabricated in a cleanroom facility by patterning a 3" silicon wafer with a photoresist using a photomask with high resolution (>5,600 dpi) as a mask. Once the mould is fabricated, it is used to cure a PDMS pre-polymer thereby forming a stamp that is negative to the master. Second, the PDMS stamp is inked with an ECM solution, dried with nitrogen, and printed onto a tissue culture substrate. The unpatterned regions are passivated with Pluronic F-127 to prevent non-specific cell attachment. Cells are then seeded onto the patterned surface and the cells that do not bind specifically to the ECM are washed away.

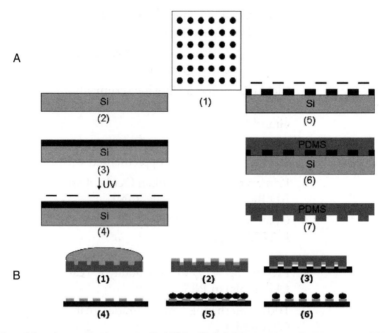

Fig. 2.1. Overview of the micro-patterning protocol of ESCs. (**A**) Soft lithographic method to produce PDMS elastomeric stamps. (1) Photomask is designed using CAD software. (2) Polished silicon wafer (Si) is cleaned with acetone and methanol. (3) SU-8 negative photoresist is spin-coated. (4) Wafer is exposed to near-UV light to crosslink-exposed regions. (5) Non-exposed regions are dissolved in SU-8 developer. (6) Poly(dimethylsiloxane) (PDMS) pre-polymer is cured onto the template mould. (7) PDMS stamp is released from the template mould. (**B**) Micro-contact printing of ECM and cell seeding. (1) PDMS stamp is inked with an ECM solution. (2) Excess solution is aspirated and the stamp is dried with N₂ gas. (3) Stamp is placed in conformal contact with the tissue-culture substrate. (5) Cells are seeded onto the surface. (6) Non-attached cells are washed away.

3.1. Fabrication of the Template Mould

1. Design a photomask using computer aided design software such as AutoCAD or Macromedia Freehand (*see* **Note 3**).

2. Print the photomask onto a high quality transparency using a laser printer at a resolution higher than 5,600 dpi. At this resolution, features sizes as small as 8–10 μm can be made reliably.

3. In a cleanroom facility, clean silicon wafers by placing them into a wafer holder and sonicating them in an ultrasonic bath for 10 min in acetone followed by 10 min in methanol. Dry softly by N₂ gas.

4. Apply SU-8 25 photoresist and spin-coat at 1,000 rpm for 60s for a target thickness of 40 μm (*see* **Note 4**).

5. Pre-bake the wafer on a contact hot plate at 65°C for 5 min followed by a soft-bake at 95°C for 15 min. Allow for the wafers to cool room temperature slowly (*see* **Note 5**).

6. Apply the photomask onto the wafer and ensure conformal contact by placing a UV transparent glass plate on top of the photomask.

7. Expose the wafer to near UV light (350–400 nm and 300 mJ/cm^2) for 30 s. Exposure time is a function of resist thickness and UV power and needs to be optimized individually.

8. Post-bake the wafer at 65°C for 5 min followed by 95°C for 10 min.

9. Place wafer in a wafer holder and into large beaker. Fill beaker with SU-8 developer to immerse the entire wafer and agitate by magnetically stirring for 15 min.

10. Optional: Hard-bake the wafer for 1 min at 50–200°C on a contact hot plate. Duration depends on thickness (*see* **Note 6**).

11. Rinse wafer with IPA and dry gently with N$_2$ gas.

12. Place wafer in a 10 cm Petri dish and store template mould in a dry environment.

3.2. Fabrication of PDMS Stamp

1. Mix PDMS pre-polymer and curing agent in the Sylgard® 184 Silicone Elastomer kit, ratio 10:1 (w/w), in a plastic cup. Mix using a glass rod for 5 min. Approximately 15–20 g of PDMS pre-polymer is used per 3" wafer.

2. De-gas mixture by repeatedly applying and removing a vacuum until all bubbles are removed.

3. Pour mixture into the Petri dish containing the wafer.

4. Cure the PDMS replica at 70°C for 4 h to overnight.

5. To release the mould, cut the edge of the Petri dish using a pair of pliers carefully. If PDMS has cured around the silicon wafer to encapsulate it, cut the residual PDMS away with box-cutters.

6. Peel the PDMS stamp off the wafer slowly and cut the stamp into the desired shape.

7. Rinse the PDMS stamp with methanol to remove hydrophobic residues.

8. After removing the PDMS stamp, rinse the template mould with hexane to remove residual PDMS.

3.3. Preparation of ECM Solution for hESC Patterning

1. To prepare PBS (pH 5.0), combine 100 mL sterile ddH$_2$O with 79.5 mg KH$_2$PO$_4$, 900 mg NaCl, and 14.4 mg Na$_2$HPO$_4$.

2. Test solution with pH meter. Adjust to pH 5.0 by addition of Na$_2$HPO$_4$ (base) and KH$_2$PO$_4$ (acid).

3. Filter sterilize PBS using a bottle top filter.

4. On ice, add 29 mL of PBS solution (pH 5.0) to 1 mL GFR-Matrigel. Diluted GFR-Matrigel can be stored at 4°C for 1–2 weeks. Alternatively, aliquot into 1–2 mL increments and store at −20°C for longer storage.

3.4. Preparation of ECM Solution for mESC Patterning

1. Prepare 50 μg/mL gelatin solution in ddH$_2$O and autoclave.
2. Combine 1 mL of 0.1% fibronectin to 39 mL of gelatin solution. Final concentration of the ECM solution will be 25 μg/mL of fibronectin and 50 μg/mL of gelatin. Aliquot in 6–7 mL increments and store at −20°C until use.

3.5. Micro-contact Printing of ECM

1. Sterilize PDMS stamps by submerging in 70% ethanol solution for 1 h to overnight.
2. If using tissue culture slides, sterilize silicone gaskets by UV for 10 min on each side.
3. Dry stamps with N$_2$ gas (30–40 psi) in a biosafety cabinet. Dry each stamp individually. Use sterile tweezers when handling stamps.
4. Place stamps in sterile dish, stamp-side up.
5. Dispense 0.5–0.75 mL of ECM solution onto the center of each stamp and gently pipette up and down to spread.
6. Incubate inked stamps for 1 h. Matrigel-coated stamps should be kept at 4.0°C while fibronectin-gelatin inked stamps can be left at room temperature.
7. Turn on humidifier in humidity chamber. Ideal relative humidity should be 55–70%.
8. Carefully aspirate ECM solution off the stamp.
9. Rinse stamp thoroughly with autoclaved ddH$_2$O.
10. Dry stamp with N$_2$ (*see* **Note 7**).
11. Place stamp gently onto slide or culture dish and ensure conformal contact by pressing lightly.
12. Place patterned slides in humidity chamber for at least 10 min.
13. If using tissue culture slides, attach adhesive silicone gaskets around the patterned region. Press firmly to ensure no leakage from the gasket.
14. Passivate slides with 5% w/v Pluronic F-127 in ddH$_2$O for 1 h at 4°C.
15. After printing, PDMS stamps can be cleaned and reused by sonicating the stamps in Fisherbrand ultrasonic cleaning solution for 10 min, followed by sonication in 95% ethanol for 10 min, and then rinsed with ddH$_2$O. PDMS stamps need to be sterilized using 70% ethanol before being used again (*see* **Note 8**).

3.6. Seeding mESCs onto Patterned Substrates

1. Before patterning, mESCs can be maintained using standard mESC culture protocols consisting of sub-culturing mESCs every 2–3 days on gelatinized flasks using mESC culture media that contains FBS and LIF.

2. Rinse cells with PBS to remove traces of serum.
3. Trypsinize cells in 0.25% Trypsin–EDTA for 3 min at 37°C.
4. Add serum-based media to quench trypsin and pipette repeatedly to mechanically dissociate mESCs into single cells.
5. To ensure a single cell suspension, filter cells through a 40 μm filter (optional).
6. Centrifuge at 1,000 rpm for 5 min and re-suspend in mESC-SFM at a concentration of 200–500 cells/μL (*see* **Note 9**).
7. Seed cells onto the patterned slide or culture dish for 4 h.
8. Wash slide 3× with media to remove non-attached cells.

3.7. Seeding hESCs onto Patterned Substrates

1. Before patterning, hESCs can be maintained using standard hESC culture protocols using mouse embryonic feeder (MEF) layers, MEF-conditioned media, or serum-free systems.
2. Rinse cells with PBS to remove traces of serum.
3. Add TrypleE Express to cells for 10 min at 37°C.
4. Add PBS to the cells to dilute TrypleE Express and pipette repeatedly to mechanically dissociate hESCs into single cells.
5. To ensure a single cell suspension, filter cells through a 40 μm filter (optional).
6. Centrifuge at 1,000 rpm for 5 min and re-suspend in hESC-SFM at a concentration of 1,000–1,500 cells/μL.
7. Seed cells onto the patterned slide or culture dish overnight.
8. Wash slide 3× with media to remove non-attached cells. Bright field images of patterned hESCs are shown in **Fig. 2.2**.

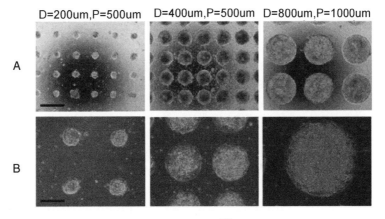

Fig. 2.2. H9 hESC colonies cultured on patterned GFR-Matrigel™ at different diameters. (**A**) Images taken at 4× magnification. Scale bar represents 625 μm. (**B**) Images taken at 10× magnification. Scale bar represents 250 μm. D: diameter, P: pitch (distance between colonies).

3.8. Immunocytochemistry of Patterned ESCs

1. Dissociate cultures into single cells using protocols described above (see **Sections 3.6** and **3.7**)
2. Transfer $0.5–1.0 \times 10^6$ cells into 1.5 mL Eppendorf Tube. Centrifuge at 3,000 rpm for 3 min.
3. Re-suspend 1×10^6 cells in 100 µL of Intraprep Reagent 1 with 50 µl of 2% HF for 15 min at room temperature.
4. Wash 2× with 1 mL of 2% HF.
5. Dilute primary mouse Oct-4 antibody (1:100) in 100 µL of Intraprep Reagent 2 with 5 µL of 5% BSA. Incubate with cells for 30 min at room temperature.
6. Wash 2× with 1 mL of 2% HF.
7. Dilute secondary goat anti-mouse IgG-FITC (1:100) in 100 µL of 2% HF with 5 µL of 5% BSA. Incubate with cells for 20 min at room temperature.
8. Wash 2× with 1 mL of 2% HF.
9. Re-suspend cells in 500 µL 2% into round-bottom FACS tubes for flow cytometry.

3.9. Immunohistochemistry of Patterned ESCs

1. Wash cells once with PBS.
2. Fix cells in 3.7% formaldehyde in PBS for 15 min at 37°C.
3. Wash 3× in 0.5% FBS/PBS.
4. Permeabilize cells in 100% methanol for 2 min at room temperature.
5. Wash 3× in 0.5% FBS/PBS.
6. Block overnight with 10% FBS/PBS at 4°C.
7. Dilute primary mouse Oct-4 antibody (1:100) in 10% FBS/PBS. Incubate with cells overnight.
8. Wash 3× in 0.5% FBS/PBS.
9. Dilute secondary goat anti-mouse IgG Alexafluor 488 (1:200) and Hoechst 33342 (1:200) in 10%FBS/PBS. Incubate with cells for 1 h.
10. Wash 3× in 0.5% FBS/PBS.
11. Cells can now be imaged in a fluorescent microscope. Excitation at 488 nm induces fluorescence of Alexafluor488 (green emission), while excitation at 365 nm induces Hoechst 33342 (blue emission).
12. If a high throughput automated fluorescence microscope such as the Cellomics™ Arrayscan V^{TI} platform is available, single-cell fluorescence measurements can be made. Using the Target Activation and Morphology Explorer algorithms, individual cells and colonies can be identified and the colony diameter, area, and number of cells per colony can be measured (see **Fig. 2.3** and Color Plate 2).

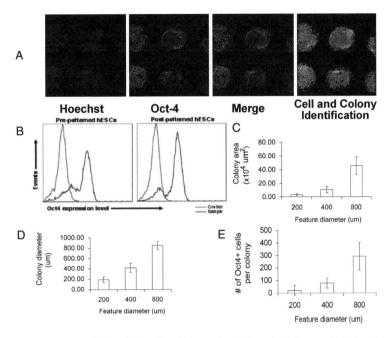

Fig. 2.3. Analysis of patterned H9 hESC colonies (D = 400 μm, P = 500 μm). (**A**) Immunohistochemistry to identifying Oct4+ cells (*green*)), all nuclei by Hoechst 33342 (*blue*). Individual Oct4+ cells (*light green*) and Oct4- cells (*dark blue*) are identified by the Target Activation algorithm of the Cellomics™ ArrayscanV^TI software by drawing overlay masks around the nuclei. (**B**) Representative flow cytometry plots of H9 hESCs before and after patterning by immunocytochemistry. Data demonstrating how the Morphology Explorer algorithm of the Cellomics™ Arrayscan V^TI software can be used to identify cells and colonies, (**C**) calculate their area, (**D**) diameter, and (**E**) the number of Oct4+ cells in the colony. Error bars represent the standard deviation over 150 colonies. (*See* Color Plate 2)

4. Notes

1. Typical cleanroom facilities are equipped for soft lithography. This equipment includes an ultrasonic bath, spin coater, programmable convection oven, programmable hot plate, and a mask aligner.

2. Depending on the desired surface area, different silicone gaskets can be purchased from Grace Bio-Labs. JTR8R-A-2.5 gaskets fit 8-wells per microscope slide with a diameter of 9 mm and a height of 1 mm. JTR20-A-2.5 gaskets fit 2-wells per microscope slide with a diameter of 20 mm and a height of 2.5 mm. For very large surfaces, 60 mm tissue culture-treated dishes can be purchased. If analyses of the cultures are done by immunohistochemistry, then minimizing surface area is desired to reduce costs of antibodies and other reagents. If immunocytochemistry is being used, then more cells are necessary and thus require a larger surface.

3. Design of the photomask must incorporate knowledge of the photoresist that will be used. If using a negative photoresist, e.g. SU-8, protected areas of the mask (dark) will not be

crosslinked by UV light and thus will be removed by the developer giving the topography of the template mould to be the negative of the mask. When PDMS is cured on top of the template mould, the PDMS replica will be a positive of the original photomask.

4. Thickness of the resist layer depends on photoresist viscosity, rotor speed, and spin coat time. All the parameters given in the soft lithography section were optimized empirically or taken from product manuals of the photoresist used. Calibration and standardization are required for consistent and optimal results. Referring to the manufacturer's product information for the photoresist is also recommended. Target thickness should be approximately equal to the feature size to maximize the moment of inertia of the stamp, i.e. if the feature is 25 μm in diameter, the thickness of the resist should be 25 μm as well. To obtain thicker layers between 50 and 120 μm, SU-8 50 (Microchem Corp.) can be used.

5. All bake times (pre-, soft-, post-, and hard-bake) are dependent on the thickness of the resist layer and need to be empirically optimized for a given thickness.

6. Strength of the resist layer and adhesion to the Si wafer is dependent on bake and exposure times. Allowing the wafers to cool slowly will prevent cracking and strengthen the resist. Increasing post- and hard-baking times after developing will also improve adhesion and strength.

7. After drying the stamp, it is recommended to print onto the substrate as soon as possible. Leaving the stamp dry for over 1 min results in poor transfer of the ECM to the substrate and protein denaturation.

8. PDMS stamps can also be cleaned by etching the surface with a plasma-cleaner (Model #: PDC-32G, Harrick Plasma, Ithaca, NY).

9. Volume to seed the substrate depends on the size of the well. For 9 mm gaskets with 1 mm height, 100–150 μL is required. Similarly for 20 mm gaskets with 2.5 mm height, 750 μL is required and for 6 mm dishes, 4 mL is required.

Acknowledgments

R.P. is funded by an Ontario Graduate Scholarship and C.B. by an Ontario Graduate Scholarship in Science and Technology. This work was supported by CIHR stem cell new emerging technologies grant. PWZ is the Canada Research Chair in Stem Cell Bioengineering.

References

1. Flaim, C.J., S. Chien, and S.N. Bhatia, (2005) *An extracellular matrix microarray for probing cellular differentiation.* Nat Methods, 2(2): p. 119–25.
2. McBeath, R., et al., (2004) *Cell shape, cytoskeletal tension, and RhoA regulate stem cell lineage commitment.* Dev Cell, 6(4): p. 483–95.
3. Soen, Y., et al., (2006) *Exploring the regulation of human neural precursor cell differentiation using arrays of signaling microenvironments.* Mol Syst Biol, 2: p. 37.
4. Tan, J.L., et al., (2004) *Simple approach to micropattern cells on common culture substrates by tuning substrate wettability.* Tissue Eng, 10(5–6): p. 865–72.
5. Chen, C.S., et al., (1997) *Geometric control of cell life and death.* Science, 276(5317): p. 1425–8.
6. Anderson, D.G., S. Levenberg, and R. Langer, (2004) *Nanoliter-scale synthesis of arrayed biomaterials and application to human embryonic stem cells.* Nat Biotechnol, 22(7): p. 863–6.
7. Rettig, J.R. and A. Folch, *Large-scale single-cell trapping and imaging using microwell arrays.* Anal Chem, 2005. 77(17): p. 5628–34.(2004) *Molded polyethylene glycol microstructures for capturing cells within microfluidic channels.* Lab Chip, 4(5): p. 425–30.
9. Folch, A., et al., (1999) *Molding of deep polydimethylsiloxane microstructures for microfluidics and biological applications.* J Biomech Eng, 121(1): p. 28–34.
10. Huang, Y., et al., (2001) *Electric manipulation of bioparticles and macromolecules on microfabricated electrodes.* Anal Chem, 73(7): p. 1549–59.
11. Kane, R.S., et al., (1999) *Patterning proteins and cells using soft lithography.* Biomaterials, 20(23–24): p. 2363–76.

Chapter 3

Efficient Gene Knockdowns in Human Embryonic Stem Cells Using Lentiviral-Based RNAi

Asmin Tulpule and George Q. Daley

Abstract

Human embryonic stem cells (hESCs) represent a powerful platform to study human development and its dysfunction in human disease. However, certain biological properties have hampered the application of standard gain of function and loss of function tools to these cells. For example, while traditional gene knockouts by homologous recombination (HR) have been reported, the low cloning efficiency of hESCs has made HR a lengthy and laborious undertaking. An alternative method of achieving loss of function is the use of small interfering RNAs (siRNAs) that can be introduced either as pre-synthesized duplexed oligonucleotides or via lentiviral vector. The use of a lentiviral vector to deliver siRNAs has proven to be a rapid and specific way to achieve highly efficient and persistent gene knockdowns in hESCs. In this chapter, we will summarize the key requirements for the successful application of lentiviral RNAi in hESCs.

Key words: RNAi, RNA-interference, shRNA, siRNA, human embryonic stem cell, hESC, lentivirus.

1. Introduction

Human embryonic stem cells (hESCs) present unique opportunities to study the aspects of human development that are unapproachable in any other model systems. An important approach for such studies is the ability to affect loss (and gain) of function for key genes to understand their role in the maintenance of the pluripotent state and in cell lineage specification during differentiation. Gene knockout by homologous recombination is widely practiced in mouse ESCs and has been described in hESCs *(1)*. However, hESCs are relatively resistant to electroporation as a mean of introducing exogenous gene sequences, and their low cloning efficiency makes selection for genetically modified clones, a challenging and time-intensive proposition. Thus, while HR remains the gold standard for achieving genetic loss of function, more rapid techniques with increased ease of use have emerged.

Lentiviral-based RNAi in hESCs provides several key advantages over other loss-of-function techniques. First, the application to hESCs can be achieved much more rapidly than HR. Second, lentiviral RNAi allows for stable, specific, high-efficiency knockdown, and through the use of a selectable marker can generate a pure population of knockdown cells. Finally, conditional lentiviral siRNA systems *(2)* allow for the interrogation of genes that adversely affect the pluripotent state of hESCs – something that could not be achieved via HR because of the likely impossibility of generating a clone where the knocked out gene is necessary for hESC self-renewal. In our experience, lentiviral RNAi has proven to be an effective and rapid way to interrogate loss of gene function in hESCs *(3)*.

2. Materials

2.1. Oligos

We ordered 100 nmol of DNA oligos with PAGE purification and 5′-phosphorylation from Integrated DNA Technologies (www.idtdna.com).

2.2. Annealing Buffer

The annealing buffer to form the siRNA is 100 mM of potassium acetate (Sigma P1190-100G), 30 mM of HEPES (pH 7.4; Sigma H4034-25G) using potassium hydroxide (Sigma P5958-250G), and 2.0 mM of magnesium acetate (Sigma M2545-250G).

2.3. F12 hESC Media

F12 hESC media are prepared by combining 200 mL of DMEM/F12 media (Invitrogen 11330-032), 50 mL of knockout serum replacement (KOSR; Invitrogen 10828-028), 2.5 mL of l-glutamine (Invitrogen 25030-081), 3.55 µL of β-mercaptoethanol (Sigma 7522), 2.5 mL of non-essential amino acids (Invitrogen 11140-050), and 500 µL of basic FGF (from a 10 ng/µL stock; Invitrogen 13256-029).

3. Methods

3.1. Oligo Design and Virus Production

1. Currently, there are many available siRNA expression systems; of which, we have used the Lentilox *(4)* and Lentihair *(5)* systems with good success in hESCs. Both these systems utilize a RNA polymerase III-based U6 promoter to transcribe a short hairpin RNA (shRNA) and contain an independently

transcribed selectable marker (GFP and puromycin, respectively). The shRNA is a transcript consisting of both the sense and the antisense siRNA as well as an intervening non-annealing loop sequence. The transcript self-anneals into a stem loop structure that is then processed into its active double-stranded RNA form.

2. The design of the oligonucleotides to be expressed from the U6 promoter is:

   Sense oligonucleotide: 5′-T(GN$_{18}$)(TTCAAGAGA)(N*$_{18}$C)TTTTTT-3′

   Antisense oligonucleotide: Complement of the overall sense oligonucleotide

   GN$_{18}$ represents the siRNA that we will select against the target gene. The loop sequence shown earlier (5′-TTCAAGAGA-3′) is taken from Brummelkamp et al., 2002 *(6)*. N*$_{18}$C is the complement of the GN$_{18}$ sequence The 3′-poly-thymidine sequence (at least four nucleotides) is the terminator signal.

3. We utilize the siRNA selection Program at the Whitehead Institute for Biomedical Research (jura.wi.mit.edu/siRNAext/) to identify transcript sequences appropriate for siRNA targeting. While entering into transcript sequence information into the selection program, it is important to include both the coding region and the 3′-untranslated region, as we have found siRNA's targeting the 3′ untranslated region to be especially efficacious. As detailed earlier, the sequence of the final core siRNA duplex should be AAGN$_{18}$TT (by selecting that option from the menu, one can restrict the search to siRNAs that fit this pattern). We generally select at least three siRNA sequences for each target, if possible.

4. The sense and antisense oligonucleotides are ordered from Integrated DNA Technologies (idtdna.com) with 5′ phosphorylation and PAGE purification. Appropriate modifications to the sense and antisense oligonucleotides must be made to facilitate the cloning of the duplex into the lentiviral vector of choice.

5. To form the duplex, the sense and antisense oligonucleotides must be annealed by adding each oligonucleotide to a final concentration of 6 μM (60 pmol/μl) in annealing buffer. Heat mixed oligonucleotides to 95°C for 10 min, tap to mix, and allow to cool slowly to room temperature (~30 min).

6. The oligonucleotide duplex should then be ligated into the lentiviral vector of choice and used to transform bacteria. Positive colonies should be screened by restriction digest and reamplified to obtain sufficient starting material for virus production. It is highly recommended that maxi prep DNA be sequenced prior to its use in virus production, as mutations can be especially problematic in the context of siRNAs.

7. We will not revisit the production and harvesting of lentiviral particles here as the specific details of such protocols vary among laboratories, and many protocols suffice. In our laboratory, we transfect lentiviral DNA with the cloned shRNA into 293T cells along with DNA encoding the other viral proteins and the appropriate coat protein (such as glycoprotein of vesicular stomatitis virus or VSV-g) on separate plasmids. The use of separate plasmids for transfection creates non-replication competent viral particles that do not contain the DNA that encodes the viral envelope and accessory proteins. However, because of the slight chance of recombination during transfection, the lack of replication competence must always be verified experimentally.

8. Three days after transfection, the supernatant from the 293T cells is then harvested and concentrated via ultracentrifugation. Aliquots of virus are used directly for viral titering and biosafety tests to rule out generation of replication competent virus (see Tiscornia et al., 2006 for details, ref. *(7)*), while the rest is stored in small aliquots at −80°C.

3.2. Infection Protocol

3.2.1. Preparing Mouse Embryonic Feeder Conditioned Media

1. Two days prior to infection, plate 3 wells (of a 6 well plate) of CF1 mouse embryonic feeders (MEFs; Chemicon PMEF-CF) per lentiviral siRNA at a density of 1 million cells/100 cm^2. Each well is coated with 0.1% gelatin for 30 min at 37°C. Aspirate gelatin and plate feeders at appropriate density. Allow incubation overnight.

2. The next day, check to make sure feeders have appropriately attached. Wash well 1× with PBS and add 3 mL of F12 hESC media per well.

3. On infection day, harvest the MEF-conditioned media (MEF-CM) from the MEFs and add bFGF to a concentration of 20 ng/mL. Add 3 mL of F12 media per well for the following day.

3.2.2. Preparing Matrigel Plates

1. Two days prior to infection, thaw a 10 mL Matrigel (BD Bioscience 354234) bottle on ice at 4°C overnight.

2. Rapidly aliquot the Matrigel into cryovials (we usually use 0.5 mL aliquots). This step must be performed rapidly as Matrigel will solidify at room temperature. Freeze all aliquots at −20°C.

3. The night before infection, place sufficient aliquots of Matrigel at 4°C on ice overnight for use tomorrow. Matrigel is provided at different concentrations in the 10 mL bottle. In order to standardize, 0.33 mg is used to coat one well. The infections will require one Matrigel well per gene being targeted, plus one or two control wells (*see* **Note 1**).

4. On the day of the infection, determine the number of wells needed and add 1 mL ice-cold DMEM/F12 media (no KOSR or serum) for each calculated well to a 50 mL conical tube.

5. Rapidly add 0.33 mg Matrigel for each calculated well to the 50 mL conical tube.

6. Mix and aliquot 1 mL of the diluted Matrigel to each well. Make sure the surface is completely covered and kept at room temperature for 2 h.

7. Wash wells gently 1× with PBS and add 2 mL MEF-CM + 20 ng/mL bFGF to each well.

3.2.3. Preparing the Cells for Infection

Prepare one confluent well of hESCs on MEFs per lentiviral siRNA for testing (*see* **Note 2**).

1. Wash well 1× with PBS.

2. Incubate the hESCs with 1 mL of 200 U/mL Collagenase IV (Invitrogen 17104-019) for 10 min. The edges of the colonies should curl up slightly after treatment.

3. Remove the collagenase IV and add 2 mL F12 media to each well.

4. Cut up the hESC colonies by running a glass pipette vertically and horizontally across the surface of the well. Pieces of the colonies should come off the plate into the supernatant.

5. Transfer the supernatant to a 15 mL conical tube and add 2 mL F12 media to the well.

6. Wash the well and remove residual cell clumps, then transfer to the same 15 mL conical tube.

7. Spin cells down for 3 min at 1,000 rpm in a hanging bucket centrifuge.

8. Aspirate the supernatant and resuspend the pellet in 3 mL MEF-CM + bFGF. Pipette up and down at least ten times such that the cell clumps are mostly between 50 and 100 cells when viewed under the microscope. See **Note 3** for a discussion of clump size.

9. Distribute the hESCs evenly between the three Matrigel coated wells, shake to distribute, and place in incubator overnight.

3.2.4. Infection Protocol

1. After 8–12 hr, check that the hESCs have plated onto the Matrigel. There will be some clumps that will not attach, which are to be expected.

2. Prepare MEF-CM + bFGF as described previously and feed each well with 2 mL.

3. Add protamine sulfate (Sigma P3369) at a final concentration of 6 μg/mL to each well.

4. All steps from this step onwards, the material should be treated as "Biosafety Level 2 plus" (BL-2+). Appropriate precautions should be employed (e.g., double gloves, bleaching of waste, etc.).

5. Add knockdown virus at approximately MOI of 100. Since plating efficiencies can vary significantly from passage to passage, we calculate an approximate MOI using the fact that one confluent well contains about 900,000 cells. From a 1:3 split and an average plating efficiency of 66%, there are roughly 200,000 cells per Matrigel well. Therefore, to achieve an MOI of 100, infect each well with 2×10^7 infectious units of virus. Always remember to include a well infected with a control virus and an uninfected well.

6. The following day, remove the supernatant from each infected well and place into a 50 mL conical tube for treatment with bleach.

7. Wash the infected wells 2× with PBS and transfer washes to 50 mL conical tubes.

8. Feed each well with 3 mL of MEF-CM + bFGF.

9. Add bleach to 50 mL of conical tubes to neutralize the virus. Allow to keep for 20 min, then transfer to autoclave bags for autoclaving/hazardous waste disposal.

10. Infected hESCs now no longer need to be treated as BL-2+, provided the viruses have tested negative for replication competence.

3.3. Verification of Gene Knockdown/ Generating Pure Populations

1. Infected hESCs should then be expanded by serial passage on Matrigel or by transfer to MEFs. Provided that the lentiviral vector has a GFP marker, infection efficiency can be determined by subjecting a small aliquot to flow cytometry analysis. In general, we have observed efficiencies between 15 and 30%. hESC colonies tend to be infected in a patchwork fashion, as shown in **Fig. 3.1A** and Color Plate 3.

2. To test the effectiveness of a gene knockdown, we recommend expanding the infected hESCs to sufficient numbers to allow for fluorescent-activated cell sorting (FACS) of infected cells and immediate lysis to analyze protein levels via western blotting and RNA levels by quantitative real-time PCR. In general, we have seen knockdown levels in the range of 70–100% in a well-infected GFP+ population (*see* **Note 4**).

3. To generate a pure or enriched knockdown population, we recommend FACS of infected cells into 96 well dishes (if clonal populations are desired) or into a 6 well plate, followed by repeated expansion and FACS to isolate a stable knockdown population. Due to the poor replating efficiency of hESCs

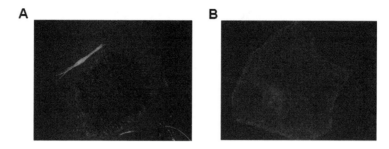

Fig. 3.1. Infection of hESCs with GFP marked lentiviral RNAi vectors. (**A**) hESC colony infected with a GFP marked lentivirus displays a patchwork of infected cells. (**B**) hESC colony after enrichment of GFP+ cells by FACS. (*See* Color Plate 3)

following FACS, we have found that at least 500 cells/96 well and 50,000 cells/6 well are required to generate colonies after sorting (*see* **Note 5**). Even after one or two rounds of serial sorting, hESC cultures that are greater than 90% GFP+ may appear to have significant variability of GFP levels within individual colonies (**Fig. 3.1B**). This must be considered in analyzing the effect of a given knockdown.

4. Notes

1. There is always concern about off-target effects in any RNAi experiments. We recommend generating a few control viruses: (a) a non-specific RNAi directed against a gene not present in hESCs, such as Luciferase and (b) siRNAs where the target sequence has one or two mismatches. hESCs infected with these control viruses, and not uninfected hESCs, should serve as an appropriate positive control in any experiment.

2. At times, we have generated a "mixed-bag" lentiviral siRNA preparation, where three to four siRNAs against the same gene are transfected into 293Ts together and the viral preparation contains virions expressing all three siRNAs. We recommend this approach for initial testing of the efficacy of siRNA knockdown.

3. Clump size during passaging has a significant effect on both replating efficiency and infection efficiency. Clumps with less than 50 cells frequently will not replate or the colonies will immediately differentiate upon replating. However, larger clumps (>200 cells) are often poorly infected due to physical limitations of infecting the center of the clump. Therefore, there is a crucial balance that needs to be struck to optimize both replating and infection efficiencies.

4. We have generally had success in targeting a wide variety of genes with siRNA. However, some siRNAs have proven to be

less efficient than others and consequently required increased MOIs to achieve satisfactory knockdown. It should be noted that higher MOIs do carry the risk of increased off-target effects as the concentration of dsRNA rises.

5. The isolation of pure populations of hESCs which are knocked down for a specific gene could be problematic, if the gene of interest is necessary for hESC self-renewal or survival. In these situations, a conditional RNAi approach is suggested.

References

1. Zwaka TP and JA Thomson. (2003). Homologous recombination in human embryonic stem cells. Nat Biotechnol. 21, 319–21.
2. Szulc J, M Wiznerowicz, MO Sauvain, D Trono and P Aebischer. (2006) A versatile tool for conditional gene expression and knockdown. Nat Methods. 3(2), 109–16.
3. Zaehres H, MW Lensch, L Daheron, SA Stewart, J Itskovitz-Eldor and GQ Daley. (2005). High-efficiency RNA interference in human embryonic stem cells. Stem Cells. 23, 299–305.
4. Rubinson DA, CP Dillon, AV Kwiatkowski, C Sievers, L Yang, J Kopinja, DL Rooney, MM Ihrig, MT McManus, FB Gertler, ML Scott and L Van Parijs. (2003). A lentivirus-based system to functionally silence genes in primary mammalian cells, stem cells and transgenic mice by RNA interference. Nat Genet. 33, 401–6.
5. Stewart SA, DM Dykxhoorn, D Palliser, H Mizuno, EY Yu, DS An, DM Sabatini, IS Chen, WC Hahn, PA Sharp, RA Weinberg and CD Novina. (2003). Lentivirus-delivered stable gene silencing by RNAi in primary cells. RNA. 9, 493–501.
6. Brummelkamp TR, R Bernards and R Agami. (2002). A system for stable expression of short interfering RNAs in mammalian cells. Science. 296, 550–3.
7. Tiscornia G, O Singer and IM Verma. (2006). Production and purification of lentiviral vectors. Nature Protocols. 1– 241–5.

Chapter 4

Measurement of Cell-Penetrating Peptide-Mediated Transduction of Adult Hematopoietic Stem Cells

Aziza P. Manceur and Julie Audet

Abstract

The ability of cell-penetrating peptides (CPPs) to cross cell membranes and transport cargo into cells makes them an attractive tool for the molecular engineering of stem cells. Even though the exact mechanism of transduction remains unclear, their potential has been demonstrated for diverse applications, including hematopoietic stem cell expansion and the generation of islets cells from embryonic stem cells. Several parameters can affect the intracellular delivery of CPP-based constructs. Those include the type of cells targeted, the type of CPP used, and the properties of the cargo. For this reason, it is important to have a means to quantitatively assess the transduction efficiency of specific constructs in the cell type of interest in order to select the best vector for a specific application. In this chapter, we describe a method to measure the uptake of HIV transactivator of transcription (TAT) and the homeobox protein Antennapedia (Antp) constructs in primary hematopoietic progenitor cells and hematopoietic cell lines. This method is useful to compare, select, and optimize different strategies to deliver CPP-based constructs into a given cell type.

Key words: Hematopoietic stem cell, cell-penetrating peptides, protein transduction domains, TAT, antennapedia, intracellular delivery, biological transport, flow cytometry, confocal microscopy.

1. Introduction

Several different strategies have been developed to introduce a variety of molecules into cells. However, except for lentiviral transfection, most techniques are unsuited for stem cell research as they cause a high level of cell death (1, 2). Cell-penetrating peptides (CPP), also known as protein transduction domains (PTDs), have the ability of crossing cell-membranes (3). The exact mechanism of action is still unclear, but their potential as vectors has been demonstrated numerous times. Peptides, proteins, oligonucleotides, and even nanoparticles have been transduced efficiently into different cell types (3–5). From past studies, it is now clear that the uptake depends on the CPP and the cell type used (6). Other important factors include the concentration of the CPP and the incubation time. Since there are many parameters to optimize, to maximize the cellular uptake, flow cytometry with fluorescently-labeled constructs is an attractive approach because of its high-throughput and multiplex capabilities. However, to obtain meaningful uptake measurements using this approach, it is necessary to minimize non-specific fluid phase uptake and eliminate surface-bound, non-internalized constructs (7). This chapter describes a protocol to quantitatively assess the uptake of transactivator of transcription (TAT)- or homeobox protein Antennapedia (Antp)-based constructs into Sca-1$^+$Lin$^-$ mouse bone marrow cells and human hematopoietic cell lines using flow cytometry. In addition, this chapter describes a method to visualize the intracellular localization of TAT and Antp constructs by confocal microscopy and a procedure to assess the toxicity of different CPPs using the AlamarBlue^{TM} cell proliferation assay.

2. Materials

2.1. Cell Isolation and Culture

1. Iscove's modified Dulbecco's medium (IMDM) (Gibco) supplemented with 10% fetal bovine serum (FBS) (Gibco).
2. Scissors, tweezers, ethanol, alcohol swabs, 23 gauge needle, and 5 cc syringe.
3. Potassium-buffered saline (PBS) (Gibco) supplemented with 2% FBS.
4. NH$_4$Cl (StemCell Technologies).
5. Fc receptor blocker: 1 μg rat serum (StemCell Technologies).
6. Rat phycoerythrin (PE)-conjugated anti-Sca-1 antibody (Invitrogen).
7. Biotin-labeled antibody cocktail of lineage-specific [CD5 (Ly-*1*), CD11b (Mac-1), CD45R/B220, Ly-6G/C (Gr-1), Neutrophils (*7-4*), TER119] (StemCell Technologies) (8).
8. Streptavidin-fluorescein isothiocyanate (SA-FITC; Invitrogen).
9. 7-Amino actinomycin D (7-ADD) buffer or propidium iodide (PI) buffer (Invitrogen) (*see* **Note 1**).

2.2. Sample Preparation

1. Stock solutions (250 μM) of TAT (YGRKKRRQRRR)- or Antp (RQIKIWFQNRRMKWKK)-conjugated constructs labeled with FITC, as well as a peptide derived from myelin basic protein (APRTPGGRR) labeled with FITC as a negative control (Anaspec) (*see* **notes 2** and **3**).

2. Potassium-buffered saline (PBS) (Gibco).
3. PBS containing 10% fetal bovine serum (FBS) (Gibco).
4. PBS containing 2% FBS.
5. Trypsin solution (Sigma) diluted to 1 mg/mL in PBS.
6. 1.5 mL low-retention microcentrifuge tube (Eppendorff) and pipette tips to minimize adsorption of the peptides on the surface of the tube (Fisher Scientific).

2.3. Flow Cytometry

1. BD falcon tubes (BD Biosciences).
2. Sheath fluid (BD FACSFlow Sheath Fluid, BD Biosciences).
3. 7-Amino actinomycin D (7-ADD) buffer or propidium iodide (PI) buffer (Invitrogen). Dilute 7-AAD or PI to a concentration of 1 µg/mL using PBS + 2% FBS.
4. Trypan blue (Merck KgaA) diluted to 1 mg/mL in PBS (*see* **Note 4**).

2.4. Confocal Microscopy

1. Glass-bottom petri dishes coated with poly-l-lysine to improve cell adherence (Mat Tek corporation) (*see* **note 5**).
2. PBS containing 2% FBS.
3. Nuclear stain: Combine Hoechst 33342 (Molecular Probes) with PI buffer (Invitrogen). Dilute in PBS + 2% FBS so that you get a concentration of 1 µg/mL for both PI and Hoechst (*see* **Note 6**).

2.5. Cell Viability Assay

1. IMDM supplemented with 10% FBS (Gibco).
2. AlamarBlue™ (Biosource).
3. 96-well plate.

3. Methods

The uptake measurements are based on changes in the median cellular fluorescence of the cell sample following incubation with FITC-labeled CPPs constructs. Cells are first exposed to CPP constructs at an extracellular concentration of 5 µM for 5 min at 4°C; this incubation temperature is used in order to minimize the contribution of non-specific fluid-phase endocytosis during the first step of the loading procedure, when the cells are exposed to a high CPP concentration. This first step is followed with a washing procedure which decreases the concentration of CPPs in the fluid phase surrounding the cells. Finally, cells are incubated for 10 min at 37°C to restore membrane fluidity and cellular activity and complete the process of internalization. This incubation is performed in the presence of trypsin which removes cell surface-bound peptides/proteins before measuring the uptake into live cells with flow cytometry. The efficacy of the trypsin treatment for removing surface-bound molecules can be evaluated using trypan blue to quench the cell-surface FITC emission *(9, 10)*. Following the loading procedure, the distribution and localization of the CPP constructs in live cells can be observed with confocal microscopy using Hoechst (blue) and PI (red) for nuclear and viability staining, respectively. In addition a 96-well plate cell toxicity assay based on AlamarBlue^{TM} can be used determine suitable maximum CPP-construct concentrations for a given cell type; AlamarBlue^{TM} is reduced by metabolically active cells only, which leads to a change in color (from an oxidized indigo state to a reduced fluorescent pink form).

3.1. Cell Isolation and Culture

1. Maintain the K562 cell line (ATCC CCL-243) in IMDM supplemented with 10% FBS (*see* **Note 7**).

2. Before the isolation of mouse bone marrow Sca-1 Lin⁻ cells, euthanize the animals by cervical dislocation or CO_2 asphyxiation (*see* **Note 8**).

3. Prepare a 50 mL tube filled with ethanol to rinse scissors and tweezers.

4. Lay the mice on their back and thoroughly spray the mice with ethanol.

5. Make a small incision in the fur in the middle of the stomach and pull on both sides to expose the legs.

6. Cut the bone just below the knee and hold the extremity attached to the femur with tweezers.

7. Using scissors, remove as much flesh as possible around the femur. Cut the femur as high as possible, just below the hip.

8. Clean up the femur to remove all the flesh. Use ethanol swabs if necessary.

9. Keep the femurs on ice until you are ready to flush the bone marrow (*see* **Note 9**).

10. Prepare 2 mL PBS + 2% FBS in a 5 mL tube, a 23 gauge needle and a 5 cc syringe.

11. Cut the ends of the femurs. Hold the bone with the tweezers above the 5 mL tube and flush the bone marrow into the tube using the syringe. You might have to re-do it several times to get all the bone marrow out. Gently aspirate up and down to get a single-cell suspension.

12. Spin down the bone marrow and resuspend in 5 mL clean PBS + 2% FBS. Add 1–2 mL NH_4Cl to lyse the red blood cells, vortex, and incubate on ice for 5–10 min.

13. Wash two to three times to remove the NH_4Cl.

14. You can freeze the cells in IMDM + 10% FBS + 10% DMSO. However, a lot of cells will die during thawing. Alternatively, you can sort them right away using a fluorescence-activated cell sorter (FACS) (go to step 20).
15. Thaw cells quickly at 37°C (until there is a bit of ice left).
16. Gently transfer cells into a 50 mL tube.
17. Slowly add 15 mL of PBS without Ca^{2+} or Mg^{2+} dropwise while gently swirling tube.
18. Fill tube to 50 mL and gently invert tube to mix and spin down.
19. Discard supernatant, flick tube gently and resuspend cells in 5 mL HBS + 2% FBS.
20. Cell staining for sorting: Resuspend at 5×10^5 cells/100 μL; 5×10^6–10×10^6 cells/mL is suitable for staining for flow cytometry.
21. Add 100 μL of cells to the following Eppendorffs (for a total of 5×10^5 cells per Eppendorff) and bring volume to 200 μL:
    a. Unstained
    b. 7AAD compensation control (or PI) (not always necessary)
    c. PE compensation control (Sca-PE)
    d. FITC compensation control (lin-FITC)
    e. Sca-PE isotype control (not always necessary)
    f. Samples (Sca-PE + 7AAD) – about 5 mL at 5×10^6 to 10×10^6 cells/mL per tube
22. Incubate cells with Fc Block for 10 min on ice (1:100 = 1 μg/10^6 cells i.e. 2 μL to PE and FITC compensation control tubes and 50 μL to sample tube). Do not wash the cells.
23. Add antibodies and isotype controls one after the other:
    a. Sca-PE first (incubate for 20 min on ice) dilute 1:200 (1 μL to the PE compensation control tube and 25 μL to the sample tube). Incubate for 20 min on ice
    b. Wash
    c. Lin$^+$-biotin (16 μL/50×10^6 cells). Incubate for 20 min on ice.
    d. Wash
    e. FITC-Streptavidin 1:100 (2 μL to the FITC compensation control tube and 50 μL to the sample tube). Incubate for 20 min on ice.
    f. Wash (twice) (see **Note 10**).
24. Resuspend unstained, PE compensation control, and Sca-PE isotype control tubes in 400 μl PBS + 2% FCS.
25. Resuspend the 7AAD compensation control in 400 μl 7AAD buffer (1 μg/mL).

26. Resuspend samples in 2–5 mL of 7AAD buffer (so that you get 10 millions cells/mL).
27. Transfer to polystyrene FACS tubes (*see* **Note 11**).
28. Sort for Sca-1$^+$Lin$^-$ cells (*see* **Note 12**).

3.2. Sample Preparation

1. Count cells.
2. Transfer 50,000–100,000 cells per 1.5 mL microcentrifuge tube (Eppendorff). Spin down and resuspend in 200 μL PBS + 10% FBS (*see* **Note 13**).
3. Incubate cells on ice for 10 min.
4. Add 4 μL of the CPP stock solution for a final concentration of 5 μM. Vortex immediately and incubate on ice for 5 min.
5. Spin down and resuspend in 200 μL PBS + 10% FBS – incubate for 15 min on ice.
6. Spin down and resuspend in 200 μL PBS to remove the FBS.
7. Spin down and resuspend in 200 μL pre-warmed trypsin solution (1 mg/mL) for 10 min at 37°C (*see* **Note 14**).
8. Add 200 μL PBS + 10% FBS to neutralize the trypsin and spin down.
9. Resuspend in 250–500 μL PI or 7AAD buffer and analyze cells by flow cytometry (*see* **Section 3.3**).

3.3. Flow Cytometry

Keep the cells on ice during analysis.
1. Prepare three control tubes for compensation (unstained, PI only, and FITC only) (*see* **Note 15**).
2. Read the tubes. **Figure 4.1** shows an example of the histograms obtained with Antp-treated K562 cells (**Fig. 4.1A**) and TAT-treated Sca-1$^+$Lin$^-$ cells (**Fig. 4.1B**).

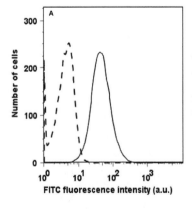 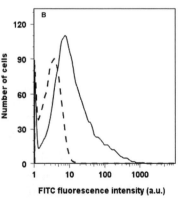

Fig. 4.1. Fluorescence intensity of cells exposed to 5 μM FITC-labeled CPPs for 5 min on ice. (**A**) Antp-treated K562 cells. (**B**) TAT-treated Sca-1$^+$Lin$^-$ mouse bone marrow cells. The dotted lines correspond to the untreated control sample.

3. For comparison purposes, normalize your values by dividing the FITC median of your sample, by the FITC median of the unstained cells. A summary of the uptake of TAT and Antp compared to the negative control MBP for both cell types is presented in **Fig. 4.2**. For both CPPs and both cell types, the uptake was significantly greater than the uptake of MBP (ANOVA, $\alpha=0.05$). **Figure 4.3** shows the results obtained with K562 cells loaded with Antp conjugated to a 15-amino acid FITC-labeled peptide. The uptake of the construct was significant compared to MBP but lower than that of FITC-Antp.

4. To confirm the completeness of the surface-bound peptide/protein removal, add a 1:1 volume of trypan blue (1 mg/mL) to the left-over CPP-treated cell sample directly in the BD falcon FACS tube.

5. Incubate on ice for 5 min.

6. Read again using the flow cytometer. The FITC signal should remain the same before and after quenching with trypan blue.

3.4. Confocal Microscopy

1. Prepare the sample as in steps 1–8 of **Section 3.2**. After neutralizing the trypsin, resuspend the cells in nuclear stain buffer (PI and Hoechst together) (*see* **Note 16**). Leave the cells on ice for 15 min.

2. Spin the cells down and resuspend in PBS + 2% FBS.

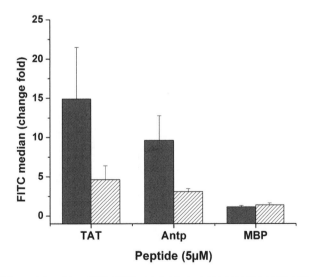

Fig. 4.2. Comparison of the uptake of two CPPs (TAT and Antp), and one negative control (MBP) into K562 cells and Sca1$^+$Lin$^-$ bone marrow cells. The results are expressed as fold changes in the median fluorescence of the samples relative to a control sample not exposed to fluorescent peptides (changes in median ± SD). K562 (n=3): ▬ Sca1$^+$Lin$^-$ cells (n=3): ▨.

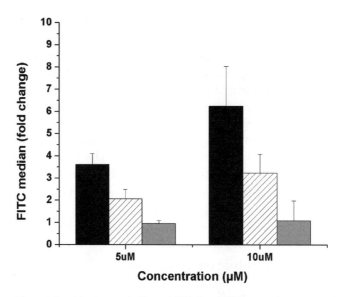

Fig. 4.3. Comparison of the uptake of two concentrations of FITC-Antp ■, a small peptide conjugated to Antp ▨, and MBP ▨ into K562 cells.

3. Deposit a small volume of cells (100–200 μL) in the middle of the poly-l-lysine coated petri dish. Allow the cells to settle down.
4. Add 1 mL PBS around the cells to minimize evaporation.
5. Image (see **Fig. 4.4**) using the appropriate filter sets and lasers (see **Note 17**).

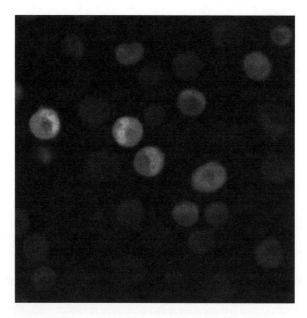

Fig. 4.4. Confocal microscopy pictures of live K562 cells exposed to 5 μM FITC-TAT for 5 min.

3.5. Toxicity Assay

1. First, prepare a calibration curve which depicts AlamarBlue™ fluorescence as a function of cell number. The optimal cell density to be used for the assay is the concentration that corresponds to the steepest region of the curve, before saturation of the signal occurs.

2. In the case of K562 cells, we found the optimal cell density to be around 150,000 cells (**Fig. 4.5**). Therefore, seed

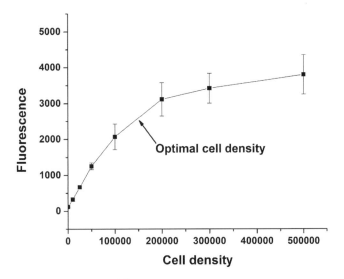

Fig. 4.5. Calibration curve showing AlamarBlue™ fluorescence as a function of cell number (K562 cells).

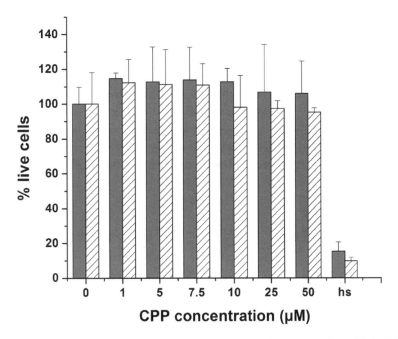

Fig. 4.6. Cell viability of K562 cells after 5 min incubation on ice with increasing concentrations of Antp. Cells were heat-shocked (hs) as a negative control. TAT: ▨ Antp: ▨

150,000 K562 cells/well in 200 μL PBS + 10% FBS in a 96-well plate.
3. Perform steps 3–8 of **Section 3.2**, using the appropriate CPP concentration.
4. Resuspend the cells in 180 μL IMDM + 10% FBS.
5. Add 20 μL AlamarBlue™ to each well (final concentration = 10%) (*see* **Note 18**).
6. Incubate the cells at 37°C with 5% CO_2 for 3 h.
7. Read the plate with a spectrophotometer (excitation $\lambda = 530$–560 nm and emission $\lambda = 590$ nm). Express the results by dividing the value obtained for your sample by the value of the non-treated cells, and multiply by 100. **Figure 4.6** shows the cell viability of K562 cells treated with increasing concentrations of TAT and Antp, compared with cells subjected to a heat shock (*see* **Note 18**). No sign of toxicity were observed up to 50 μM (ANOVA = 0.05).

4. Notes

1. The fluorophores are light sensitive. Keep the antibody and the samples in the dark.
2. It was reported that fluorescent dyes alter the intracellular distribution of CPPs *(12)*. Therefore, make sure that you use the same fluorophore when comparing different CPPs or CPP constructs.
3. The FITC fluororophore is light and pH sensitive. Keep the samples in the dark and at the same pH to avoid changes in fluorescence intensity.
4. We selected the trypan blue distributed by Merck (Merck KgaA) as it was reported to have a better quenching efficiency than other brands of trypan blue *(13)*.
5. It is preferable to perform live-cell imaging as cell fixation may promote the redistribution of soluble proteins in the cell *(7)*.
6. Ensure that UV excitation is available on the confocal microscope when using Hoechst staining.
7. When using cell lines, use cells while they are in the exponential growth phase.
8. Two femurs from 1 mouse = 10×10^6 to 15×10^6 cells.
9. You can also flush the bone marrow from the tibias to increase the yield.

10. You can add Sca-PE and Lin⁺-biotin at the same time, but not Lin⁺-biotin and FITC-streptavidin.
11. Make sure that cells did not clump. You can use polystyrene tubes with cell-strainer cap (BD).
12. Two to five percent of the cells are Sca-1⁺Lin⁻.
13. The cell density and the volume used to assess the transduction efficiency were reported to influence the results *(14)*. For comparison, always use the same cell density in the same volume of solution during the experiments.
14. The trypsin solution was previously warmed up in a 37°C water bath. Use a shorter incubation time than 10 min for sensitive cells.
15. PI and FITC overlap and require compensation. For the FITC compensation control tube, you can use unwashed TAT-treated cells (cells that have been incubated with TAT for 5 min, spun down, and resuspended in PBS + 2% FBS).
16. Hoechst stains the DNA of all cells and PI only stains dead cells. Keep cells on ice during analysis to reduce cell death.
17. It might be difficult to image the FITC-CPP if the uptake is too low. TF-1 cells treated with FITC-TAT present a high uptake and can be used as a positive control. TF-1 cells are human hematopoietic progenitor cells which are grown in RPMI + 10% FBS, supplemented with 5 ng/mL human granulocyte macrophage-colony-stimulating factor (GM-CSF) (R&D Systems). When using cell lines, perform experiments during the exponential growth phase.
18. As a control, add a few wells with dead cells. You can heat-shock cells by incubating the cells in a 60°C water bath for 15 min.

References

1. Green, I., Christison, R., Voyce, C. J., Bundell, K. R., and Lindsay, M. A. (2003) Protein transduction domains: are they delivering? *Trends Pharmacol Sci* **24**, 213–215.
2. Kobayashi, N., Rivas-Carrillo, J. D., Soto-Gutierrez, A., Fukazawa, T., Chen, Y., Navarro-Alvarez, N., and Tanaka, N. (2005) Gene delivery to embryonic stem cells. *Birth Defects Res C Embryo Today* **75**, 10–18.
3. Dietz, G. P., and Bahr, M. (2004) Delivery of bioactive molecules into the cell: the Trojan horse approach. *Mol Cell Neurosci* **27**, 85–131.
4. Gupta, B., Levchenko, T. S., and Torchilin, V. P. (2005) Intracellular delivery of large molecules and small particles by cell-penetrating proteins and peptides. *Adv Drug Deliv Rev* **57**, 637–651.
5. Noguchi, H., and Matsumoto, S. (2006) Protein transduction technology: a novel therapeutic perspective. *Acta Med Okayama* **60**, 1–11.
6. Fischer, R., Fotin-Mleczek, M., Hufnagel, H., and Brock, R. (2005) Break on through

7. Richard, J. P., Melikov, K., Vives, E., Ramos, C., Verbeure, B., Gait, M. J., Chernomordik, L. V., and Lebleu, B. (2003) Cell-penetrating peptides. A reevaluation of the mechanism of cellular uptake. *J Biol Chem* **278**, 585–590.
8. Thomas, T. E., Miller, C. L., and Eaves, C. J. (1999) Purification of hematopoietic stem cells for further biological study. *Methods* **17**, 202–218.
9. Henriques, S. T., Costa, J., and Castanho, M. A. (2005) Re-evaluating the role of strongly charged sequences in amphipathic cell-penetrating peptides: a fluorescence study using Pep-1. *FEBS Lett* **579**, 4498–4502.
10. Letoha, T., Gaal, S., Somlai, C., Venkei, Z., Glavinas, H., Kusz, E., Duda, E., Czajlik, A., Petak, F., and Penke, B. (2005) Investigation of penetratin peptides. Part 2. In vitro uptake of penetratin and two of its derivatives. *J Pept Sci* **11**, 805–811.
11. Miller, C. L., Audet, J., and Eaves, C. J. (2002) Ex vivo expansion of human and murine hematopoietic stem cells, in Hematopoietic Stem Cell Protocols (Klug, A.A. and Jordan, C.T., eds.), Humana Press, Totowa, New Jersey, pp.189–208.
12. Szeto, H. H., Schiller, P. W., Zhao, K., and Luo, G. (2005) Fluorescent dyes alter intracellular targeting and function of cell-penetrating tetrapeptides. *Faseb J* **19**, 118–120.
13. Van Amersfoort, E. S., and Van Strijp, J. A. (1994) Evaluation of a flow cytometric fluorescence quenching assay of phagocytosis of sensitized sheep erythrocytes by polymorphonuclear leukocytes. *Cytometry* **17**, 294–301.
14. Hallbrink, M., Oehlke, J., Papsdorf, G., and Bienert, M. (2004) Uptake of cell-penetrating peptides is dependent on peptide-to-cell ratio rather than on peptide concentration. *Biochim Biophys Acta* **1667**, 222–228.

Chapter 5

Stem Cell Sources for Regenerative Medicine

Ali M. Riazi, Sarah Y. Kwon, and William L. Stanford

Abstract

Tissue-resident stem cells or primitive progenitors play an integral role in homeostasis of most organ systems. Recent developments in methodologies to isolate and culture embryonic and somatic stem cells have many new applications poised for clinical and preclinical trials, which will enable the potential of regenerative medicine to be realized. Here, we overview the current progress in therapeutic applications of various stem cells and discuss technical and social hurdles that must be overcome for their potential to be realized.

Key words: Stem cells, cell therapy, regeneration.

1. Stem Cells and Mechanisms of Regeneration

The term, 'regenerative medicine', coined by William A. Haseltine in 2000, is now widely used to describe biomedical approaches to heal the body by stimulation of endogenous cells to repair damaged tissues, or transplantation of cells or engineered tissues to replace diseased or injured ones. A number of tissues in mammals including epithelia, bone marrow, and muscle undergo regeneration to keep tissue integrity after cell loss due to normal cell turnover or injury. Organ regeneration in adult mammals, however, has only been demonstrated in deer antler (1). In other mammalian species including human, tissue injuries are normally repaired by formation of fibrous scar tissue (2, 3). This is unlike some metazoans, such as planarians, uredale amphibians, and zebrafish that can regenerate body organs and appendages (4). Hence, unraveling the mechanisms of regeneration may eventually help in designing methods for cell therapy and tissue regeneration in humans. The underlying molecular mechanisms of regeneration are not completely elucidated; nonetheless, two general mechanisms mediated by either stem cells or dedifferentiation of lineage-specified cells play roles in regeneration (5). In the freshwater planarian *S. mediterranea*, pluripotent, stem-like cells called neoblasts distributed throughout the body divide and differentiate into the approximately 40 cell types necessary to regenerate the planarian body from small fragments (6). Similarly, activation of preexisting stem cells in mammals is responsible for the limited regeneration seen in tissues such as muscle and blood (see sections below). In other cases, such as epimorphic regeneration of limbs in newts, muscle, bone, cartilage, nerve sheath, and connective tissue cells at the site of amputation dedifferentiate and form a pool of proliferating progenitor cells in a microstructure called regeneration blastema. A similar structure is believed to be formed in zebrafish at the site of fin or heart amputation (7). The identity and origin of cells within the blastema are still controversial and a recent study has indicated that some of the proliferating cells within the blastema of regenerating salamander limbs are derived from a population of skeletal muscle satellite (stem) cells rather than cell dedifferentiation (8). Cells within blastema redifferentiate into multiple cell types to regenerate the limb. Interestingly, the extracts of the newt regenerating limb (9) or the ectopic expression of the transcription factor Msx1 (10) render dedifferentiation of multinucleated mouse myotubes into mononucleated myoblasts in vitro suggesting that some of the pathways involved in regeneration are conserved throughout the animal kingdom. The identification of genes involved in blastema formation and dedifferentiation is being actively pursued (reviewed in (11)). Upregulation of signaling molecules, such as platelet-derived growth factors (PDGF-a and PDGF-b) (12), fibroblast growth factors (FGFs) (13), bone morphogenic protein-4 (BMP4) (14), Notch1 (14), and matrix metalloproteinases (15) have been implicated in heart, fin, and limb regeneration in zebrafish and newts. The serendipitous finding that a particular strain of mice (MRL) confers the ability to regenerate, resembling that of amphibians, raises the possibility that humans may also have untapped regeneration capacity (16). However, the evolution of immune system and ability to regenerate may be inversely related (17). The inhibition of specific proinflammatory cytokines or decreased expression of genes involved in inflammation may be the basis for regeneration ability in MRL mice (18).

While regeneration biologists strive to understand the developmental biology of regeneration to harness our inherent regenerative ability, stem cell biologists study the molecular pathways involved in self-renewal and lineage-specific differentiation of mouse and human embryonic and somatic stem cells to generate specific cell types required for clinical applications. This has initiated controversy over the approach or source of cells to be used in cell-based therapies. Here we describe some of the characteristics of embryonic and somatic stem cells, their applications in various diseases, and the relevant technical and ethical issues.

2. Embryonic Stem Cells in Regenerative Medicine

Embryonic stem cells (ESCs) are pluripotent cells derived from the preimplantation embryo. Mouse ESCs can generate all cell types in the body including the germline following reintroduction into the preimplantation embryo by blastocyst injection (19) or morula aggregation (20). ESCs have great promise in regenerative medicine as they provide a source of cells with the potential to differentiate into various cell types to repair injured tissues. The therapeutic capacity of ESCs was further extended by demonstration that ESC injection into blastocysts of mice harboring mutations in the inhibitor of differentiation (Id) genes, resulted in correction of cardiac abnormalities and rescue of embryonic lethality through a non-cell autonomous pathway (21). This suggested that ESCs can also exert their function by cell and gene therapy to induce regeneration.

The ability to differentiate ESCs into derivatives of all three primary germ layers in vitro is a powerful approach to study the molecular pathways involved in pluripotency and cell lineage specification during early embryogenesis (22). Understanding the molecular basis of ESC self-renewal is also vital for their expansion and therapeutic application. Mouse ESCs can be maintained in culture, in an undifferentiated state, indefinitely in the presence of leukemia inhibitory factor (LIF) (23). LIF activates gp130 signaling through binding to its cognate receptor, LIFRβ, which then dimerizes with gp130 and transduces the signal through the JAK-STAT pathway. While STAT3 plays an important role in self-renewal of ESCs, $Stat3^{-/-}$ embryos can undergo gastrulation suggesting the existence of a STAT3-independent pathway for ESC self-renewal (24). It was later found that BMP2, BMP4, or GDF6 (normally present in bovine serum) and activation of Id genes are also critical for clonal propagation of ESCs (25). Despite similarities to their mouse counterpart, human ESCs require bFGF and not LIF for maintaining their undifferentiated state (26). However, self-renewal of both human and mouse ESCs requires a critical level of transcription factors OCT3/4 (27),

NANOG *(28, 29)*, and SOX2 *(30)*. Protein interactions and transcriptional targets of these essential self-renewal genes have been studied and shown to co-occupy a large portion of target genes *(31–34)*.

Differentiation of both mouse and human ESCs into ectodermal, mesodermal, and endodermal derivatives have been documented and culture conditions for differentiating ESCs into cardiomyocytes *(35, 36)*, smooth muscle cells (*(37)*, **Chapter 22**), hematopoietic cells *(38, 39)*, neural tissue (*(40, 41)*, **Chapters 11** and **13**), insulin-secreting cells *(42)*, endothelial cells (*(43, 44)*, **Chapter 21**), osteoblasts *(45, 46)*, and hepatocytes *(47)* have been determined. This volume has updated protocols for guiding differentiation of ESCs into many of these cell fates.

3. Challenges for Therapeutic Application of ESCs

Clinical application of human ESCs is complicated by three formidable challenges: the ethical and social issues associated with the derivation of ESCs from spare embryos generated during in vitro fertilization (IVF), cell graft rejection due to unmatched histocompatibility antigens on ESC-derived cells, and the possibility of teratoma formation after ESC-derived cells are transplanted into patients (reviewed in *(48, 49)*). Derivation of ESCs from IVF blastocysts has been opposed by some religious authorities and policy makers. Their arguments are based on defining blastocyst-stage embryos as sentient human beings, hence their destruction has been deemed unethical (reviewed in *(50)*). A presidential edict in the United States has restricted federal research funding to only 21 ESC lines created before August 9, 2001. However, since continued culturing of these cells may lead to genetic changes and chromosomal aberrations, derivation of new cell lines is necessary *(51)*. Legislation in other countries varies from permissive (UK) to restrictive (Germany). Canada has recently given conditional permission for using fresh or frozen human embryos for stem cell research with parental consent (Canadian Stem Cell Network and *(52)*). A recent achievement to derive human ESCs without destroying embryos *(53)* may dissipate some of the ethical concerns, however, it is unlikely to answer all the criticisms since it is argued that removing cells from embryos may lower their chance of implantation *(54)*. The ethical debate surrounding ESCs is closely linked to current controversies over somatic cell nuclear transfer (SCNT) for reproductive and therapeutic cloning. SCNT for reproductive means is considered unethical by essentially all policy makers and scientific organizations, while SCNT for the isolation of human ESC lines (see section below) is illegal in most countries and debated by the scientific community.

The derivation of ESCs by SCNT is one strategy to circumvent graft rejection. Another possibility to address the immunological barrier of transplantation of ESC-derived cells is to establish an ESC bank comprising lines with an extensive variety of human leukocyte antigen (HLA) alleles *(55, 56)*. While it may be possible to generate ESC lines for about 60% of the population by carefully selecting fertilized eggs prescreened for HLA types *(57)*, it would not be feasible to use this strategy to generate lines for all haplotypes considering the great number of different HLA alleles in the population. More creative alternative strategies, including genetic engineering, will need to be developed to overcome the limitations of immunological tolerance.

The third formidable hurdle in the therapeutic application of ESC-derived cells is the possibility of teratoma formation by contaminating undifferentiated ESCs. Petersen and colleagues showed that NOD/scid diabetic mice can be rescued by transplantation of ESC-derived insulin producing cells, but therapy failed due to immature formation of teratomas *(58)*. Other studies have generated similar results. For instance, transplantation of human ESCs into immune-deficient mice led to formation of multiple teratomas *(59)*. This is likely due to the high degree of heterogeneity and proliferation of the transplanted ESCs. Further understanding of the molecular pathways that underlie differentiation of ESCs into specific lineages may allow genetic manipulation or modification of culture conditions to force differentiation of ESCs into a particular lineage, and therefore, exclude the teratoma-forming undifferentiated ESCs from the culture.

4. Somatic Stem Cells in Regenerative Medicine

Stem cells can be distinguished from progenitor cells by their ability to robustly contribute to the maintenance of a tissue for an extended period rather than a short time (in other words, stem cells possess long-term self-renewal while progenitors do not). However, the cell transplantation assay required to differentiate stem cells from progenitors has not been adequately developed for a number of tissues and thus it is difficult to distinguish between progenitors and stem cells for a number of organ systems. For ease of understanding in this chapter, we will not dwell on the evidence of whether or not there is evidence to distinguish between progenitors and stem cells in a particular cell fraction. Instead, we will use the terminology utilized by the authors of the studies that we are discussing.

The identification of stem or primitive progenitor cell activity and enrichment of these cells from a wide range of adult tissues, including the brain *(60–62)*, skin *(63–65)*, skeletal muscle *(66, 67)*, bone marrow *(68–72)*, and liver *(73, 74)* propelled regenerative medicine into the scientific and public spotlight. Unfortunately, not all tissues have demonstrated regenerative potential or somatic stem cells contributing to homeostasis. Of the tissues with reported stem or progenitor activity, the only one used extensively in cell-based therapies is bone marrow—the source of hematopoietic stem cells and mesenchymal progenitors – which has been used to treat osteogenesis imperfecta through bone marrow transplantation *(75)* and is currently being used in a number of clinical trials. The ready access to bone marrow-derived stem cells through bone marrow aspirates, mobilized peripheral blood, or umbilical cords, and the ability to transplant these cells helped propel the use of hematopoietic and mesenchymal stem and progenitor cells to the clinic (although mesenchymal progenitor cells poorly engraft using current protocols). Numerous technical hurdles must be overcome for the wide use of non-bone marrow-derived tissue-resident stem cells in regenerative therapies, including isolation, directed expansion or differentiation in culture, and transplantation. Thus, the initial reports of cells from the bone marrow capable of non-hematopoietic and non-connective tissue differentiation were greeted by skeptical optimism by the scientific community and enthusiasm by the public. While little enthusiasm still remains in the scientific community, groups opposed to ESC research point to bone marrow plasticity as a reason why ESC research is not needed.

5. Cloning and Nuclear Reprograming

Briggs and King's success in generating live *Xenopus* frogs in 1952 *(76)* by transplanting nuclei from frog blastula was the first report of animal cloning. Subsequently, experiments by John Gurdon showed that nuclei from differentiated intestinal cells of tadpoles could be reprogramed by transplantation into oocytes to create live frogs *(77, 78)*. These experiments laid the foundation for nuclear reprograming research. It was not until 1997, however, with the cloning of the sheep Dolly that reprograming of adult mammalian nuclei was achieved *(79)*. The success rate of animal cloning by nuclear transfer (NT) remains very low and most cloned animals are born with abnormal growth or die prematurely (reviewed in *(80, 81)*). The recent achievements in nuclear reprograming of differentiated cells have led to the possibility of generating autologous cells by NT using donated oocytes and deriving patient-specific, pluripotential stem cells for cell therapy, a process dubbed by some as therapeutic cloning *(82)*. Rideout et al. *(83)* reported the first successful case of treatment of a genetic defect in mice by NT-derived ESCs. Although possible in principle, SCNT to generate ESCs has not yet been successful for human cells *(84)*. One hurdle in using SCNT is the need for egg donation, an invasive and costly procedure. However, NT into mouse zygotes temporarily arrested in mitosis has been recently reported *(85)*. Such procedures may eventually alleviate the need for human donor eggs in SCNT. Other methods to reprogram somatic nuclei have been pursued such as transfer of human fibroblast-derived nuclei into rabbit oocytes *(86)*. Nuclear reprograming can also be achieved in cell hybrids made between somatic cells with mouse or human ESCs *(87, 88)*. Human fibroblast fusion with ESCs elicits reprograming events in the somatic cell DNA indicated by expression of pluripotency markers including OCT4, NANOG, REX1, and TDGF1 *(89)*. Exposure of differentiated cells to extracts of embryonic carcinoma cells or ESCs has also been reported to trigger expression of genes representing pluripotent stem cells *(90)*. However, whether the detected pluripotency marker resulted from resetting the genome or cell extracts used in experiments has been questioned by some investigators *(81)*. The recent finding that combined expression of only four transcription factors, OCT3/4, SOX2, c-MYC, and KLF4, is sufficient to convert mouse fibroblasts into ES-like cells *(91)* further raises hopes that stem cells may be produced by reprograming but without the controversial use of oocytes or preimplantation embryos.

Incomplete reversal of the epigenetic state of donor nuclei appears to be the underlying reason for low efficiency of nuclear reprograming, regardless of method used. The epigenetic state of donor nuclei influences the success rate of SCNT, as cloning efficiency using nuclei from ESCs and neural stem cells was shown to be higher than that from the fibroblasts *(92)*. The nuclei in ESCs undergo epigenetic changes in prolonged culture and as cells differentiate. The main epigenetic landmarks include DNA methylation and histone methylation and acetylation (reviewed in *(93)*). Methylation of CpG islands is associated with repression of gene expression and found in maternally and paternally imprinted genes (reviewed in *(94)*). Actively transcribed genes are typically acetylated on lysine residues by histone acetyltransferases (HATs), while histones associated with silent genes are deacetylated leading to methylation of histones *(95)*. Trimethylation of histone H3 on lysine 27, in particular, is important for ESCs to retain their pluripotent state by repressing cell lineage-specific genes *(96)*. Methylation of H3 is believed to recruit polycomb group proteins to the promoter region of downstream genes —a process that leads to gene silencing *(96, 97)*. Further understanding of the mechanisms of epigenetic reprograming is critical for the production of patient-derived ESCs for cell therapy.

6. Therapeutic Applications of Stem Cells

The greatest limitation for all cell- and organ-based therapies is the tissue source, whether it is HLA matching non-autologous tissues, isolation of autologous stem cells, or engineering tissues from ESCs. Once a suitable cell source is identified, another daunting limitation arises: the low survival of cells grafted into the recipient *(98)*. That is likely due to several factors including an inflammatory response associated with transplantation and a loss of the normal interaction between the engrafted cells and their natural microenvironment. Here we describe the current progress and challenges of cell-based therapies for the following systems: hematopoietic, nervous, musculoskeletal, and cardiovascular. In addition, advancements for liver and pancreas renewal are discussed. Although regeneration of the epidermis and vascular system are not discussed in this chapter, progress and methods are described in Parts IV and VI of this book.

6.1. Hematopoietic System

The cardiovascular system, including the blood, vasculature, and heart, is the first organ system to develop, and is required to carry oxygen to the developing embryo. Blood cells first emerge in the yolk sac around day 7.5 of mouse embryogenesis *(99)* and day 16 of human development (reviewed in *(100)*). Within blood islands, hematopoietic cells arise in the vicinity of endothelial cells. Various studies in avian and murine species have demonstrated existence of a common mesodermal progenitor, so-called hemangioblast, which can produce cells from both endothelial and hematopoietic lineages *(101–103)*. Definitive hematopoietic stem cells (HSCs) and progenitors arise independently in the region of aorta-gonads-mesonephros (AGM), which then seed the fetal liver around mid-gestation. The fetal liver remains the primary site of hematopoiesis until the bone marrow takes over this role shortly after birth (reviewed in *(104)*).

6.1.1. Somatic-Derived Hematopoietic Stem Cells

Hematopoietic cells have a high turnover rate and are constantly replaced by progenitor cells originating from hematopoietic stem cells. Thus, homeostasis of the blood system is driven by the HSC. The technical ease of bone marrow transplantation in the mouse has greatly facilitated analysis and isolation of HSCs. The assay of the HSC is the stringent competitive transplantation assay *(105)*, and serial transplantation is used as the gold standard to assay self-renewal. Using a combination of flow cytometric sorting and bone marrow transplantation, HSCs were the first somatic stem cell to be highly enriched *(69)*. In the mouse, HSCs can be isolated to near purity (1:2–1:3 cells that is long-term repopulating) using several different combinations of cell surface antigens and multi-color cell sorting. Furthermore, various restricted multilineage progenitors can also be isolated by multicolor cell sorting. In **Chapter 6**, Williams and Nilsson describe methods to isolate HSCs from the central and endosteal regions of the bone marrow and analyze HSC homing to their niche. The analysis of human HSCs has been greatly advanced by the immunocompromised mouse transplantation model *(106)* including the enrichment of HSCs from progenitors and differentiated cells. Unfortunately, because some of the markers that have been so useful in the purification of mouse HSCs are not expressed in the same manner in the human hematopoietic system, human HSCs have not yet been isolated to near purity.

Allogeneic bone marrow (HSC) transplantation has become standard practice for the treatment of blood malignancies and autoimmune disorders (reviewed in *(107)*), while autologous bone marrow transplantation is now being used in conjunction with high-dose chemotherapy to aid in the treatment of non-hematopoietic cancers. As with all allogeneic transplantation procedures, allogeneic bone marrow (HSC) transplantations are limited by the availability of HLA matched donors and complicated by graft-versus-host disease (GVHD) and infections *(108)*. The advent of umbilical cord blood (UCB) banking has the potential to provide a rich resource of HLA-typed sources of HSCs with no risk to the donor. The use of UCB in pediatric patients with blood disorders has demonstrated full reconstitution of the hematopoietic system with lower GVHD compared to bone marrow transplantation *(109–111)*. It is also rapidly becoming the preferred HSC source for treatment of child blood diseases worldwide *(112)*. Using UCB in adolescents and adults, however, is not effective since the cell dose required for transplantation (1.5×10^7/kg) often exceeds the number of nucleated cells in UCB *(113–115)*, although higher rates of nucleated cells from UCB have been recently achieved by a modified collection method *(116)*. Thus, a significant amount of work has been invested to develop methods to expand HSCs. Most of this effort has been performed by altering cytokine cocktails including more sophisticated factorial design strategies *(117, 118)*; however, the best cytokine cocktails only preserve the number of stem cells for one or two divisions rather than expand the number of HSCs. Some recent success has been achieved by genetic overexpression of HoxB4 or transduction of HOXB4 protein into HSCs using cell-penetrating peptides *(119, 120)*. **Chapter 4** by Manceur and Audet is devoted to methods for use in optimizing the delivery of proteins by cell-penetrating peptides. Another strategy demonstrating promise is a closed system bioreactor that combines growth of enriched stem cells in optimal growth factor combinations with the removal of differentiating cells that secrete factors that can induce differentiation of HSCs. This bioprocess has demonstrated greater than five-fold expansion of NOD/SCID repopulating cells (HSCs) *(121, 122)*.

While much of the focus in hematopoietic-related bioengineering and regenerative medicine has centered around the expansion of HSCs, the transplantation (or transfusion) of differentiated blood lineages (notably blood and platelet rich plasma) is much more commonplace than bone marrow transplants. In **Chapters 7** and **8**, protocols for megakaryocyte expansion and platelet production and red blood cell production, respectively, are detailed.

6.1.2. Embryonic Stem Cell-Derived Hematopoietic Lineages

Human *(123)* and mouse *(124)* ESCs can be readily directed to differentiate into hematopoietic cell lineages either using embryoid body (EB) differentiation or co-culture with OP9 stromal cells (reviewed in *(39)*). Using lineage-specific markers, the generation of erythroid and myeloid lineages has been shown. The ESC-derived blood precursors have been invaluable in understanding how hematopoietic and endothelial lineages are formed which led to the identification of blast-colony-forming cells (BL-CFCs), progenitor cells that give rise to hematopoietic and endothelial cells. BL-CFCs have identical characteristics to yolk sac hemangioblasts *(102, 125)*. While cultured ESCs give rise to proliferating hematopoietic cells, they have limited ability to reconstitute the hematopoietic system in lethally irradiated mice, suggesting that HSCs are not produced by ESCs in vitro using standard differentiation protocols. However, ectopic expression of HOXB4 *(126)*, STAT5 *(127)*, and CDX4 *(128)* in mouse ESCs leads to the generation of hematopoietic cells capable of engrafting irradiated hosts, suggesting that it might be possible to engineer definitive HSCs from ESCs. In contrast to mouse ESCs, expression of HOXB4 in human ESCs fails to promote generation of repopulating HSCs *(129)* and efficient generation of human ESC-derived HSCs still remains to be shown.

6.2. Nervous System

The loss of neurons and glia results in various neurological diseases such as Parkinson's, Huntington's, and multiple sclerosis.

Similarly, spinal cord lesions interrupt axonal pathways, leading to the loss of neurons and glia. In each of these disorders, a different spectrum of cells is damaged, requiring different neuronal cell types for cell replacement *(130)*. Neurons are highly specialized and terminally differentiated cells that cannot divide postnatally. The recent advances in methods to derive neuronal cells from stem cells have fuelled interest in the development of stem cell-based therapies for treatment of neurological conditions. Although the central nervous system (CNS) has endogenous stem cells that persist in the adult CNS, these cells are unable to confer an adequate regenerative response for unknown reasons *(131)*. Therefore, stem cell therapy has been suggested as an approach to replace the lost neuronal cells. In addition to stem cell therapy, neuronal cell transplantation therapy has been found to be achievable in humans to treat diseases. It was shown that symptoms of Parkinson's disease were reversed in patients by grafting human fetal dopaminergic neurons into the host brain *(132, 133)*. Within Part III of this book, methods are given for the isolation and manipulation of neural stem cells and skin-derived precursors (SKPs) from mice using a neurosphere or neurosphere-like culture system. Methods are also given for the transplantation of embryonic stem cell-derived neurons and neural stem cells into hosts, such as chick, mouse, and monkey.

6.2.1. Neural Stem/Progenitor Cells

Although neither purified nor studied in vivo to the extent that HSCs have been, neural stem/progenitor cells (NSCs) are a heterogeneous population of cells that can self-renew and have the potential to generate neural and glial lineages in developing and adult CNS *(61, 134)*. Human NSCs have been isolated and cultured from brain tissue of aborted fetuses *(135)* and deceased individuals *(136)* or alternatively derived from ESCs *(137, 138)*. In adult brain, NSCs are concentrated in the subventricular and subgranular zones *(61, 139)*. Upon transplantation into the brains of neonatal mice, NSCs can engraft and differentiate into cells from neuronal and astrocytic lineages *(135)*. To study the clinical potential of NSCs, animal models of neurodegenerative and spinal cord injuries *(140, 141)* are commonly used. NSCs can engraft and survive in injured brain or spinal cord microenvironments and exert functional recovery. However, the putative mechanism(s) by which NSCs can sustain their repair capabilities is controversial *(131)*. Most investigators have reported an absence of NSC differentiation into neurons in injured CNS and shown that the majority of grafted cells give rise to astrocyte and oligodendrocyte lineages *(141, 142)*. Conversely, a more recent report indicates differentiation of human NSCs isolated from fetal brain into neurons and oligodendrocytes upon transplantation into a spinal cord injury model *(143)*. In order to fully determine their repair capabilities, further experiments expanding and transplanting

NSCs must be performed. **Chapters 11–13** provide the state-of-the-art protocols for transplantation and analysis in the chick, mouse, and non-human primate nervous systems. Since isolating human neural stem cells from CNS of patients is not feasible with current technology, the potential of other stem cell types is being evaluated.

6.2.2. Stem Cells from Non-neuronal Tissues

The mammalian dermis is a source of neural crest-derived multipotent cells called skin-derived precursor cells (SKPs) *(63)*. SKPs differentiate into cells expressing markers of peripheral neurons (e.g., βIII tubulin) and Schwann cells (e.g., myelin basic protein) *(144)* (*see* **Chapter 10**). They are an attractive source of cells as they can be obtained from patients with relative ease and transplanted autologously. In wild-type or *shiverer* mice (genetically deficient mice in myelin basic protein), the ability of SKPs or SKP-derived Schwann cells to myelinate was shown *(145)*. In **Chapter 10**, Fernandes and Miller describe protocols using mouse skin to culture and neuronally differentiate SKPs as well as analyze the cultures using immunocytochemistry. As at least a fraction of mesenchymal progenitors are neural crest-derived, it is not surprising that the differentiation of bone marrow mesenchymal progenitors into astrocytes and neurons, both in vivo following transplantation *(146)* and in vitro *(147)*, has been reported. Similarly, a fraction of human non-hematopoietic UCB mononucleated cells demonstrate neuronal, astrocyte, and oligodendrocyte differentiation potential *(148, 149)*. Transplantation of bone marrow cells into the spinal cord of patients with spinal cord injury *(150)* or amyotrophic lateral sclerosis (ALS) *(151)* has been associated with positive outcome. Similarly, injection of UCB-derived cells in animal models of neurodegeneration *(152)* has resulted in functional improvement; however, the mechanism underlying functional improvement in these studies is unknown.

6.2.3. ESC-Derived Neurons and Glial Cells

Neurons can be derived from mouse and human ESCs in the absence of LIF and treatment with retinoic acid *(153)*. Alternatively, neural precursors and dopaminergic neurons have been generated from mouse ESCs by addition of FGF2 in serum-free medium *(154, 155)*. Expression of a selectable marker under the control of neuron or glial specific promoters, such as T-alpha-1 tubulin *(156)* and CNP *(157)*, has been employed to increase the yield of ESC-derived neurons, glial cells, or their progenitors. It is expected that one of the first uses of ESC-derived cells in the clinic will be neuronal cells for spinal cord repair and treatment of Parkinson's disease, because the immune response is expected to be minimal within the spinal cord and brain. In **Chapter 11**, Wichterle and co-authors describe the derivation of motor neurons from ESCs by xenotransplantation of ESCs into the developing spinal cord. Also, in **Chapter 13**, Takahashi and co-authors provide methods to culture and differentiate monkey ESCs into dopaminergic neurons and transplant them into MPTP-treated monkeys to treat symptoms of Parkinson's. Such methods are crucial in understanding the molecular cues necessary to guide differentiation of human ESCs into specific neuronal lineages.

6.2.4. Stem Cell-Based Delivery of Therapeutic Substances

The application of stem cells is not confined to differentiated cell therapy; rather, stem cells may be used as a vehicle to deliver therapeutic substances to specific sites in the brain. Several reports have shown that endogenous or transplanted NSCs are attracted to lesions within the brain, including glioblastoma *(158)* and degenerating motor neurons, in a mouse model of ALS syndrome *(159)*. The tropic property of NSCs may be exploited to deliver drugs to kill tumors or secrete survival/growth factors to halt degeneration (reviewed in *(160)*).

6.3. Musculoskeletal System

Musculoskeletal disorders such as Duchenne muscular dystrophy and bone defects have long been therapeutic targets for cell-based therapies. Finding stem cell sources that can give rise to muscle, bone, and cartilage cells are integral parts of any regenerative strategy for musculoskeletal disorders.

6.3.1. Skeletal Muscle-Derived Stem Cells

In mammalian embryos, skeletal muscle is formed from the dorsal part of the somite, called dermomyotome *(161)*. Muscle precursor cells migrate to somites under the regulation of the Pax3 transcription factor and activate the myogenic determination pathway by expressing the MRF (myogenic regulatory factor) transcription factors, Myf5 and MyoD (reviewed in *(162)*). The postnatal growth and repair of skeletal muscle is conferred mainly by a population of muscle-resident cells known as satellite cells (reviewed in *(163)*). These cells reside between the basal lamina and the sarcolema of muscle fibers and are normally quiescent, but are activated and proliferate during postnatal growth or in response to injury *(164)*. Within **Chapter 20**, Collins and Zammit describe a method to isolate satellite cells and their associated anatomical niche from single muscle fibers to analyze them in culture as well as transplant the muscle fiber into mouse muscles to observe satellite cell behavior in vivo.

The expression of MRFs in muscle satellite cells largely recapitulates the hierarchical expression of these genes during development. Satellite cells express specific markers, including hepatocyte growth factor receptor (c-met), M-cadherin, and transcription factors, such as Pax7, Myf5, and MyoD (reviewed in *(165)*). The crucial role of Pax7 in the development of satellite cells was established in *Pax7* null mice which lack satellite cells and have disrupted skeletal muscle growth *(166)*. Recent studies have demonstrated that satellite cells divide asymmetrically when activated, indicating their self-renewal ability *(167)* (as described in **Chapter 19** by Tajbakhsh and colleagues).

Although less well characterized, an additional population of skeletal muscle-resident cells that shows stem cell characteristics has been identified. These so-called skeletal muscle-derived stem cells (MDSC) that exclude Hoechst dye 33342 are often called side population (SP) cells due to their appearance in flow cytometry profiles. MDSCs are multipotent, express stem cell markers *(168–170)*, and are a heterogeneous population. The CD45⁻ subpopulation of MDSCs is capable of differentiation into myogenic lineages both in vivo and in vitro *(170)* as well as into other lineages, such as blood *(171)*. Since the SP compartment of skeletal muscle can give rise to satellite cells and contribute to muscle regeneration after intramuscular and intravenous transplantation, they have been proposed to contain precursors for muscle satellite cells *(163)*. It has been reported that the transplantation of MDSCs in a mouse model of muscle degeneration (*mdx* mice) leads to a higher frequency of muscle fibers formed in comparison with satellite cells *(163)* and therefore, MDSCs may be a more favorable myogenic cell source.

6.3.2. Mesenchymal Stem/Progenitor Cells and Bone Repair

Stromal populations of various sources, including the bone marrow, that have the capacity to form myocytes, osteoblasts, chondrocytes, and adipocytes are often termed mesenchymal stem cells (MSCs) *(172)*. These populations, however, are very heterogeneous and contain differentiated cells; uni-, bi-, and tripotent progenitors; as well as primitive, self-renewing stem-like cells. Hence, two white papers have been published recently by *The International Society for Cellular Therapy* stating that the term "Mesenchymal Stem Cell" should not be used unless stem cell properties of the cell type have been demonstrated and suggest that the term "Mesenchymal Stromal Cell" (also abbreviated MSC) should be used in its place *(173, 174)*. **Chapter 18** by Gnecchi and Melo provide protocols to analyze the properties of MSCs derived from bone marrow. A number of surface markers have been reported to be expressed by MSCs, such as CD105, CD73, CD106, CD54, CD29, CD44, CD90, and STRO-1 *(70, 172, 175)*; however, these markers do not appear to be able to fractionate more primitive from differentiated cells. The hematopoietic markers, CD45 and CD34, are normally not expressed in MSCs. A recent study has identified ESC-specific antigen, SSEA-4 as a marker for human MSCs in the bone marrow *(176)*. In **Chapter 16**, Short and colleagues describe protocols to isolate MSCs expressing Stem Cell Antigen-1 (Sca-1) from the mouse compact bone, a particularly rich source of multilineage mesenchymal cells. MSCs have primarily been studied on the basis of their in vitro differentiation capacity with little known about their role in tissue homeostasis. Our laboratory generated Sca-1-deficient mice and demonstrated these animals exhibit decreased multipotent mesenchymal progenitor (i.e., MSC) self-renewal resulting in premature type II (age-related) osteoporosis *(177, 178)*. Thus, we have demonstrated that MSCs are required for the maintenance of bone mineral density during aging.

Other tissue sources for MSCs include adipose tissue (fat) *(179)*, umbilical cord *(180)*, and hair follicles *(181)*. The human umbilical cord in particular, is a non-invasive and rich source of MSCs; **Chapter 17** describes the protocols for isolating and analyzing MSCs (more specifically, human umbilical cord perivascular cells or HUCPVCs) from this source. In addition, circulating MSCs may be isolated from peripheral blood *(182)*.

MSCs are already being used clinically to repair bone defects, with numerous trials ongoing. This strategy often combines expanded autologous MSCs with biodegradable scaffolds for treatment of difficult-to-heal large bones (e.g., *see (183)*). This strategy does not require that the MSC population contain self-renewing stem cells; instead, osteoprogenitors, which have been found within the "MSC" population, are sufficient. In contrast, bone diseases such as osteogenesis imperfecta (OI) and osteoporosis could be treated by systemic transplantation of true self-renewing (i.e., true stem cell) MSCs. OI results from mutations in either of the two genes encoding type I collagen, COL1A1 or COL1A2, the major structural protein in bone. Severe forms of OI are due to dominant mutations that affect the assembly of wild-type and mutant collagen bundles, which leads to bony deformities, excessive bone fragility with fracturing, and reduced bone growth and stature in these patients. Systemic delivery of MSCs has been used to treat children with severe forms of OI. Despite only 1.5–2.0% MSC/osteoblast engraftment into five of the six patients treated in two clinical studies, the engrafted patients demonstrated increased bone mineral content, growth in body length, and reduced fractures *(75, 184, 185)*.

More recently, allogeneic HLA-mismatched male fetal MSCs were transplanted into a female fetus with multiple intrauterine fractures during the 32nd week of gestation *(186)*. At 9 months of age, the analysis demonstrated that the patient had as much as 7.4% of the bone cells derived from the donor and demonstrated no lymphocyte activation in response to the donor cells. Only three fractures were apparent in the first two years of life, while the patient grew appreciably, suggesting a successful transplantation *(186)*. While protocols have been developed to differentiate ESCs into osteoprogenitors, it is unlikely that ESCs will make an important clinical impact in bone repair because MSCs are being shown to have a clinical impact. Widespread clinical use of MSCs is hindered by their poor homing and engraftment, as noted in the transplantation of childhood OI patients and confirmed in animal models *(187)*. In addition, the development of strategies to engender more robust osteogenic differentiation would benefit bone repair strategies.

6.3.3. Regeneration of Cartilage

Articular cartilage defects and osteoarthritis (OA) are among the most abundant orthopedic impairments *(188)*. Chondrocytes, the cellular component of cartilage, can be isolated and cultured in vitro. Autologous chondrocyte transplantation has been attempted; however, the long-term efficacy of chondrocyte transplantation remains unsatisfactory *(189, 190)*. In vitro expansion of chondrocytes leads to dedifferentiation and a loss-of-function. Moreover, normal chondrocytes are not available in sufficient numbers for transplantation *(188)* and therefore, alternative cell sources are required.

Chondrogenic potential of marrow-derived MSCs has been established by demonstrating expression of chondrocyte-specific genes: type II collagen, type X collagen, aggrecan, and proteoglycan *(191, 192)*. Differentiation of MSCs into chondrocytes is dependent on cell seeding density, the extent of culture expansion *(193)*, and is also enhanced by hypoxia *(194)*. The potential of MSCs to repair cartilageous tissues in degenerative discs and regeneration of nucleus pulposus has been demonstrated in a rabbit model *(195)*. Skeletal MDSCs can also differentiate into chondrocytes. When transplanted into articular cartilage in a rabbit model, these cells improved healing with efficiency similar to chondrocytes *(196)*.

However, a clear limitation of MSCs (and possibly MDSCs) is that these cells are programed to differentiate to hypertrophic chondrocytes (rather than articular chondrocytes) and form cartilage that can mineralize. Although methods to prevent this process have been identified, such as treating the cells with parathyroid hormone-related protein (PTHrP), it is not known whether this will be effective when the cells are implanted in vivo *(197)*. Thus, ESCs are an attractive potential source for engineering articular cartilage since the developmental programs controlling articular chondrocyte differentiation should be accessible to manipulation.

6.4. Cardiovascular System

The cardiovascular system remains one of the primary targets of regenerative disease. Despite progress in the treatment and management of heart disease, heart-related problems are still the leading cause of hospitalization in the world *(198)*. Since cardiomyocytes are post-mitotic shortly after birth, any substantial loss of cells (i.e., such as what occurs during myocardial infarction) cannot be repaired *(3)*. Cell-based therapies for myocardial disease are significantly challenged by the lack of appropriate cells that can engraft and differentiate into functional cardiomyocytes in adequate numbers. The following are potential cell sources that are currently being investigated in various laboratories.

6.4.1. Cardiac Stem/Progenitor Cells

Myocardial cells are derived from the mesoderm and emerge from the primitive streak during gastrulation. Cardiac progenitor cells migrate and become committed to the cardiac fate within the primary and the secondary (anterior) heart fields. Myocardial cells within the heart tube then acquire different lineage characteristics as cardiac chambers are formed *(199, 200)*. Recent studies have suggested the existence of a pool of cardiac stem cells (CSCs) in the adult heart, isolated based on efflux of Hoechst 33342 dye, and expression of Sca-1 *(201)*, c-kit *(202)*, and Abcg2 *(203)* cell surface markers. CSCs have been shown to be multipotent, differentiating into cardiomyocytes, endothelial, and smooth muscle cell lineages *(202, 204)*.

An apparently distinct population of cardiac progenitor cells that expresses the LIM-homeodomain transcription factor, Islet-1 (Isl-1), has been identified in embryonic and postnatal hearts. The Isl-1+ cells make a substantial contribution to the embryonic heart *(205)*, while their numbers progressively decrease during development. The postnatal Isl-1+ cardiac progenitors appear to be the remnants of the fetal progenitor population that remain undifferentiated and adopt a fully differentiated cardiomyocyte phenotype and express cardiomyocyte markers when induced in vitro *(206)*. The relationship between these various populations remains unclear. Recent studies have indicated the existence of multipotent Isl-1+ and Flk-1+ cells capable of differentiation into cardiac, endothelial, and smooth muscle cell lineages *(207–209)*.

The CSC niches found at the heart apex, atria, and base midregion of the ventricle are partly defined. CSCs are believed to be surrounded by lineage-committed cells and adhered to the matrix by $\alpha_4\beta_1$ integrin *(210)*. Although the role of CSCs in myocardial homeostasis is not clear, analysis of c-Kit mutant mice demonstrates that heart-resident c-Kit+ CSCs do not function in cardiac repair or remodeling following myocardial infarction (MI) *(211)*. Thus, the potential of CSCs in regenerative medicine is dependent on developing strategies to activate CSCs in situ to mediate repair or successfully isolate and propagate CSCs in vitro for tissue engineering or cell transplantation.

6.4.2. Embryonic Stem Cell-Derived Cardiomyocytes

Differentiation of mouse and human ESCs into cardiomyocytes has been well documented *(212)*. Cardiomyocytes arise from a population of mesodermal cells that express Brachyury and Flk-1 *(207)*. Further differentiation of this mesodermal cell population into cardiomyocytes is associated with downregulation of Brachyury and upregulation of cardiac markers, such as Nkx2-5 and α-myosin heavy chain (MHC) *(213)*. Morphological and electrophysiological studies of ESC-derived cardiomyocytes (ES-CMs) have revealed similarities to pacemaker cells at earlier stages of differentiation, while cells resembling atrial, ventricular, and sinus nodal cells are generated upon further differentiation (*(214)*, reviewed in *(215)* and *(213)*). The percentage of cardiomyocytes arising in differentiated ESC cultures ranges from 1% to 5% depending on the parental ESC line *(216)*. The commonly used R1 ESC line generates a relatively low percentage of cardiomyocytes (Riazi & Stanford, unpublished data), while D3 cells normally differentiate into cardiomyocytes at a higher frequency *(217)*. Expression of a green fluorescent protein or an antibiotic resistance gene has been utilized in various studies in order to enrich the population of ES-CMs. Klug and colleagues used such a strategy to enrich a cardiomyocyte population in differentiating ESCs that were genetically modified to express neo^R driven by the αMHC promoter *(218)*. Similar selection schemes using α-actin, Mlc2v, and Ncx1 promoters have been employed to enrich for populations of ES-CMs in tissue culture *(219–221)*.

Injection of human and mouse undifferentiated ESCs or derived cardiomyocytes into injured murine heart has demonstrated the potential of ESCs to engraft and improve cardiac function in animal models *(222, 223)*. The uncontrolled differentiation into non-cardiac cells, however, restricts the clinical use of ESCs.

6.4.3. Other Sources of Stem Cells

Transplantation of exogenous adult stem cells capable of differentiation into contractile myocytes is an alternative cell therapy approach for treatment of myocardial injuries. The potential of various cell types to replace cardiomyocytes lost by MI and/or to prevent MI-induced hypertrophy and remodeling has been studied (reviewed in *(224)*). Progenitor cell populations derived from skeletal muscle *(225)*, bone marrow *(226)*, and endothelial progenitor cells (EPCs) *(227)* have received the most attention and have been shown to improve heart function following their transplantation into the hearts of cardiac injury models *(228)*. This improvement in heart function, however, has been largely attributed to the paracrine effect of the cells injected and not to transdifferentiation of the cells into cardiomyocytes (*(229)* – see **Chapter 18**) *(224, 230, 231)*. Injected skeletal myoblasts form multinucleated myotubes that do not make electrochemical couplings with host cardiomyocytes and therefore could be arrhythmogenic *(231)*. Identification of non-satellite skeletal muscle cells, a skeletal-based precursor of cardiomyocytes (Spoc cells), which can differentiate into beating cardiomyocyte-like cells *(232)* suggests a potential therapeutic utility for these cells. Further analysis of their origin and the markers specifically expressed by these cells is necessary for their purification *(233)*.

The possibility that bone marrow-derived cells can acquire a cardiac phenotype has also been largely questioned *(234)*. Transplantation of Lin-c-kit+ bone marrow stem cells (BMSCs) from αMHC-nlacZ into the heart in a mouse injury model did not result in any lacZ+ cells in the heart *(235)* refuting the previous reports of transdifferentiation *(226, 236)*. Some of the reports of stem cell plasticity may be attributed to cell heterogeneity or cell–cell fusion (reviewed in *(234)*). A recent large clinical trial using bone marrow cells in MI patients did not show an improvement of global left ventricular function, while it affected infarct remodeling *(237)*, further complicating the search for an optimal cell source for cardiac cell therapy. Cardiac transplantation of EPCs has been performed *(227, 238)*. As expected, EPCs improve the cardiac function by neovascularization and not by differentiating into cardiac myocytes *(228)*.

6.4.4. Stem Cells for Bioengineering Heart Valve and Blood Vessels

Other important applications of stem cells in regenerative medicine for the cardiovascular system include heart valve replacement and tissue-engineered blood vessels. Heart valve replacement is relatively common with approximately 100,000 related operations performed every year in the USA alone *(239)*. Mechanical valve replacements are complicated by the requirement of lifelong anticoagulant and limited durability *(240, 241)*. Early attempts to engineer heart valve leaflets using synthetic scaffolds seeded with endothelial cells and fibroblasts were promising but these valves often failed in vivo due to calcification and deformation during heart remodeling *(242)*. Heart valves are composed of endocardial endothelial, valvular interstitial, cardiac muscle, and smooth muscle cells (SMC) *(243)*. Therefore, the notion of using stem cells that can differentiate into these different cell types is an attractive strategy. Several studies have indicated construction of a heart valve substitute using ovine *(244)* and human *(245)* bone marrow-derived cells seeded onto a biodegradable scaffold. Once transplanted into the heart, these biosynthetic valves remodeled into structures with cell composition very similar to a native valve *(246)*.

Construction of tissue-engineered vascular grafts has also been actively pursued. Cell types including EPCs and mesenchymal, hematopoetic, and embryonic stem cells have been used (*see (247)* for a recent review).

6.5. Liver and Pancreas

The liver is developed as an out-pouching of the gut tube during embryogenesis. Under the influence of FGFs produced by the adjacent cardiac mesoderm, the cells in the liver bud are induced toward hepatic development at embryonic day 8.5 in mice and the fourth week of development in humans. In the absence of FGF signaling, the foregut endoderm is destined to make pancreas precursor cells *(248)*. Hepatoblasts formed during development diverge along two lineages, hepatocytes and cholangiocytes (bile duct epithelial cells), under the influence of hepatocyte growth factor and Notch signaling pathways, respectively (reviewed in *(249)*). Liver can regenerate after partial hepatectomy and removal of up to two-thirds of its mass by hepatocyte hyperplasia, while liver repopulation after acute liver failure such as viral hepatitis is dependent on stem or progenitor cells *(250)*. The shortage of liver for transplantation to treat acute liver disease has made cell therapy—using human hepatocytes or stem or progenitor cells—a necessary alternative *(251)*.

6.5.1. Adult Liver Stem/Progenitor Cells

The identity and origin of cell types involved in liver regeneration are poorly understood, but the contribution of both mature hepatocytes and resident stem cells in liver regeneration has been shown *(249)*. Adult hepatocytes are capable of proliferation and repopulation in mouse liver injury models *(252, 253)*. Furthermore, hepatocytes have been shown to differentiate into cholangiocytes where there is a massive bile duct injury *(254)*, suggesting that hepatocytes may be able to dedifferentiate into bipotent progenitors. A recent report by Koenig et al. *(255)* demonstrated that treatment of hepatocytes in culture containing mitogens can stimulate expression of biliary as well as extrahepatic progenitor markers.

The oval cell (*see* **Chapter 24**) is one of several liver stem/progenitor cell populations that has been reported. These cells are located in the terminal bile ducts and are activated in the face of liver injury. They have been shown to possess the ability to produce both hepatocytes and cholangiocytes in vitro *(249)*. Oval cells express markers of immature and mature hepatocytes (α-fetoprotein and albumin), biliary epithelial cells (OV-6, cytokeratin-7, and -19) *(256, 257)*, as well as hematopoietic stem cell markers, such as c-kit *(258)* and/or Sca-1 *(259)*. In **Chapter 24**, Petersen and colleagues provide protocols for the isolation and characterization of oval cells using the oval cell surface marker, Thy-1 and magnetic sorting. Importantly, liver repopulation following oval cell transplantation has been demonstrated by several investigators. Song et al. *(260)* FAC-sorted and transplanted oval cells from GFP transgenic mice into liver after partial hepatectomy and demonstrated that approximately 50% of the regenerated liver was GFP-positive. Thus, oval cells are a candidate for cell transplantation to cure liver disease.

Another recently reported population of resident stem/progenitor cells are human liver stem cells (HLSC), which are seemingly different from oval cells. These cells express mesenchymal, but not hematopoietic or oval cell markers; can undergo hepatic, osteogenic, and endothelial differentiation; and acquire insulin-producing islet-like structures *(261)*.

6.5.2. Targeting Diabetes: Stem Cells to Replenish Pancreatic Islet Cells

Replacement of pancreatic β-cells to maintain blood insulin homeostasis is a promising approach for curing diabetes. Isolation and transplantation of human islets using the Edmonton Protocol to treat type I diabetes has generated positive results. It is still complicated, however, by side effects and the shortage of donor organs *(262)*, emphasizing the importance of finding cell sources that can substitute β-cells *(263)*. The existence of an islet stem cell population in adult pancreas has been controversial. Lineage tracing experiments involving mice expressing tamoxifen-dependent Cre under an insulin promoter bred with reporter mice revealed that existing pancreatic β-cells, not a stem cell population, are the source of new β-cells in normal or regenerating pancreas *(264)*. However, it has been proposed that putative pancreatic stem cells may reside in the epithelium of pancreatic ducts *(265)*, or inside islets *(266)*, or both in ducts and islets *(267)*. Others have proposed that during β-cell neogenesis, ductal epithelial cells *(268)* or acinar cells *(269)* dedifferentiate into progenitor cells capable of producing new islets and acini *(263)*. In **Chapter 23**, Hanley and Rosenberg describe protocols to isolate adult human pancreatic islet cells, dedifferentiate them into proliferative duct-like structures, and treat them with a peptide (INGAP) to form islet-like structures that resemble freshly isolated islets. Other cell sources that can give rise to β-cells are also being pursued.

Differentiation of mouse ESCs into insulin-producing cells has been established *(270)* at low frequency. The generation of insulin-secreting cells from human ESCs has also recently been reported *(271)*. Constitutive expression of pancreatic and duodenal homeobox factor-1 (Pdx1) in human ESCs enhances the expression of various pancreatic markers but not that of insulin, suggesting that complete pancreatic differentiation requires other unknown factors *(272)*. Therefore, application of ESCs to generate functional β-cells requires a better understanding of islet cell development and the genes involved.

As the pancreas shares developmental origin with endodermal organs, such as liver and intestine, transdifferentiation of hepatocytes and intestinal cells may also be considered as a source for generating β-cells. In **Chapter 25**, Thowfeequ and co-authors present protocols designed to induce transdifferentiation of hepatocytes to pancreatic cell types by the overexpression of pancreatic transcription factors in human and rat cells. In a study by Zalzman and colleagues, co-expression of telomerase gene (hTERT) and Pdx1 in fetal human liver cells resulted in insulin-producing cells that could reverse diabetes in immunodeficient diabetic mice when transplanted *(273)*. Similarly, persistent expression of Pdx1 in oval cells appears to cause transdifferentiation into insulin-producing cells *(274)*. Uncontrolled differentiation and/or cell division after gene modification leading to cancer remain a concern with these current approaches requiring further studies before genetically manipulated β-cells can be applied to patients.

7. Conclusion

The ability to direct embryonic stem cells or stem cells from different adult tissues to differentiate into specific lineages will have a tremendous impact on future therapeutic approaches. The field of regenerative medicine is still in its infancy and a great deal of experimentation and optimization is needed to refine the conditions for derivation of different cell lineages. The question of which stem cell source will be optimal for each disease condition remains to be answered. Also, the issue of host-rejection continues to be a hurdle in using ESC-derived cells. Eventually, it should be possible to use SCNT or other methods of reprograming to generate individualized ESCs that can be directed to differentiate into tissue progenitor cells or mature cells for transplantation. On the way to the clinic, a number of obstacles must be overcome, but when they are, the true potential of stem cell therapies will be realized.

References

1. Price, J., Faucheux, C., and Allen, S. (2005) Deer antlers as a model of Mammalian regeneration. *Currs Top Dev Biol* **67**, 1–48
2. Klussmann, S., and Martin-Villalba, A. (2005) Molecular targets in spinal cord injury. *J Mol Med* **83**, 657–671
3. Sun, Y., and Weber, K. T. (2000) Infarct scar: a dynamic tissue. *Cardiovasc Res* **46**, 250–256
4. Brockes, J. P., and Kumar, A. (2005) Appendage regeneration in adult vertebrates and implications for regenerative medicine. *Science NY* **310**, 1919–1923
5. Odelberg, S. J. (2004) Unraveling the molecular basis for regenerative cellular plasticity. *PLoS Biol* **2**, E232
6. Sanchez Alvarado, A., and Kang, H. (2005) Multicellularity, stem cells, and the neoblasts of the planarian Schmidtea mediterranea. *Exp Cell Res* **306**, 299–308
7. Poss, K. D., Keating, M. T., and Nechiporuk, A. (2003) Tales of regeneration in zebrafish. *Dev Dyn* **226**, 202–210
8. Morrison, J. I., Loof, S., He, P., and Simon, A. (2006) Salamander limb regeneration involves the activation of a multipotent skeletal muscle satellite cell population. *J Cell Biol* **172**, 433–440
9. McGann, C. J., Odelberg, S. J., and Keating, M. T. (2001) Mammalian myotube dedifferentiation induced by newt regeneration extract. *Proc Natl Acad Sci USA* **98**, 13699–13704
10. Odelberg, S. J., Kollhoff, A., and Keating, M. T. (2000) Dedifferentiation of mammalian myotubes induced by msx1. *Cell* **103**, 1099–1109
11. Odelberg, S. J. (2005) Cellular plasticity in vertebrate regeneration. *Anat Rec B New Anat* **287**, 25–35
12. Lien, C. L., Schebesta, M., Makino, S., Weber, G. J., and Keating, M. T. (2006) Gene Expression Analysis of Zebrafish Heart Regeneration. *PLoS Biol* **4**, E260
13. Lee, Y., Grill, S., Sanchez, A., Murphy-Ryan, M., and Poss, K. D. (2005) Fgf signaling instructs position-dependent growth rate during zebrafish fin regeneration. *Development* **132**, 5173–5183
14. Beck, C. W., Christen, B., and Slack, J. M. (2003) Molecular pathways needed for regeneration of spinal cord and muscle in a vertebrate. *Dev Cell* **5**, 429–439
15. Vinarsky, V., Atkinson, D. L., Stevenson, T. J., Keating, M. T., and Odelberg, S. J. (2005) Normal newt limb regeneration requires matrix metalloproteinase function. *Dev Biol* **279**, 86–98
16. Leferovich, J. M., and Heber-Katz, E. (2002) The scarless heart. *Semin Cell Dev Biol* **13**, 327–333

17. Harty, M., Neff, A. W., King, M. W., and Mescher, A. L. (2003) Regeneration or scarring: an immunologic perspective. *Dev Dyn* **226**, 268–279
18. Li, X., Mohan, S., Gu, W., and Baylink, D. J. (2001) Analysis of gene expression in the wound repair/regeneration process. *Mamm Genome* **12**, 52–59
19. Bradley, A., Evans, M., Kaufman, M. H., and Robertson, E. (1984) Formation of germ-line chimaeras from embryo-derived teratocarcinoma cell lines. *Nature* **309**, 255–256
20. Nagy, A., Rossant, J., Nagy, R., Abramow-Newerly, W., and Roder, J. C. (1993) Derivation of completely cell culture-derived mice from early-passage embryonic stem cells. *Proc Natl Acad Sci USA* **90**, 8424–8428
21. Fraidenraich, D., Stillwell, E., Romero, E., Wilkes, D., Manova, K., Basson, C. T., and Benezra, R. (2004) Rescue of cardiac defects in id knockout embryos by injection of embryonic stem cells. *Science NY* **306**, 247–252
22. Beddington, R. S., and Robertson, E. J. (1989) An assessment of the developmental potential of embryonic stem cells in the midgestation mouse embryo. *Development* **105**, 733–737
23. Nichols, J., Evans, E. P., and Smith, A. G. (1990) Establishment of germ-line-competent embryonic stem (ES) cells using differentiation inhibiting activity. *Development* **110**, 1341–1348
24. Takeda, K., Noguchi, K., Shi, W., Tanaka, T., Matsumoto, M., Yoshida, N., Kishimoto, T., and Akira, S. (1997) Targeted disruption of the mouse Stat3 gene leads to early embryonic lethality. *Proc Natl Acad Sci USA* **94**, 3801–3804
25. Ying, Q. L., Nichols, J., Chambers, I., and Smith, A. (2003) BMP induction of Id proteins suppresses differentiation and sustains embryonic stem cell self-renewal in collaboration with STAT3. *Cell* **115**, 281–292
26. Xu, R. H., Peck, R. M., Li, D. S., Feng, X., Ludwig, T., and Thomson, J. A. (2005) Basic FGF and suppression of BMP signaling sustain undifferentiated proliferation of human ES cells. *Nat Methods* **2**, 185–190
27. Niwa, H., Miyazaki, J., and Smith, A. G. (2000) Quantitative expression of Oct-3/4 defines differentiation, dedifferentiation or self-renewal of ES cells. *Nat Genet* **24**, 372–376
28. Mitsui, K., Tokuzawa, Y., Itoh, H., Segawa, K., Murakami, M., Takahashi, K., Maruyama, M., Maeda, M., and Yamanaka, S. (2003) The homeoprotein Nanog is required for maintenance of pluripotency in mouse epiblast and ES cells. *Cell* **113**, 631–642
29. Chambers, I., Colby, D., Robertson, M., Nichols, J., Lee, S., Tweedie, S., and Smith, A. (2003) Functional expression cloning of Nanog, a pluripotency sustaining factor in embryonic stem cells. *Cell* **113**, 643–655
30. Avilion, A. A., Nicolis, S. K., Pevny, L. H., Perez, L., Vivian, N., and Lovell-Badge, R. (2003) Multipotent cell lineages in early mouse development depend on SOX2 function. *Genes Dev* **17**, 126–140
31. Boyer, L. A., Lee, T. I., Cole, M. F., Johnstone, S. E., Levine, S. S., Zucker, J. P., Guenther, M. G., Kumar, R. M., Murray, H. L., Jenner, R. G., Gifford, D. K., Melton, D. A., Jaenisch, R., and Young, R. A. (2005) Core transcriptional regulatory circuitry in human embryonic stem cells. *Cell* **122**, 947–956
32. Boiani, M., and Scholer, H. R. (2005) Regulatory networks in embryo-derived pluripotent stem cells. *Nat Rev Mol Cell Biol* **6**, 872–884
33. Okumura-Nakanishi, S., Saito, M., Niwa, H., and Ishikawa, F. (2005) Oct-3/4 and Sox2 regulate Oct-3/4 gene in embryonic stem cells. *J Biol Chem* **280**, 5307–5317
34. Orkin, S. H. (2005) Chipping away at the embryonic stem cell network. *Cell* **122**, 828–830
35. Kehat, I., Khimovich, L., Caspi, O., Gepstein, A., Shofti, R., Arbel, G., Huber, I., Satin, J., Itskovitz-Eldor, J., and Gepstein, L. (2004) Electromechanical integration of cardiomyocytes derived from human embryonic stem cells. *Nat Biotechnol* **22**, 1282–1289
36. He, J. Q., Ma, Y., Lee, Y., Thomson, J. A., and Kamp, T. J. (2003) Human embryonic stem cells develop into multiple types of cardiac myocytes: action potential characterization. *Circ Res* **93**, 32–39
37. Sinha, S., Wamhoff, B. R., Hoofnagle, M. H., Thomas, J., Neppl, R. L., Deering, T., Helmke, B. P., Bowles, D. K., Somlyo, A. V.,

and Owens, G. K. (2006) Assessment of contractility of purified smooth muscle cells derived from embryonic stem cells. *Stem cells (Dayton, Ohio)* **24**, 1678–1688

38. Chadwick, K., Wang, L., Li, L., Menendez, P., Murdoch, B., Rouleau, A., and Bhatia, M. (2003) Cytokines and BMP-4 promote hematopoietic differentiation of human embryonic stem cells. *Blood* **102**, 906–915

39. Olsen, A. L., Stachura, D. L., and Weiss, M. J. (2006) Designer blood: creating hematopoietic lineages from embryonic stem cells. *Blood* **107**, 1265–1275

40. Reubinoff, B. E., Itsykson, P., Turetsky, T., Pera, M. F., Reinhartz, E., Itzik, A., and Ben-Hur, T. (2001) Neural progenitors from human embryonic stem cells. *Nat Biotechnol* **19**, 1134–1140

41. Tabar, V., Panagiotakos, G., Greenberg, E. D., Chan, B. K., Sadelain, M., Gutin, P. H., and Studer, L. (2005) Migration and differentiation of neural precursors derived from human embryonic stem cells in the rat brain. *Nat Biotechnol* **23**, 601–606

42. Assady, S., Maor, G., Amit, M., Itskovitz-Eldor, J., Skorecki, K. L., and Tzukerman, M. (2001) Insulin production by human embryonic stem cells. *Diabetes* **50**, 1691–1697

43. Levenberg, S., Golub, J. S., Amit, M., Itskovitz-Eldor, J., and Langer, R. (2002) Endothelial cells derived from human embryonic stem cells. *Proc Natl Acad Sci USA* **99**, 4391–4396

44. Bautch, V. L. (2002) Embryonic stem cell differentiation and the vascular lineage. *Methods Mol Biol* **185**, 117–125

45. Sottile, V., Thomson, A., and McWhir, J. (2003) In vitro osteogenic differentiation of human ES cells. *Cloning Stem Cells* **5**, 149–155

46. Bielby, R. C., Boccaccini, A. R., Polak, J. M., and Buttery, L. D. (2004) In vitro differentiation and in vivo mineralization of osteogenic cells derived from human embryonic stem cells. *Tissue Eng* **10**, 1518–1525

47. Shirahashi, H., Wu, J., Yamamoto, N., Catana, A., Wege, H., Wager, B., Okita, K., and Zern, M. A. (2004) Differentiation of human and mouse embryonic stem cells along a hepatocyte lineage. *Cell Transplant* **13**, 197–211

48. Lerou, P. H., and Daley, G. Q. (2005) Therapeutic potential of embryonic stem cells. *Blood Rev* **19**, 321–331

49. Vogel, G. (2005) Stem cells. Deriving 'controversy-free' ES cells is controversial. *Science NY* **310**, 416–417

50. Daar, A. S., and Sheremeta, L. (2003) The science of stem cells: ethical, legal and social issues. *Exp Clin Transplant* **1**, 139–146

51. Cowan, C. A., Klimanskaya, I., McMahon, J., Atienza, J., Witmyer, J., Zucker, J. P., Wang, S., Morton, C. C., McMahon, A. P., Powers, D., and Melton, D. A. (2004) Derivation of embryonic stem-cell lines from human blastocysts. *N Engl J Med* **350**, 1353–1356

52. Reicin, C., and McMahon, E. (2005) Stem cell research in Canada: business opportunities for U.S. companies. *J Biolaw Bus* **8**, 61–64

53. Klimanskaya, I., Chung, Y., Becker, S., Lu, S. J., and Lanza, R. (2006) Human embryonic stem cell lines derived from single blastomeres. *Nature* **444**, 481–485

54. Pearson, H. (2006) Early embryos can yield stem cells... and survive. *Nature* **442**, 858

55. Trounson, A. O. (2001) The derivation and potential use of human embryonic stem cells. *Reprod Fertil Dev* **13**, 523–532

56. Snodgrass, H. R., Graham, D. K., Stanford, W. L., and Licato, L. L. (1993) Embryonic stem cells: research and clinical potentials. In *Peripheral Blood Stem Cells* (Smith, D. M., Sacher, R. A., and Jefferies, L. C., eds.), American Association of Blood Banks, Bethesda, MD, pp. 65–83

57. Taylor, C. J., Bolton, E. M., Pocock, S., Sharples, L. D., Pedersen, R. A., and Bradley, J. A. (2005) Banking on human embryonic stem cells: estimating the number of donor cell lines needed for HLA matching. *Lancet* **366**, 2019–2025

58. Fujikawa, T., Oh, S. H., Pi, L., Hatch, H. M., Shupe, T., and Petersen, B. E. (2005) Teratoma formation leads to failure of treatment for type I diabetes using embryonic stem cell-derived insulin-producing cells. *Am J Pathol* **166**, 1781–1791

59. Przyborski, S. A. (2005) Differentiation of human embryonic stem cells after transplantation in immune-deficient mice. *Stem cells (Dayton, Ohio)* **23**, 1242–1250

60. Temple, S. (2001) The development of neural stem cells. *Nature* **414**, 112–117

61. Gage, F. H. (2000) Mammalian neural stem cells. *Science NY* 287, 1433–1438
62. Nyfeler, Y., Kirch, R. D., Mantei, N., Leone, D. P., Radtke, F., Suter, U., and Taylor, V. (2005) Jagged1 signals in the postnatal subventricular zone are required for neural stem cell self-renewal. *Embo J* 24, 3504–3515
63. Toma, J. G., Akhavan, M., Fernandes, K. J., Barnabe-Heider, F., Sadikot, A., Kaplan, D. R., and Miller, F. D. (2001) Isolation of multipotent adult stem cells from the dermis of mammalian skin. *Nat Cell Biol* 3, 778–784
64. Tumbar, T., Guasch, G., Greco, V., Blanpain, C., Lowry, W. E., Rendl, M., and Fuchs, E. (2004) Defining the epithelial stem cell niche in skin. *Science NY* 303, 359–363
65. Germain, L., Auger, F. A., Grandbois, E., Guignard, R., Giasson, M., Boisjoly, H., and Guerin, S. L. (1999) Reconstructed human cornea produced in vitro by tissue engineering. *Pathobiology* 67, 140–147
66. Shi, X., and Garry, D. J. (2006) Muscle stem cells in development, regeneration, and disease. *Genes Dev* 20, 1692–1708
67. Mauro, A. (1961) Satellite cell of skeletal muscle fibers. *J Biophys Biochem Cytol* 9, 493–495
68. Bryder, D., Rossi, D. J., and Weissman, I. L. (2006) Hematopoietic stem cells: the paradigmatic tissue-specific stem cell. *Am J Pathol* 169, 338–346
69. Spangrude, G. J., Heimfeld, S., and Weissman, I. L. (1988) Purification and characterization of mouse hematopoietic stem cells. *Science NY* 241, 58–62
70. Jiang, Y., Jahagirdar, B. N., Reinhardt, R. L., Schwartz, R. E., Keene, C. D., Ortiz-Gonzalez, X. R., Reyes, M., Lenvik, T., Lund, T., Blackstad, M., Du, J., Aldrich, S., Lisberg, A., Low, W. C., Largaespada, D. A., and Verfaillie, C. M. (2002) Pluripotency of mesenchymal stem cells derived from adult marrow. *Nature* 418, 41–49
71. Owen, M., and Friedenstein, A. J. (1988) Stromal stem cells: marrow-derived osteogenic precursors. *Ciba Found Symp* 136, 42–60
72. Dzau, V. J., Gnecchi, M., and Pachori, A. S. (2005) Enhancing stem cell therapy through genetic modification. *J Am Coll Cardiol* 46, 1351–1353
73. Oh, S. H., Hatch, H. M., and Petersen, B. E. (2002) Hepatic oval 'stem' cell in liver regeneration. *Semin Cell Dev Biol* 13, 405–409
74. Herrera, M. B., Bruno, S., Buttiglieri, S., Tetta, C., Gatti, S., Deregibus, M. C., Bussolati, B., and Camussi, G. (2006) Isolation and characterization of a stem cell population from adult human liver. *Stem cells (Dayton, Ohio)* 24, 2840–2850
75. Horwitz, E. M., Gordon, P. L., Koo, W. K., Marx, J. C., Neel, M. D., McNall, R. Y., Muul, L., and Hofmann, T. (2002) Isolated allogeneic bone marrow-derived mesenchymal cells engraft and stimulate growth in children with osteogenesis imperfecta: Implications for cell therapy of bone. *Proceedings of the National Academy of Sciences of the United States of America* 99, 8932–8937
76. Briggs, R., and King, T. J. (1952) Transplantation of Living Nuclei From Blastula Cells into Enucleated Frogs' Eggs. *Proc Natl Acad Sci USA* 38, 455–463
77. Gurdon, J. B. (1962) The developmental capacity of nuclei taken from intestinal epithelium cells of feeding tadpoles. *J Embryol Exp Morphol* 10, 622–640
78. Gurdon, J. B. (2006) From Nuclear Transfer to Nuclear Reprogramming: The Reversal of Cell Differentiation. *Annu Rev Cell Dev Biol* 22, 1–22
79. Wilmut, I., Schnieke, A. E., McWhir, J., Kind, A. J., and Campbell, K. H. (1997) Viable offspring derived from fetal and adult mammalian cells. *Nature* 385, 810–813
80. Jouneau, A., Zhou, Q., Camus, A., Brochard, V., Maulny, L., Collignon, J., and Renard, J. P. (2006) Developmental abnormalities of NT mouse embryos appear early after implantation. *Development* 133, 1597–1607
81. Hochedlinger, K., and Jaenisch, R. (2006) Nuclear reprogramming and pluripotency. *Nature* 441, 1061–1067
82. Meissner, A., and Jaenisch, R. (2006) Mammalian nuclear transfer. *Dev Dyn* 235, 2460–2469
83. Rideout, W. M., 3rd, Hochedlinger, K., Kyba, M., Daley, G. Q., and Jaenisch, R. (2002) Correction of a genetic defect by nuclear transplantation and combined cell and gene therapy. *Cell* 109, 17–27

84. Kennedy, D. (2006) Editorial retraction. *Science (New York, N.Y* **311**, 335
85. Egli, D., Rosains, J., Birkhoff, G., and Eggan, K. (2007) Developmental reprogramming after chromosome transfer into mitotic mouse zygotes. *Nature* **447**, 679–685
86. Chen, Y., He, Z. X., Liu, A., Wang, K., Mao, W. W., Chu, J. X., Lu, Y., Fang, Z. F., Shi, Y. T., Yang, Q. Z., Chen da, Y., Wang, M. K., Li, J. S., Huang, S. L., Kong, X. Y., Shi, Y. Z., Wang, Z. Q., Xia, J. H., Long, Z. G., Xue, Z. G., Ding, W. X., and Sheng, H. Z. (2003) Embryonic stem cells generated by nuclear transfer of human somatic nuclei into rabbit oocytes. *Cell Res* **13**, 251–263
87. Tada, M., Morizane, A., Kimura, H., Kawasaki, H., Ainscough, J. F., Sasai, Y., Nakatsuji, N., and Tada, T. (2003) Pluripotency of reprogrammed somatic genomes in embryonic stem hybrid cells. *Dev Dyn* **227**, 504–510
88. Strelchenko, N., Kukharenko, V., Shkumatov, A., Verlinsky, O., Kuliev, A., and Verlinsky, Y. (2006) Reprogramming of human somatic cells by embryonic stem cell cytoplast. *Reprod Biomed Online* **12**, 107–111
89. Cowan, C. A., Atienza, J., Melton, D. A., and Eggan, K. (2005) Nuclear reprogramming of somatic cells after fusion with human embryonic stem cells. *Science NY* **309**, 1369–1373
90. Taranger, C. K., Noer, A., Sorensen, A. L., Hakelien, A. M., Boquest, A. C., and Collas, P. (2005) Induction of dedifferentiation, genomewide transcriptional programming, and epigenetic reprogramming by extracts of carcinoma and embryonic stem cells. *Mol Biol Cell* **16**, 5719–5735
91. Takahashi, K., and Yamanaka, S. (2006) Induction of pluripotent stem cells from mouse embryonic and adult fibroblast cultures by defined factors. *Cell* **126**, 663–676
92. Blelloch, R., Wang, Z., Meissner, A., Pollard, S., Smith, A., and Jaenisch, R. (2006) Reprogramming efficiency following somatic cell nuclear transfer is influenced by the differentiation and methylation state of the donor nucleus. *Stem cells (Dayton, Ohio)* **24**, 2007–2013
93. Czermin, B., and Imhof, A. (2003) The sounds of silence–histone deacetylation meets histone methylation. *Genetica* **117**, 159–164
94. Jaenisch, R. (1997) DNA methylation and imprinting: why bother? *Trends Genet* **13**, 323–329
95. Armstrong, L., Lako, M., Dean, W., and Stojkovic, M. (2006) Epigenetic modification is central to genome reprogramming in somatic cell nuclear transfer. *Stem cells (Dayton, Ohio)* **24**, 805–814
96. Boyer, L. A., Plath, K., Zeitlinger, J., Brambrink, T., Medeiros, L. A., Lee, T. I., Levine, S. S., Wernig, M., Tajonar, A., Ray, M. K., Bell, G. W., Otte, A. P., Vidal, M., Gifford, D. K., Young, R. A., and Jaenisch, R. (2006) Polycomb complexes repress developmental regulators in murine embryonic stem cells. *Nature* **441**, 349–353
97. Papp, B., and Muller, J. (2006) Histone trimethylation and the maintenance of transcriptional ON and OFF states by trxG and PcG proteins. *Genes Dev* **20**, 2041–2054
98. Muller-Ehmsen, J., Krausgrill, B., Burst, V., Schenk, K., Neisen, U. C., Fries, J. W., Fleischmann, B. K., Hescheler, J., and Schwinger, R. H. (2006) Effective engraftment but poor mid-term persistence of mononuclear and mesenchymal bone marrow cells in acute and chronic rat myocardial infarction. *J Mol Cell Cardiol* **41**, 876–884
99. Moore, M. A., and Metcalf, D. (1970) Ontogeny of the haemopoietic system: yolk sac origin of in vivo and in vitro colony forming cells in the developing mouse embryo. *Br J Haematol* **18**, 279–296
100. Tavian, M., and Peault, B. (2005) Embryonic development of the human hematopoietic system. *Int J Dev Biol* **49**, 243–250
101. Haar, J. L., and Ackerman, G. A. (1971) A phase and electron microscopic study of vasculogenesis and erythropoiesis in the yolk sac of the mouse. *Anat Rec* **170**, 199–223
102. Huber, T. L., Kouskoff, V., Fehling, H. J., Palis, J., and Keller, G. (2004) Haemangioblast commitment is initiated in the primitive streak of the mouse embryo. *Nature* **432**, 625–630
103. Shalaby, F., Ho, J., Stanford, W. L., Fischer, K. D., Schuh, A. C., Schwartz, L., Bernstein, A., and Rossant, J. (1997) A requirement for Flk1 in primitive and

definitive hematopoiesis and vasculogenesis. *Cell* **89**, 981–990
104. Zambidis, E. T., Oberlin, E., Tavian, M., and Peault, B. (2006) Blood-forming endothelium in human ontogeny: lessons from in utero development and embryonic stem cell culture. *Trends Cardiovasc Med* **16**, 95–101
105. Harrison, D. E. (1980) Competitive repopulation: a new assay for long-term stem cell functional capacity. *Blood* **55**, 77–81
106. Larochelle, A., Vormoor, J., Hanenberg, H., Wang, J. C., Bhatia, M., Lapidot, T., Moritz, T., Murdoch, B., Xiao, X. L., Kato, I., Williams, D. A., and Dick, J. E. (1996) Identification of primitive human hematopoietic cells capable of repopulating NOD/SCID mouse bone marrow: implications for gene therapy. *Nat med* **2**, 1329–1337
107. Weissman, I. L. (2000) Translating stem and progenitor cell biology to the clinic: barriers and opportunities. *Science NY* **287**, 1442–1446
108. Brown, J. M., and Weissman, I. L. (2004) Progress and prospects in hematopoietic stem cell expansion and transplantation. *Exp hematol* **32**, 693–695
109. Gluckman, E., Broxmeyer, H. A., Auerbach, A. D., Friedman, H. S., Douglas, G. W., Devergie, A., Esperou, H., Thierry, D., Socie, G., Lehn, P., and et al. (1989) Hematopoietic reconstitution in a patient with Fanconi's anemia by means of umbilical-cord blood from an HLA-identical sibling. *N Engl J Med* **321**, 1174–1178
110. Kurtzberg, J., Laughlin, M., Graham, M. L., Smith, C., Olson, J. F., Halperin, E. C., Ciocci, G., Carrier, C., Stevens, C. E., and Rubinstein, P. (1996) Placental blood as a source of hematopoietic stem cells for transplantation into unrelated recipients. *N Engl J Med* **335**, 157–166
111. Wagner, J. E., Kernan, N. A., Steinbuch, M., Broxmeyer, H. E., and Gluckman, E. (1995) Allogeneic sibling umbilical-cord-blood transplantation in children with malignant and non-malignant disease. *Lancet* **346**, 214–219
112. Brunstein, C. G., and Wagner, J. E. (2006) Umbilical cord blood transplantation and banking. *Annu Rev Med* **57**, 403–417
113. Gluckman, E., Rocha, V., Boyer-Chammard, A., Locatelli, F., Arcese, W., Pasquini, R., Ortega, J., Souillet, G., Ferreira, E., Laporte, J. P., Fernandez, M., and Chastang, C. (1997) Outcome of cord-blood transplantation from related and unrelated donors. Eurocord Transplant Group and the European Blood and Marrow Transplantation Group. *N Engl J Med* **337**, 373–381
114. Rubinstein, P., Carrier, C., Scaradavou, A., Kurtzberg, J., Adamson, J., Migliaccio, A. R., Berkowitz, R. L., Cabbad, M., Dobrila, N. L., Taylor, P. E., Rosenfield, R. E., and Stevens, C. E. (1998) Outcomes among 562 recipients of placental-blood transplants from unrelated donors. *N Engl J Med* **339**, 1565–1577
115. Gluckman, E., Rocha, V., and Chevret, S. (2001) Results of unrelated umbilical cord blood hematopoietic stem cell transplant. *Transfus Clin Biol* **8**, 146–154
116. Bornstein, R., Flores, A. I., Montalban, M. A., del Rey, M. J., de la Serna, J., and Gilsanz, F. (2005) A modified cord blood collection method achieves sufficient cell levels for transplantation in most adult patients. *Stem cells (Dayton, Ohio)* **23**, 324–334
117. Audet, J., Miller, C. L., Eaves, C. J., and Piret, J. M. (2002) Common and distinct features of cytokine effects on hematopoietic stem and progenitor cells revealed by dose-response surface analysis. *Biotechnol bioeng* **80**, 393–404
118. Audet, J., Miller, C. L., Rose-John, S., Piret, J. M., and Eaves, C. J. (2001) Distinct role of gp130 activation in promoting self-renewal divisions by mitogenically stimulated murine hematopoietic stem cells. *Proc Natl Acad Sci USA* **98**, 1757–1762
119. Krosl, J., Austin, P., Beslu, N., Kroon, E., Humphries, R. K., and Sauvageau, G. (2003) In vitro expansion of hematopoietic stem cells by recombinant TAT-HOXB4 protein. *Nat med* **9**, 1428–1432
120. Antonchuk, J., Sauvageau, G., and Humphries, R. K. (2002) HOXB4-induced expansion of adult hematopoietic stem cells ex vivo. *Cell* **109**, 39–45
121. Madlambayan, G. J., Rogers, I., Kirouac, D. C., Yamanaka, N., Mazurier, F., Doedens, M., Casper, R. F., Dick, J. E., and Zandstra, P. W. (2005) Dynamic changes in cellular and microenvironmental composition can be controlled to elicit in vitro human hematopoietic stem cell expansion. *Exp hematol* **33**, 1229–1239

122. Madlambayan, G. J., Rogers, I., Purpura, K. A., Ito, C., Yu, M., Kirouac, D., Casper, R. F., and Zandstra, P. W. (2006) Clinically relevant expansion of hematopoietic stem cells with conserved function in a single-use, closed-system bioprocess. *Biol Blood Marrow Transplant* **12**, 1020–1030

123. Tian, X., Morris, J. K., Linehan, J. L., and Kaufman, D. S. (2004) Cytokine requirements differ for stroma and embryoid body-mediated hematopoiesis from human embryonic stem cells. *Exp hematol* **32**, 1000–1009

124. Keller, G., Kennedy, M., Papayannopoulou, T., and Wiles, M. V. (1993) Hematopoietic commitment during embryonic stem cell differentiation in culture. *Mol Cell Biol* **13**, 473–486

125. Choi, K., Kennedy, M., Kazarov, A., Papadimitriou, J. C., and Keller, G. (1998) A common precursor for hematopoietic and endothelial cells. *Development* **125**, 725–732

126. Kyba, M., Perlingeiro, R. C., and Daley, G. Q. (2002) HoxB4 confers definitive lymphoid-myeloid engraftment potential on embryonic stem cell and yolk sac hematopoietic progenitors. *Cell* **109**, 29–37

127. Kyba, M., Perlingeiro, R. C., Hoover, R. R., Lu, C. W., Pierce, J., and Daley, G. Q. (2003) Enhanced hematopoietic differentiation of embryonic stem cells conditionally expressing Stat5. *Proc Natl Acad Sci USA* **100** Suppl 1, 11904–11910

128. Wang, Y., Yates, F., Naveiras, O., Ernst, P., and Daley, G. Q. (2005) Embryonic stem cell-derived hematopoietic stem cells. *Proc Natl Acad Sci USA* **102**, 19081–19086

129. Wang, L., Menendez, P., Shojaei, F., Li, L., Mazurier, F., Dick, J. E., Cerdan, C., Levac, K., and Bhatia, M. (2005) Generation of hematopoietic repopulating cells from human embryonic stem cells independent of ectopic HOXB4 expression. *J Exp Med* **201**, 1603–1614

130. Lindvall, O., Kokaia, Z., and Martinez-Serrano, A. (2004) Stem cell therapy for human neurodegenerative disorders-how to make it work. *Nat med* **10** Suppl, S42–50

131. Martino, G., and Pluchino, S. (2006) The therapeutic potential of neural stem cells. *Nat Rev Neurosci* **7**, 395–406

132. Piccini, P., Brooks, D. J., Bjorklund, A., Gunn, R. N., Grasby, P. M., Rimoldi, O., Brundin, P., Hagell, P., Rehncrona, S., Widner, H., and Lindvall, O. (1999) Dopamine release from nigral transplants visualized in vivo in a Parkinson's patient. *Nat Neurosci* **2**, 1137–1140

133. Piccini, P., Lindvall, O., Bjorklund, A., Brundin, P., Hagell, P., Ceravolo, R., Oertel, W., Quinn, N., Samuel, M., Rehncrona, S., Widner, H., and Brooks, D. J. (2000) Delayed recovery of movement-related cortical function in Parkinson's disease after striatal dopaminergic grafts. *Ann Neurol* **48**, 689–695

134. Reynolds, B. A., and Rietze, R. L. (2005) Neural stem cells and neurospheres–reevaluating the relationship. *Nat Methods* **2**, 333–336

135. Uchida, N., Buck, D. W., He, D., Reitsma, M. J., Masek, M., Phan, T. V., Tsukamoto, A. S., Gage, F. H., and Weissman, I. L. (2000) Direct isolation of human central nervous system stem cells. *Proc Natl Acad Sci USA* **97**, 14720–14725

136. Palmer, T. D., Schwartz, P. H., Taupin, P., Kaspar, B., Stein, S. A., and Gage, F. H. (2001) Cell culture. Progenitor cells from human brain after death. *Nature* **411**, 42–43

137. Tropepe, V., Hitoshi, S., Sirard, C., Mak, T. W., Rossant, J., and van der Kooy, D. (2001) Direct neural fate specification from embryonic stem cells: a primitive mammalian neural stem cell stage acquired through a default mechanism. *Neuron* **30**, 65–78

138. Li, X. J., and Zhang, S. C. (2006) In vitro differentiation of neural precursors from human embryonic stem cells. *Methods Mol Biol* **331**, 169–177

139. Doetsch, F. (2003) A niche for adult neural stem cells. *Curr Opin Genet Dev* **13**, 543–550

140. Svendsen, C. N., Caldwell, M. A., Shen, J., ter Borg, M. G., Rosser, A. E., Tyers, P., Karmiol, S., and Dunnett, S. B. (1997) Long-term survival of human central nervous system progenitor cells transplanted into a rat model of Parkinson's disease. *Exp Neurol* **148**, 135–146

141. Wu, S., Suzuki, Y., Kitada, M., Kitaura, M., Kataoka, K., Takahashi, J., Ide, C., and Nishimura, Y. (2001) Migration, integration, and differentiation of hippocampus-

derived neurosphere cells after transplantation into injured rat spinal cord. *Neurosci Lett* **312**, 173–176
142. Vroemen, M., Aigner, L., Winkler, J., and Weidner, N. (2003) Adult neural progenitor cell grafts survive after acute spinal cord injury and integrate along axonal pathways. *Eur J Neurosci* **18**, 743–751
143. Cummings, B. J., Uchida, N., Tamaki, S. J., Salazar, D. L., Hooshmand, M., Summers, R., Gage, F. H., and Anderson, A. J. (2005) Human neural stem cells differentiate and promote locomotor recovery in spinal cord-injured mice. *Proc Natl Acad Sci USA* **102**, 14069–14074
144. Fernandes, K. J., McKenzie, I. A., Mill, P., Smith, K. M., Akhavan, M., Barnabe-Heider, F., Biernaskie, J., Junek, A., Kobayashi, N. R., Toma, J. G., Kaplan, D. R., Labosky, P. A., Rafuse, V., Hui, C. C., and Miller, F. D. (2004) A dermal niche for multipotent adult skin-derived precursor cells. *Nat Cell Biol* **6**, 1082–1093
145. McKenzie, I. A., Biernaskie, J., Toma, J. G., Midha, R., and Miller, F. D. (2006) Skin-derived precursors generate myelinating Schwann cells for the injured and dysmyelinated nervous system. *J Neurosci* **26**, 6651–6660
146. Kopen, G. C., Prockop, D. J., and Phinney, D. G. (1999) Marrow stromal cells migrate throughout forebrain and cerebellum, and they differentiate into astrocytes after injection into neonatal mouse brains. *Proc Natl Acad Sci USA* **96**, 10711–10716
147. Sanchez-Ramos, J., Song, S., Cardozo-Pelaez, F., Hazzi, C., Stedeford, T., Willing, A., Freeman, T. B., Saporta, S., Janssen, W., Patel, N., Cooper, D. R., and Sanberg, P. R. (2000) Adult bone marrow stromal cells differentiate into neural cells in vitro. *Exp Neurol* **164**, 247–256
148. Habich, A., Jurga, M., Markiewicz, I., Lukomska, B., Bany-Laszewicz, U., and Domanska-Janik, K. (2006) Early appearance of stem/progenitor cells with neural-like characteristics in human cord blood mononuclear fraction cultured in vitro. *Exp hematol* **34**, 914–925
149. El-Badri, N. S., Hakki, A., Saporta, S., Liang, X., Madhusodanan, S., Willing, A. E., Sanberg, C. D., and Sanberg, P. R. (2006) Cord blood mesenchymal stem cells: Potential use in neurological disorders. *Stem Cells Dev* **15**, 497–506
150. Park, H. C., Shim, Y. S., Ha, Y., Yoon, S. H., Park, S. R., Choi, B. H., and Park, H. S. (2005) Treatment of complete spinal cord injury patients by autologous bone marrow cell transplantation and administration of granulocyte-macrophage colony stimulating factor. *Tissue Eng* **11**, 913–922
151. Mazzini, L., Mareschi, K., Ferrero, I., Vassallo, E., Oliveri, G., Boccaletti, R., Testa, L., Livigni, S., and Fagioli, F. (2006) Autologous mesenchymal stem cells: clinical applications in amyotrophic lateral sclerosis. *Neurol Res* **28**, 523–526
152. Vendrame, M., Cassady, J., Newcomb, J., Butler, T., Pennypacker, K. R., Zigova, T., Sanberg, C. D., Sanberg, P. R., and Willing, A. E. (2004) Infusion of human umbilical cord blood cells in a rat model of stroke dose-dependently rescues behavioral deficits and reduces infarct volume. *Stroke* **35**, 2390–2395
153. Bain, G., Kitchens, D., Yao, M., Huettner, J. E., and Gottlieb, D. I. (1995) Embryonic stem cells express neuronal properties in vitro. *Dev Biol* **168**, 342–357
154. Okabe, S., Forsberg-Nilsson, K., Spiro, A. C., Segal, M., and McKay, R. D. (1996) Development of neuronal precursor cells and functional postmitotic neurons from embryonic stem cells in vitro. *Mech Dev* **59**, 89–102
155. Lee, S. H., Lumelsky, N., Studer, L., Auerbach, J. M., and McKay, R. D. (2000) Efficient generation of midbrain and hindbrain neurons from mouse embryonic stem cells. *Nat Biotechnol* **18**, 675–679
156. Schmandt, T., Meents, E., Gossrau, G., Gornik, V., Okabe, S., and Brustle, O. (2005) High-purity lineage selection of embryonic stem cell-derived neurons. *Stem Cells Dev* **14**, 55–64
157. Glaser, T., Perez-Bouza, A., Klein, K., and Brustle, O. (2005) Generation of purified oligodendrocyte progenitors from embryonic stem cells. *Faseb J* **19**, 112–114
158. Glass, R., Synowitz, M., Kronenberg, G., Walzlein, J. H., Markovic, D. S., Wang, L. P., Gast, D., Kiwit, J., Kempermann, G., and Kettenmann, H. (2005) Glioblastoma-induced attraction of endogenous neural

precursor cells is associated with improved survival. *J Neurosci* **25**, 2637–2646
159. Chi, L., Ke, Y., Luo, C., Li, B., Gozal, D., Kalyanaraman, B., and Liu, R. (2006) Motor neuron degeneration promotes neural progenitor cell proliferation, migration, and neurogenesis in the spinal cords of amyotrophic lateral sclerosis mice. *Stem cells (Dayton, Ohio)* **24**, 34–43
160. Muller, F. J., Snyder, E. Y., and Loring, J. F. (2006) Gene therapy: can neural stem cells deliver? *Nat Rev Neurosci* **7**, 75–84
161. Buckingham, M., Bajard, L., Chang, T., Daubas, P., Hadchouel, J., Meilhac, S., Montarras, D., Rocancourt, D., and Relaix, F. (2003) The formation of skeletal muscle: from somite to limb. *J Anat* **202**, 59–68
162. Buckingham, M. (2001) Skeletal muscle formation in vertebrates. *Curr Opin Genet Dev* **11**, 440–448
163. Asakura, A. (2003) Stem cells in adult skeletal muscle. *Trends Cardiovasc Med* **13**, 123–128
164. Seale, P., and Rudnicki, M. A. (2000) A new look at the origin, function, and "stem-cell" status of muscle satellite cells. *Dev Biol* **218**, 115–124
165. Holterman, C. E., and Rudnicki, M. A. (2005) Molecular regulation of satellite cell function. *Semin Cell Dev Biol* **16**, 575–584
166. Seale, P., Sabourin, L. A., Girgis-Gabardo, A., Mansouri, A., Gruss, P., and Rudnicki, M. A. (2000) Pax7 is required for the specification of myogenic satellite cells. *Cell* **102**, 777–786
167. Shinin, V., Gayraud-Morel, B., Gomes, D., and Tajbakhsh, S. (2006) Asymmetric division and cosegregation of template DNA strands in adult muscle satellite cells. *Nat Cell Biol* **8**, 677–687
168. Gussoni, E., Soneoka, Y., Strickland, C. D., Buzney, E. A., Khan, M. K., Flint, A. F., Kunkel, L. M., and Mulligan, R. C. (1999) Dystrophin expression in the mdx mouse restored by stem cell transplantation. *Nature* **401**, 390–394
169. Asakura, A., and Rudnicki, M. A. (2002) Side population cells from diverse adult tissues are capable of in vitro hematopoietic differentiation. *Exp hematol* **30**, 1339–1345
170. Qu-Petersen, Z., Deasy, B., Jankowski, R., Ikezawa, M., Cummins, J., Pruchnic, R., Mytinger, J., Cao, B., Gates, C., Wernig, A., and Huard, J. (2002) Identification of a novel population of muscle stem cells in mice: potential for muscle regeneration. *J Cell Biol* **157**, 851–864
171. Komori, T. (2006) Regulation of osteoblast differentiation by transcription factors. *J Cell Biochem* **99**, 1233–1239
172. Pittenger, M. F., Mackay, A. M., Beck, S. C., Jaiswal, R. K., Douglas, R., Mosca, J. D., Moorman, M. A., Simonetti, D. W., Craig, S., and Marshak, D. R. (1999) Multilineage potential of adult human mesenchymal stem cells. *Science NY* **284**, 143–147
173. Horwitz, E. M., Le Blanc, K., Dominici, M., Mueller, I., Slaper-Cortenbach, I., Marini, F. C., Deans, R. J., Krause, D. S., and Keating, A. (2005) Clarification of the nomenclature for MSC: The International Society for Cellular Therapy position statement. *Cytotherapy* **7**, 393–395
174. Dominici, M., Le Blanc, K., Mueller, I., Slaper-Cortenbach, I., Marini, F., Krause, D., Deans, R., Keating, A., Prockop, D., and Horwitz, E. (2006) Minimal criteria for defining multipotent mesenchymal stromal cells. The International Society for Cellular Therapy position statement. *Cytotherapy* **8**, 315–317
175. Krampera, M., Pizzolo, G., Aprili, G., and Franchini, M. (2006) Mesenchymal stem cells for bone, cartilage, tendon and skeletal muscle repair. *Bone* **39**, 678–683
176. Gang, E. J., Bosnakovski, D., Figueiredo, C. A., Visser, J. W., and Perlingeiro, R. C. (2006) SSEA-4 identifies mesenchymal stem cells from bone marrow. *Blood* **109**, 1743–1751
177. Bonyadi, M., Waldman, S. D., Liu, D., Aubin, J. E., Grynpas, M. D., and Stanford, W. L. (2003) Mesenchymal progenitor self-renewal deficiency leads to age-dependent osteoporosis in Sca-1/Ly-6A null mice. *Proc Natl Acad Sci USA* **100**, 5840–5845
178. Holmes, C., Khan, T. S., Owen, C., Ciliberti, N., Grynpas, M. D., and Stanford, W. L. (2007) Longitudinal Analysis of Mesenchymal Progenitors and Bone Quality in the Stem Cell Antigen-1 Null Osteoporotic Mouse. *J Bone Miner Res* **22**, 1373–1386
179. Lee, R. H., Kim, B., Choi, I., Kim, H., Choi, H. S., Suh, K., Bae, Y. C., and Jung,

J. S. (2004) Characterization and expression analysis of mesenchymal stem cells from human bone marrow and adipose tissue. *Cell Physiol Biochem* **14**, 311–324
180. Sarugaser, R., Lickorish, D., Baksh, D., Hosseini, M. M., and Davies, J. E. (2005) Human umbilical cord perivascular (HUCPV) cells: a source of mesenchymal progenitors. *Stem cells (Dayton, Ohio)* **23**, 220–229
181. Shih, D. T., Lee, D. C., Chen, S. C., Tsai, R. Y., Huang, C. T., Tsai, C. C., Shen, E. Y., and Chiu, W. T. (2005) Isolation and characterization of neurogenic mesenchymal stem cells in human scalp tissue. *Stem cells (Dayton, Ohio)* **23**, 1012–1020
182. Roufosse, C. A., Direkze, N. C., Otto, W. R., and Wright, N. A. (2004) Circulating mesenchymal stem cells. *Int J Biochem Cell Biol* **36**, 585–597
183. Quarto, R., Mastrogiacomo, M., Cancedda, R., Kutepov, S. M., Mukhachev, V., Lavroukov, A., Kon, E., and Marcacci, M. (2001) Repair of large bone defects with the use of autologous bone marrow stromal cells. *N Engl J Med* **344**, 385–386
184. Horwitz, E. M., Prockop, D. J., Fitzpatrick, L. A., Koo, W. W., Gordon, P. L., Neel, M., Sussman, M., Orchard, P., Marx, J. C., Pyeritz, R. E., and Brenner, M. K. (1999) Transplantability and therapeutic effects of bone marrow-derived mesenchymal cells in children with osteogenesis imperfecta. *Nat med* **5**, 309–313
185. Horwitz, E. M., Prockop, D. J., Gordon, P. L., Koo, W. W., Fitzpatrick, L. A., Neel, M. D., McCarville, M. E., Orchard, P. J., Pyeritz, R. E., and Brenner, M. K. (2001) Clinical responses to bone marrow transplantation in children with severe osteogenesis imperfecta. *Blood* **97**, 1227–1231
186. Le Blanc, K., Gotherstrom, C., Ringden, O., Hassan, M., McMahon, R., Horwitz, E., Anneren, G., Axelsson, O., Nunn, J., Ewald, U., Norden-Lindeberg, S., Jansson, M., Dalton, A., Astrom, E., and Westgren, M. (2005) Fetal mesenchymal stem-cell engraftment in bone after in utero transplantation in a patient with severe osteogenesis imperfecta. *Transplantation* **79**, 1607–1614
187. Wang, X., Li, F., and Niyibizi, C. (2006) Progenitors systemically transplanted into neonatal mice localize to areas of active bone formation in vivo: implications of cell therapy for skeletal diseases. *Stem cells (Dayton, Ohio)* **24**, 1869–1878
188. Kuo, C. K., Li, W. J., Mauck, R. L., and Tuan, R. S. (2006) Cartilage tissue engineering: its potential and uses. *Curr Opin Rheumatol* **18**, 64–73
189. Breinan, H. A., Minas, T., Hsu, H. P., Nehrer, S., Sledge, C. B., and Spector, M. (1997) Effect of cultured autologous chondrocytes on repair of chondral defects in a canine model. *J Bone Joint Surg Am* **79**, 1439–1451
190. Sams, A. E., and Nixon, A. J. (1995) Chondrocyte-laden collagen scaffolds for resurfacing extensive articular cartilage defects. *Osteoarthritis Cartilage* **3**, 47–59
191. Yoo, J. U., Barthel, T. S., Nishimura, K., Solchaga, L., Caplan, A. I., Goldberg, V. M., and Johnstone, B. (1998) The chondrogenic potential of human bone-marrow-derived mesenchymal progenitor cells. *J Bone Joint Surg Am* **80**, 1745–1757
192. Worster, A. A., Nixon, A. J., Brower-Toland, B. D., and Williams, J. (2000) Effect of transforming growth factor beta1 on chondrogenic differentiation of cultured equine mesenchymal stem cells. *Am J Vet Res* **61**, 1003–1010
193. Sekiya, I., Larson, B. L., Smith, J. R., Pochampally, R., Cui, J. G., and Prockop, D. J. (2002) Expansion of human adult stem cells from bone marrow stroma: conditions that maximize the yields of early progenitors and evaluate their quality. *Stem cells (Dayton, Ohio)* **20**, 530–541
194. Wang, D. W., Fermor, B., Gimble, J. M., Awad, H. A., and Guilak, F. (2005) Influence of oxygen on the proliferation and metabolism of adipose derived adult stem cells. *J Cell Physiol* **204**, 184–191
195. Sakai, D., Mochida, J., Iwashina, T., Watanabe, T., Nakai, T., Ando, K., and Hotta, T. (2005) Differentiation of mesenchymal stem cells transplanted to a rabbit degenerative disc model: potential and limitations for stem cell therapy in disc regeneration. *Spine* **30**, 2379–2387
196. Adachi, N., Sato, K., Usas, A., Fu, F. H., Ochi, M., Han, C. W., Niyibizi, C., and Huard, J. (2002) Muscle derived, cell based ex vivo gene therapy for treatment

of full thickness articular cartilage defects. *J Rheumatol* **29**, 1920–1930
197. Kafienah, W., Mistry, S., Dickinson, S. C., Sims, T. J., Learmonth, I., and Hollander, A. P. (2007) Three-dimensional cartilage tissue engineering using adult stem cells from osteoarthritis patients. *Arthritis rheum* **56**, 177–187
198. Miller, L. W., and Missov, E. D. (2001) Epidemiology of heart failure. *Cardiol Clin* **19**, 547–555
199. Buckingham, M., Meilhac, S., and Zaffran, S. (2005) Building the mammalian heart from two sources of myocardial cells. *Nat Rev Genet* **6**, 826–835
200. Solloway, M. J., and Harvey, R. P. (2003) Molecular pathways in myocardial development: a stem cell perspective. *Cardiovasc Res* **58**, 264–277
201. Oh, H., Chi, X., Bradfute, S. B., Mishina, Y., Pocius, J., Michael, L. H., Behringer, R. R., Schwartz, R. J., Entman, M. L., and Schneider, M. D. (2004) Cardiac muscle plasticity in adult and embryo by heart-derived progenitor cells. *Ann N Y Acad Sci* **1015**, 182–189
202. Beltrami, A. P., Barlucchi, L., Torella, D., Baker, M., Limana, F., Chimenti, S., Kasahara, H., Rota, M., Musso, E., Urbanek, K., Leri, A., Kajstura, J., Nadal-Ginard, B., and Anversa, P. (2003) Adult cardiac stem cells are multipotent and support myocardial regeneration. *Cell* **114**, 763–776
203. Martin, C. M., Meeson, A. P., Robertson, S. M., Hawke, T. J., Richardson, J. A., Bates, S., Goetsch, S. C., Gallardo, T. D., and Garry, D. J. (2004) Persistent expression of the ATP-binding cassette transporter, Abcg2, identifies cardiac SP cells in the developing and adult heart. *Dev Biol* **265**, 262–275
204. Urbanek, K., Torella, D., Sheikh, F., De Angelis, A., Nurzynska, D., Silvestri, F., Beltrami, C. A., Bussani, R., Beltrami, A. P., Quaini, F., Bolli, R., Leri, A., Kajstura, J., and Anversa, P. (2005) Myocardial regeneration by activation of multipotent cardiac stem cells in ischemic heart failure. *Proc Natl Acad Sci USA* **102**, 8692–8697
205. Cai, C. L., Liang, X., Shi, Y., Chu, P. H., Pfaff, S. L., Chen, J., and Evans, S. (2003) Isl1 identifies a cardiac progenitor population that proliferates prior to differentiation and contributes a majority of cells to the heart. *Dev Cell* **5**, 877–889
206. Laugwitz, K. L., Moretti, A., Lam, J., Gruber, P., Chen, Y., Woodard, S., Lin, L. Z., Cai, C. L., Lu, M. M., Reth, M., Platoshyn, O., Yuan, J. X., Evans, S., and Chien, K. R. (2005) Postnatal isl1+ cardioblasts enter fully differentiated cardiomyocyte lineages. *Nature* **433**, 647–653
207. Kattman, S. J., Huber, T. L., and Keller, G. M. (2006) Multipotent flk-1+ cardiovascular progenitor cells give rise to the cardiomyocyte, endothelial, and vascular smooth muscle lineages. *Dev Cell* **11**, 723–732
208. Moretti, A., Caron, L., Nakano, A., Lam, J. T., Bernshausen, A., Chen, Y., Qyang, Y., Bu, L., Sasaki, M., Martin-Puig, S., Sun, Y., Evans, S. M., Laugwitz, K. L., and Chien, K. R. (2006) Multipotent embryonic isl1+ progenitor cells lead to cardiac, smooth muscle, and endothelial cell diversification. *Cell* **127**, 1151–1165
209. Wu, S. M., Fujiwara, Y., Cibulsky, S. M., Clapham, D. E., Lien, C. L., Schultheiss, T. M., and Orkin, S. H. (2006) Developmental origin of a bipotential myocardial and smooth muscle cell precursor in the mammalian heart. *Cell* **127**, 1137–1150
210. Urbanek, K., Cesselli, D., Rota, M., Nascimbene, A., De Angelis, A., Hosoda, T., Bearzi, C., Boni, A., Bolli, R., Kajstura, J., Anversa, P., and Leri, A. (2006) Stem cell niches in the adult mouse heart. *Proc Natl Acad Sci USA* **103**, 9226–9231
211. Ayach, B. B., Yoshimitsu, M., Dawood, F., Sun, M., Arab, S., Chen, M., Higuchi, K., Siatskas, C., Lee, P., Lim, H., Zhang, J., Cukerman, E., Stanford, W. L., Medin, J. A., and Liu, P. P. (2006) Stem cell factor receptor induces progenitor and natural killer cell-mediated cardiac survival and repair after myocardial infarction. *Proc Natl Acad Sci USA* **103**, 2304–2309
212. Mummery, C. (2007) Cardiomyocytes from human embryonic stem cells: more than heart repair alone. *Bioessays* **29**, 572–579
213. Wei, H., Juhasz, O., Li, J., Tarasova, Y. S., and Boheler, K. R. (2005) Embryonic stem cells and cardiomyocyte differentiation: phenotypic and molecular analyses. *J Cell Mol Med* **9**, 804–817.

214. Maltsev, V. A., Wobus, A. M., Rohwedel, J., Bader, M., and Hescheler, J. (1994) Cardiomyocytes differentiated in vitro from embryonic stem cells developmentally express cardiac-specific genes and ionic currents. *Circ Res* **75**, 233–244

215. Boheler, K. R., Czyz, J., Tweedie, D., Yang, H. T., Anisimov, S. V., and Wobus, A. M. (2002) Differentiation of pluripotent embryonic stem cells into cardiomyocytes. *Circ Res* **91**, 189–201

216. Zandstra, P. W., Bauwens, C., Yin, T., Liu, Q., Schiller, H., Zweigerdt, R., Pasumarthi, K. B., and Field, L. J. (2003) Scalable production of embryonic stem cell-derived cardiomyocytes. *Tissue Eng* **9**, 767–778

217. Metzger, J. M., Lin, W. I., Johnston, R. A., Westfall, M. V., and Samuelson, L. C. (1995) Myosin heavy chain expression in contracting myocytes isolated during embryonic stem cell cardiogenesis. *Circ Res* **76**, 710–719

218. Klug, M. G., Soonpaa, M. H., Koh, G. Y., and Field, L. J. (1996) Genetically selected cardiomyocytes from differentiating embronic stem cells form stable intracardiac grafts. *J Clin Invest* **98**, 216–224

219. Kolossov, E., Fleischmann, B. K., Liu, Q., Bloch, W., Viatchenko-Karpinski, S., Manzke, O., Ji, G. J., Bohlen, H., Addicks, K., and Hescheler, J. (1998) Functional characteristics of ES cell-derived cardiac precursor cells identified by tissue-specific expression of the green fluorescent protein. *J Cell Biol* **143**, 2045–2056

220. Meyer, N., Jaconi, M., Landopoulou, A., Fort, P., and Puceat, M. (2000) A fluorescent reporter gene as a marker for ventricular specification in ES-derived cardiac cells. *FEBS Lett* **478**, 151–158

221. Fijnvandraat, A. C., van Ginneken, A. C., Schumacher, C. A., Boheler, K. R., Lekanne Deprez, R. H., Christoffels, V. M., and Moorman, A. F. (2003) Cardiomyocytes purified from differentiated embryonic stem cells exhibit characteristics of early chamber myocardium. *J Mol Cell Cardiol* **35**, 1461–1472

222. Singla, D. K., Hacker, T. A., Ma, L., Douglas, P. S., Sullivan, R., Lyons, G. E., and Kamp, T. J. (2006) Transplantation of embryonic stem cells into the infarcted mouse heart: formation of multiple cell types. *J Mol Cell Cardiol* **40**, 195–200

223. Leor, J., Gerecht-Nir, S., Cohen, S., Miller, L., Holbova, R., Ziskind, A., Shachar, M., Feinberg, M. S., Guetta, E., and Itskovitz-Eldor, J. (2007) Human embryonic stem cell transplantation to repair the infarcted myocardium. *Heart* **93**, 1173–1174

224. Murry, C. E., Field, L. J., and Menasche, P. (2005) Cell-based cardiac repair: reflections at the 10-year point. *Circulation* **112**, 3174–3183

225. Menasche, P., Hagege, A. A., Scorsin, M., Pouzet, B., Desnos, M., Duboc, D., Schwartz, K., Vilquin, J. T., and Marolleau, J. P. (2001) Myoblast transplantation for heart failure. *Lancet* **357**, 279–280

226. Orlic, D., Kajstura, J., Chimenti, S., Jakoniuk, I., Anderson, S. M., Li, B., Pickel, J., McKay, R., Nadal-Ginard, B., Bodine, D. M., Leri, A., and Anversa, P. (2001) Bone marrow cells regenerate infarcted myocardium. *Nature* **410**, 701–705

227. Badorff, C., Brandes, R. P., Popp, R., Rupp, S., Urbich, C., Aicher, A., Fleming, I., Busse, R., Zeiher, A. M., and Dimmeler, S. (2003) Transdifferentiation of blood-derived human adult endothelial progenitor cells into functionally active cardiomyocytes. *Circulation* **107**, 1024–1032

228. Siepe, M., Heilmann, C., von Samson, P., Menasche, P., and Beyersdorf, F. (2005) Stem cell research and cell transplantation for myocardial regeneration. *Eur J Cardiothorac Surg* **28**, 318–324

229. Gnecchi, M., He, H., Liang, O. D., Melo, L. G., Morello, F., Mu, H., Noiseux, N., Zhang, L., Pratt, R. E., Ingwall, J. S., and Dzau, V. J. (2005) Paracrine action accounts for marked protection of ischemic heart by Akt-modified mesenchymal stem cells. *Nat med* **11**, 367–368

230. Fazel, S., Cimini, M., Chen, L., Li, S., Angoulvant, D., Fedak, P., Verma, S., Weisel, R. D., Keating, A., and Li, R. K. (2006) Cardioprotective c-kit+ cells are from the bone marrow and regulate the myocardial balance of angiogenic cytokines. *J Clin Invest* **116**, 1865–1877

231. Reinecke, H., Poppa, V., and Murry, C. E. (2002) Skeletal muscle stem cells do not transdifferentiate into cardiomyocytes after cardiac grafting. *J Mol Cell Cardiol* **34**, 241–249

232. Winitsky, S. O., Gopal, T. V., Hassanzadeh, S., Takahashi, H., Gryder, D., Rogawski,

M. A., Takeda, K., Yu, Z. X., Xu, Y. H., and Epstein, N. D. (2005) Adult murine skeletal muscle contains cells that can differentiate into beating cardiomyocytes in vitro. *PLoS Biol* **3**, e87

233. Chien, K. R. (2005) Alchemy and the new age of cardiac muscle cell biology. *PLoS Biol* **3**, e131

234. Wagers, A. J., and Weissman, I. L. (2004) Plasticity of adult stem cells. *Cell* **116**, 639–648

235. Murry, C. E., Soonpaa, M. H., Reinecke, H., Nakajima, H., Nakajima, H. O., Rubart, M., Pasumarthi, K. B., Virag, J. I., Bartelmez, S. H., Poppa, V., Bradford, G., Dowell, J. D., Williams, D. A., and Field, L. J. (2004) Haematopoietic stem cells do not transdifferentiate into cardiac myocytes in myocardial infarcts. *Nature* **428**, 664–668

236. Jackson, K. A., Majka, S. M., Wang, H., Pocius, J., Hartley, C. J., Majesky, M. W., Entman, M. L., Michael, L. H., Hirschi, K. K., and Goodell, M. A. (2001) Regeneration of ischemic cardiac muscle and vascular endothelium by adult stem cells. *J Clin Invest* **107**, 1395–1402

237. Janssens, S., Dubois, C., Bogaert, J., Theunissen, K., Deroose, C., Desmet, W., Kalantzi, M., Herbots, L., Sinnaeve, P., Dens, J., Maertens, J., Rademakers, F., Dymarkowski, S., Gheysens, O., Van Cleemput, J., Bormans, G., Nuyts, J., Belmans, A., Mortelmans, L., Boogaerts, M., and Van de Werf, F. (2006) Autologous bone marrow-derived stem-cell transfer in patients with ST-segment elevation myocardial infarction: double-blind, randomised controlled trial. *Lancet* **367**, 113–121

238. Kalka, C., Masuda, H., Takahashi, T., Kalka-Moll, W. M., Silver, M., Kearney, M., Li, T., Isner, J. M., and Asahara, T. (2000) Transplantation of ex vivo expanded endothelial progenitor cells for therapeutic neovascularization. *Proc Natl Acad Sci USA* **97**, 3422–3427

239. Vesely, I. (2005) Heart valve tissue engineering. *Circ Res* **97**, 743–755

240. Dalrymple-Hay, M. J., Pearce, R., Dawkins, S., Haw, M. P., Lamb, R. K., Livesey, S. A., and Monro, J. L. (2000) A single-center experience with 1,378 Carbo-Medics mechanical valve implants. *Ann Thorac Surg* **69**, 457–463

241. Schoen, F. J., and Levy, R. J. (1999) Founder's Award, 25th Annual Meeting of the Society for Biomaterials, perspectives. Providence, RI, April 28-May 2, 1999. Tissue heart valves: current challenges and future research perspectives. *J Biomed Mater Res* **47**, 439–465

242. Schoen, F. J., and Levy, R. J. (2005) Calcification of tissue heart valve substitutes: progress toward understanding and prevention. *Ann Thorac Surg* **79**, 1072–1080

243. Durbin, A. D., and Gotlieb, A. I. (2002) Advances towards understanding heart valve response to injury. *Cardiovasc Pathol* **11**, 69–77

244. Sutherland, F. W., Perry, T. E., Yu, Y., Sherwood, M. C., Rabkin, E., Masuda, Y., Garcia, G. A., McLellan, D. L., Engelmayr, G. C., Jr., Sacks, M. S., Schoen, F. J., and Mayer, J. E., Jr. (2005) From stem cells to viable autologous semilunar heart valve. *Circulation* **111**, 2783–2791

245. Hoerstrup, S. P., Kadner, A., Melnitchouk, S., Trojan, A., Eid, K., Tracy, J., Sodian, R., Visjager, J. F., Kolb, S. A., Grunenfelder, J., Zund, G., and Turina, M. I. (2002) Tissue engineering of functional trileaflet heart valves from human marrow stromal cells. *Circulation* **106**, I143–150

246. Neuenschwander, S., and Hoerstrup, S. P. (2004) Heart valve tissue engineering. *Transpl Immunol* **12**, 359–365

247. Riha, G. M., Lin, P. H., Lumsden, A. B., Yao, Q., and Chen, C. (2005) Review: application of stem cells for vascular tissue engineering. *Tissue Eng* **11**, 1535–1552

248. Deutsch, G., Jung, J., Zheng, M., Lora, J., and Zaret, K. S. (2001) A bipotential precursor population for pancreas and liver within the embryonic endoderm. *Development* **128**, 871–881

249. Shafritz, D. A., Oertel, M., Menthena, A., Nierhoff, D., and Dabeva, M. D. (2006) Liver stem cells and prospects for liver reconstitution by transplanted cells. *Hepatology* **43**, S89–98

250. Fausto, N., Campbell, J. S., and Riehle, K. J. (2006) Liver regeneration. *Hepatology* **43**, S45–53

251. Nussler, A., Konig, S., Ott, M., Sokal, E., Christ, B., Thasler, W., Brulport, M., Gabelein, G., Schormann, W., Schulze, M., Ellis, E., Kraemer, M., Nocken, F., Fleig, W., Manns, M., Strom, S. C., and

Hengstler, J. G. (2006) Present status and perspectives of cell-based therapies for liver diseases. *J Hepatol* **45**, 144–159

252. Rhim, J. A., Sandgren, E. P., Degen, J. L., Palmiter, R. D., and Brinster, R. L. (1994) Replacement of diseased mouse liver by hepatic cell transplantation. *Science NY* **263**, 1149–1152

253. Overturf, K., al-Dhalimy, M., Ou, C. N., Finegold, M., and Grompe, M. (1997) Serial transplantation reveals the stem-cell-like regenerative potential of adult mouse hepatocytes. *Am J Pathol* **151**, 1273–1280

254. Michalopoulos, G. K., Barua, L., and Bowen, W. C. (2005) Transdifferentiation of rat hepatocytes into biliary cells after bile duct ligation and toxic biliary injury. *Hepatology* **41**, 535–544

255. Koenig, S., Krause, P., Drabent, B., Schaeffner, I., Christ, B., Schwartz, P., Unthan-Fechner, K., and Probst, I. (2006) The expression of mesenchymal, neural and haematopoietic stem cell markers in adult hepatocytes proliferating in vitro. *J Hepatol* **44**, 1115–1124

256. Evarts, R. P., Nagy, P., Marsden, E., and Thorgeirsson, S. S. (1987) A precursor-product relationship exists between oval cells and hepatocytes in rat liver. *Carcinogenesis* **8**, 1737–1740

257. Evarts, R. P., Nagy, P., Nakatsukasa, H., Marsden, E., and Thorgeirsson, S. S. (1989) In vivo differentiation of rat liver oval cells into hepatocytes. *Cancer Res* **49**, 1541–1547

258. Fujio, K., Evarts, R. P., Hu, Z., Marsden, E. R., and Thorgeirsson, S. S. (1994) Expression of stem cell factor and its receptor, c-kit, during liver regeneration from putative stem cells in adult rat. *Lab Invest* **70**, 511–516

259. Nierhoff, D., Ogawa, A., Oertel, M., Chen, Y. Q., and Shafritz, D. A. (2005) Purification and characterization of mouse fetal liver epithelial cells with high in vivo repopulation capacity. *Hepatology* **42**, 130–139

260. Song, S., Witek, R. P., Lu, Y., Choi, Y. K., Zheng, D., Jorgensen, M., Li, C., Flotte, T. R., and Petersen, B. E. (2004) Ex vivo transduced liver progenitor cells as a platform for gene therapy in mice. *Hepatology* **40**, 918–924

261. Herrera, M. B., Bruno, S., Buttiglieri, S., Tetta, C., Gatti, S., Deregibus, M. C., Bussolati, B., and Camussi, G. (2006) Isolation and Characterization of a Stem Cell Population from Adult Human Liver. *Stem cells (Dayton, Ohio)* **24**, 2840–2850

262. Robertson, R. P. (2004) Islet transplantation as a treatment for diabetes – a work in progress. *N Engl J Med* **350**, 694–705

263. Bonner-Weir, S., and Weir, G. C. (2005) New sources of pancreatic beta-cells. *Nat Biotechnol* **23**, 857–861

264. Dor, Y., Brown, J., Martinez, O. I., and Melton, D. A. (2004) Adult pancreatic beta-cells are formed by self-duplication rather than stem-cell differentiation. *Nature* **429**, 41–46

265. Bonner-Weir, S., Baxter, L. A., Schuppin, G. T., and Smith, F. E. (1993) A second pathway for regeneration of adult exocrine and endocrine pancreas. A possible recapitulation of embryonic development. *Diabetes* **42**, 1715–1720

266. Zulewski, H., Abraham, E. J., Gerlach, M. J., Daniel, P. B., Moritz, W., Muller, B., Vallejo, M., Thomas, M. K., and Habener, J. F. (2001) Multipotential nestin-positive stem cells isolated from adult pancreatic islets differentiate ex vivo into pancreatic endocrine, exocrine, and hepatic phenotypes. *Diabetes* **50**, 521–533

267. Seaberg, R. M., Smukler, S. R., Kieffer, T. J., Enikolopov, G., Asghar, Z., Wheeler, M. B., Korbutt, G., and van der Kooy, D. (2004) Clonal identification of multipotent precursors from adult mouse pancreas that generate neural and pancreatic lineages. *Nat Biotechnol* **22**, 1115–1124

268. Bonner-Weir, S., Toschi, E., Inada, A., Reitz, P., Fonseca, S. Y., Aye, T., and Sharma, A. (2004) The pancreatic ductal epithelium serves as a potential pool of progenitor cells. *Pediatr Diabetes* **5** Suppl 2, 16–22

269. Baeyens, L., De Breuck, S., Lardon, J., Mfopou, J. K., Rooman, I., and Bouwens, L. (2005) In vitro generation of insulin-producing beta cells from adult exocrine pancreatic cells. *Diabetologia* **48**, 49–57

270. Rajagopal, J., Anderson, W. J., Kume, S., Martinez, O. I., and Melton, D. A. (2003) Insulin staining of ES cell progeny from insulin uptake. *Science NY* **299**, 363

271. Baharvand, H., Jafary, H., Massumi, M., and Ashtiani, S. K. (2006) Generation of insulin-secreting cells from human embryonic stem cells. *Dev Growth Differ* **48**, 323–332
272. Lavon, N., Yanuka, O., and Benvenisty, N. (2006) The effect of overexpression of Pdx1 and Foxa2 on the differentiation of human embryonic stem cells into pancreatic cells. *Stem cells (Dayton, Ohio)* **24**, 1923–1930
273. Zalzman, M., Gupta, S., Giri, R. K., Berkovich, I., Sappal, B. S., Karnieli, O., Zern, M. A., Fleischer, N., and Efrat, S. (2003) Reversal of hyperglycemia in mice by using human expandable insulin-producing cells differentiated from fetal liver progenitor cells. *Proc Natl Acad Sci USA* **100**, 7253–7258
274. Yang, L. J. (2006) Liver stem cell-derived beta-cell surrogates for treatment of type 1 diabetes. *Autoimmun Rev* **5**, 409–413

Part II

Regeneration of the Blood System

Chapter 6

Investigating the Interactions Between Haemopoietic Stem Cells and Their Niche: Methods for the Analysis of Stem Cell Homing and Distribution Within the Marrow Following Transplantation

Brenda Williams and Susan K. Nilsson

Abstract

Interactions between haemopoietic stem cells (HSC) and their microenvironment serve multiple functions including the attraction to and retention and regulation in the bone marrow HSC niche. However, the cell adhesion molecules involved, their HSC receptors and the mechanisms underpinning these processes remain poorly understood. An ability to thoroughly investigate the roles of specific molecules in this process relies on a variety of in vitro and in vivo assays including the assessment of a HSC ability to home to the bone marrow and analysis of its lodgement within the bone marrow.

Key words: Niche, haemopoietic stem cells, bone marrow, homing.

1. Introduction

The engraftment of HSC post-transplantation is a multi-step process involving homing, trans-marrow migration and lodgement of HSC within a bone marrow (BM) niche. Homing is the specific recruitment of HSC to the bone marrow and involves the recognition of HSC by the bone marrow microvascular endothelium and trans-endothelial cell migration into the haemopoietic space. Lodgement is defined as the selective migration of HSC to a suitable haemopoietic microenvironment within the extravascular compartment. It is this microenvironment, in which HSC become anchored that ultimately determine their fate.

Encapsulating the concept of highly specific, local interactions regulating haemopoiesis, Schofield (1) formulated the 'niche' hypothesis in which it was suggested that the most primitive haemopoietic cell exists in association with one or more other supporting cells and would therefore, in essence, be fixed tissue cells. These microenvironmental cells were postulated to form a specific niche which, when in close association with the HSC, confer upon it the HSC attribute of indefinite self-renewal capacity, while effectively inhibiting differentiation and maturation of the cell.

We as well as others have demonstrated that the HSC niche resides at the endosteal region of the BM *(2–5)*. Furthermore, we have recently demonstrated that HSC isolated from the endosteal region have an increased ability to home to the bone marrow, lodge at the endosteum and have greater haemopoietic potential *(6)*. Herein, we describe methodologies for the isolation of HSC from the central and endosteal regions of the BM, and murine models for the in vivo assessment of their ability to home to the BM post-transplantation as well as the analysis of their site of lodgement within the bone marrow extravascular space. It has been through the use of methodologies such as these that we have been able to demonstrate that molecules, such as hyaluronic acid and osteopontin are critical components of the HSC niche *(7, 8)*.

2. Materials

2.1. Isolation of Bone Marrow

1. Adult C57BL/6 J (Ly5.2) mice, 7–10 week old (*see* **Note 1**).
2. Sterile #11 surgical blade and #3 handle.
3. Phosphate-buffered saline (PBS): pH 7.2, 310 mOsm (*see* **Note 2**) supplemented with 2% serum: Defined bovine calf serum, iron supplemented (Se, Hyclone, Utah).
4. Murine stem cell isolation kit (Chemicon #SCR051-S), containing a mortar and pestle, the required digestive enzymes and cell strainers to filter bone marrow cells after isolation.
5. One millilitre syringes attached to 23- and 21-gauge needles to flush marrow from bones.
6. Fifty millilitre conical tubes for collection of bone marrow and flushed bones.
7. 37°C orbital shaker: We use an Eppendorf Thermomixer comfort model #5355 000.011.
8. Hemocytometer and microscope equipped with phase contrast or an automated cell counter. We use a Sysmex model KX-21 N.

2.2. Density Gradient Separation

1. Nycoprep™ 1.077Animal (Axis-Shield, Oslo, Norway).
2. Cannulas (Unomedical, #50011012) attached to 20 ml syringes.

2.3. HSC Pre-Enrichment Immunomagnetic Cell Separation

1. Lineage depletion antibody cocktail: a mixture of purified rat anti-mouse antibodies recognising the cell surface antigens: B220, GR-1, MAC-1 and TER119 (Pharmingen) (see **Note 3**).
2. Antibody dilution buffer: PBS supplemented with 2% serum.
3. PBS, 310 mOsm (see **Note 2**) supplemented with 2 mM EDTA and 0.1% (w/v) fraction V bovine serum albumen (BSA; pH 7.4).
4. Dynabeads for magnetic labelling of the cells: Sheep anti-rat IgG beads- 4.5 μm diameter, 4×10^8 beads/ml (Dynal Biotech ASA, Oslo, Norway).
5. Magnets: Dynal MPC-MPC-S for 20 μl–2 ml samples, MPC-L for a 1–8 ml sample.
6. Suspension mixer: allowing both tilting and rotation at 4–8°C for Dynabead incubation step. (We use a Ratek suspension mixer (RSM6) in cold room.)
7. Falcon polypropylene round bottom tubes: 5 ml tubes (12 × 75 mm style-352063) and 14 ml tubes (17 × 100 mm style-352059).

2.4. HSC Fluorescence Activated Cell Sorting

1. Fluorescent conjugated antibodies: a mixture of purified fluorescein isothiocyanate (FITC) conjugated rat anti-mouse stem cell antigen 1 (Sca-1) and phycoerythrin (PE) conjugated rat anti-mouse c-kit (Pharmingen).
2. Antibody dilution buffer: PBS (pH 7.2), 310 mOsm (see **Note 2**) supplemented with 2% serum.
3. 5 ml polystyrene, round-bottom tube with 35 μm cell strainer cap (Becton Dickinson) (see **Note 4**).
4. Dual-laser flow cytometer with sorting capability. We use a Becton Dickinson FACSVantage serum with DIVA option.
    a. Primary laser is a Sapphire 488-200 CDRH (DPSS) laser tuned to emit 200 mW of 488 nm laser light (Coherent Inc., Santa Clara, CA).
    b. Secondary laser is a 170c-Spectrum MLUV (DPSS) laser tuned to emit 50 mW of 350.7–356.4 nm laser light (Coherent Inc.).
    c. We use an 80 μm nozzle and sort at 30 psi, drop delay frequency of 61 kHz for HSC sorting.
5. Sheath fluid: Isoton II (Beckman Coulter).
6. Collection microtube –1.7 ml, sterile (Scientific Specialties, Inc. -1210-00).
7. Hemocytometer.
8. Microscope equipped with phase contrast.

2.5. HSC Tracking

1. Carboxyfluorescein diacetate, succinimidyl ester (CFDA, SE) (Molecular Probes) (*see* **Note 5**).
2. Dimethyl sulphoxide (DMSO) AnalaR grade (MERCK).
3. PBS (pH 7.2), 310 mOsm (*see* **Note 2**) supplemented with 0.5% heat inactivated serum.
4. PBS (pH 7.2), 310 mOsm (*see* **Note 2**) supplemented with 20% heat inactivated serum.
5. PBS (pH 7.2), 310 mOsm (*see* **Note 2**).
6. Water bath maintained at 37°C.
7. Hemocytometer.
8. Microscope equipped with phase contrast.

2.6. Section Analysis

1. 2% Paraformaldehyde, 0.05% gluteraldehyde fixative made in 0.1 M Sorensen's phosphate buffer (*see* **Note 6**).
2. Perfusion apparatus (B/Braun compact S).
3. Perfusion tubing (Microtube Extrusions #PE8040).
4. Two 26-gauge needles.
5. 10 ml syringe.
6. 10% EDTA (pH 7.0) made in dH$_2$O.
7. Ratek suspension mixer (RSM6).
8. Histology department capable of embedding and sectioning paraffin blocks.
9. Vectashield mounting media (Vector laboratories H 1000).
10. Fluorescence microscope equipped with dual filter for FITC and Texas Red (green excitation at 578 nm and red excitation at 610 nm).

2.7. Homing Analysis

1. Phycoerythrin (PE) conjugated rat anti-mouse CD45 antibody (Pharmingen).
2. Flow cytometer or analyser with ability to distinguish FITC and PE. Refer to **Table 6.1**.

Table 6.1
Flow cytometry instrumentation setup

Fluorochrome label	Antibody conjugate	PMT parameter	Excitation wavelength	Optical filter
FITC	Sca-1	FL 1	488	530/30
PE	c-Kit	FL 2	488	585/42
		FL 4	488	780/60

3. Methods

3.1. Bone Marrow Harvesting

3.1.1. Central Bone Marrow

1. Kill mice by cervical dislocation and dissect femurs, tibias and iliac crests (*see* **Note 7**).
2. The metaphyseal region of each of the femur and tibia are removed and added into a 50 ml centrifuge tube containing PBS-2% serum for harvesting the endosteal sample. For the iliac crest, the acetabular notch is removed and added to the 50 ml centrifuge tube containing PBS-2% serum for harvesting the endosteal sample.
3. Using a 1 ml syringe containing PBS-2% serum attached to a 21-gauge needle insert the needle into each epiphysis of the femoral shaft and the knee epiphysis of the tibia, repeatedly flushing the marrow contents into a 50 ml centrifuge tube containing 40 ml of PBS-2% serum. To flush the iliac crest, use a 23 gauge needle attached to a 1 ml syringe containing PBS-2% serum and flush from the acetabular notch into the 50 ml centrifuge tube. Marrow is flushed equally into two collection tubes (*see* **Note 8**).
4. Wash the cells by centrifuging at 400 g for 5 min at 4°C.
5. Decant supernatant and resuspend the cell pellets in 20 ml PBS-2% serum (*see* **Note 9**).
6. Filter the cell suspension through a 40 μm nylon cell strainer into a fresh 50 ml conical tube.
7. Dilute cells to 40 ml with PBS-2% serum, store on ice for density gradient separation (*see* **Section 3.2**) and perform a cell count.

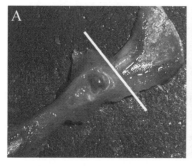 

Fig. 6.1. The iliac crest with a line through where to remove the flat triangular piece of cartilage (**A**) and where the acetabular notch is removed ready to be flushed (**B**). (*See* Color Plate 4)

3.1.2. Sampling of Endosteal Marrow

1. Kill the mice, harvest bones and flush marrow as for central marrow harvest (*see* **Section 3.1.1**), except place bones in a 50 ml centrifuge tube containing PBS-2% serum.
2. Decant the bones in buffer into a sterile mortar.
3. Grind the bones with the pestle (*see* **Note 10**).
4. Remove cell supernatant with a pipette and filter through a 40 µm nylon cell strainer into one of two 50 ml conical tubes.
5. Rinse the crushed bone fragments and filter as in step **4** to collect a total of 100 ml PBS-2% serum. Set tubes aside on ice until step 10.
6. Transfer the crushed bones into a 50 ml conical tube containing 10 ml of enzymes from stem cell isolation kit and agitate for 5 min at 37°C in an orbital shaker, 750 rpm.
7. Add 25 ml of PBS to the bone fragments and shake vigorously for 10 s.
8. Filter the cell suspension through a 40 µm nylon cell strainer into a 50 ml conical tube, prior to washing the bone fragments with 25 ml PBS-2% serum by again vigorously shaking for 10 s.
9. Filter the cell suspension through the cell strainer into the 50 ml conical tube (*see* **Note 11**).
10. Centrifuge all tubes of cell suspensions (4 × 50 ml tubes) at 400 g, 5 min, 4°C.
11. Decant supernatant (*see* **Note 9**), resuspend and pool cell pellets in 40 ml PBS-2% serum prior to storing on ice for density gradient separation (*see* **Section 3.2**). Perform a cell count.

3.2. Density Gradient Separation

1. Dilute the cell suspension to approximately 2×10^8 cells/20 ml with PBS-2% serum.
2. Divide 20 ml aliquots of cell suspension over an even number of 50 ml centrifuge tubes.
3. Underlay each gradient with 10 ml Nycoprep 1.077A using a cannula attached to a 20 ml syringe.
4. Centrifuge the gradients at 600 g for 20 min at room temperature with no de-acceleration.
5. Collect the mononuclear cells from the interface between the PBS layer and the Nycoprep solution into a 50 ml centrifuge tube using a cannula attached to a 10 ml syringe. Collect the mononuclear cells of two gradients into a 50 ml and fill the tube with PBS-2% serum.
6. Collect the mononuclear cells of two gradients into a 50 ml and fill the tube with PBS-2% serum.
7. Centrifuge the tubes at 400 g for 5 min, 4°C with no de-acceleration.

8. Decant the supernatant, and resuspend the pooled cell pellets in 50 ml PBS-2% serum.
9. Perform a cell count.

3.3. HSC Pre-enrichment Immunomagnetic Cell Separation

3.3.1. Immunolabelling Cells with a Cocktail of Lineage Antibodies

1. Pellet cells by centrifuging at 400 g for 5 min, 4°C.
2. Decant supernatants, leaving a dry pellet.
3. Resuspend cell pellets at 1×10^7 cells/100 µl in an optimally pre-titred (*see* **Note 12**) cocktail of lineage antibodies (*see* **Note 13**).
4. Incubate cells for 15 min on ice.
5. Wash labelled cells in PBS-2% serum by centrifuging at 400 g for 5 min, 4°C to remove unbound antibody (*see* **Note 14**).
6. Pipette off supernatant completely and resuspend the cell pellets in PBS supplemented with 2 mM EDTA and 0.1% BSA at a concentration of between 10^7 and 10^8 cells/ml (*see* **Note 15**).

3.3.2. Dynabeads Washing Procedure

1. Resuspend Dynabeads.
2. The optimal Dynabead to cell ratio used in this protocol has been established as half a bead per cell, with the depletion repeated with the same number of beads (*see* **Note 16**). Dispense beads for both steps into individual 1.5 ml tubes and follow the washing procedure as outlined below:
    a. Add 1 ml of 2 mM EDTA and 0.1% BSA to each tube and mix.
    b. Place the tubes in the magnet for 1 min, remove and discard supernatant.
    c. Remove tube from magnet and resuspend the Dynabeads in 1.0 ml of 2 mM EDTA and 0.1% BSA.
    d. Repeat step **b**.
3. Remove tube from the magnet and resuspend the Dynabeads in 0.25 ml of 2 mM EDTA and 0.1% BSA.

3.3.3. Immunomagnetic Separation

1. Add washed Dynabeads to the cell suspensions.
2. Incubate for 5 min at 2–8°C with gentle tilting and rotation.
3. Place tube in the magnet for 2 min.
4. Transfer supernatant containing the unbound cells to a fresh 5 ml collection tube.
5. Whilst still in the magnet, rinse the bead bound cells with 1 ml buffer.

6. Transfer the supernatant containing any residual unbound cells to the collection tube.
7. Add the second aliquot of washed Dynabeads to the cell suspension in the collection tube.
8. Incubate for 10 min at 2–8°C with gentle tilting and rotation.
9. Place tubes (A) in magnet for 2 min.
10. Transfer the supernatant containing the unbound cells to new 5 ml collection tube (B).
11. Place collection tube (B) in magnet for 2 min (*see* **Note 17**).
12. Transfer the supernatant containing unbound cells to a 10 ml polypropylene collection tube (*see* **Note 18**).
13. Make up the volume of the unbound, lineage negative cell suspension to 10 ml and count.

3.4. HSC Fluorescence-Activated Cell Sorting

3.4.1. HSC Labelling for Cell Sorting

1. Centrifuge lineage negative cells and aspirate supernatant.
2. Resuspend cell pellet at 1×10^7 cells/100 µl in an optimally pre-titred (*see* **Note 12**) antibody cocktail of anti-mouse Sca-1-FITC and anti-mouse c-kit-PE.
3. Incubate light protected on ice for 20 min.
4. Wash cells in a 10-fold volume of PBS-2% serum. Centrifuge at 400 g for 5 min, 4°C. Decant the supernatant.
5. Resuspend the cells in PBS-2% serum and filter the cell suspension through a cell strainer prior to fluorescence-activated cell sorting. A cell concentration of 10×10^6 cells/ml is considered optimal for cell sorting (*see* **Note 19**).

3.4.2. Sort Gating Strategies for Isolation of Haemopoietic Stem Cells

The set up of any flow cytometer, including biological controls for compensation, is essential for the identification and accurate sorting of specific subsets of cells. This is particularly critical when sorting very rare events such as HSC (*see* **Note 20**). Aliquots of the following cell samples are required for selection of instrument settings and fluorescence compensation.

1. Unstained bone marrow cells for setting scatter profiles and gains for background fluorescence, as well as a lymphoblastoid light scatter gate.
2. Bone marrow cells stained with CD45-FITC for FITC compensation control.
3. Bone marrow cells stained with CD45-PE for PE compensation control (*see* **Note 21**).

Both forward and orthogonal angle light-scatter signals are detected through 488 nm bandpass filters, amplified and measured

Table 6.2
Sequential gating strategy and region definitions

Region	
R1 SSC-H vs SSC-A	Doublet discrimination
AND R2 FSC vs SSC	Viable lin⁻ cells in lymphoblastoid gate
NOT R3 Sca-1 FITC (−) vs FL4	Autofluorescent cells (*see* **Note 22**)
AND R4 Sca-1 FITC (+) vs c-kit PE (+)	Sca⁺c-kit⁺ cells

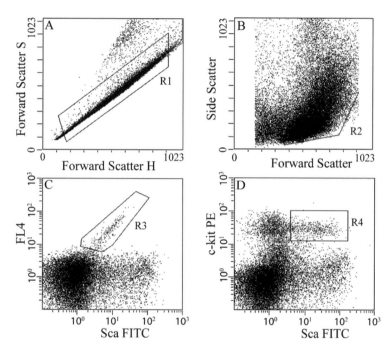

Fig. 6.2. Sequential gating for LSK cells as described in **Table 6.2**. Single cells are selected (**A**), then a blast region imposed (**B**). Autofluorescent cells are excluded (**C**) prior to LSK cells being selected (**D**).

on a linear scale. All fluorescent signals are amplified and measured over a logarithmic scale as specified in **Table 6.1**. Sequential gating strategies were used as described in **Table 6.2** and **Fig. 6.2**.

3.4.3. Cell Sorting — Cells were sorted and collected at 4°C into PBS 2% serum.

3.5. Transplantation of CFDA, SE Labelled HSC

3.5.1. Preparation of CFDA, SE Stock

1. Add 1 ml anhydrous DMSO to 25 mg vial to make a 44.8 M stock solution of CSFE.

2. In a glass vial, make a working solution of 5 mM CFDA, SE by adding a 100 μl aliquot of the 44.8 M stock solution of CFDA, SE to 796 μl of DMSO. Store the CFDA, SE at −20°C in a secondary container with desiccant.

3.5.2. CFDA, SE Staining of Cells

1. In a 5 ml polypropylene tube, resuspend cells at 5×10^6 cells/ml in PBS supplemented with 0.5% heat inactivated serum (*see* **Note 23**).
2. Pre-warm cells in a 37°C water bath for 2 min.
3. Prepare a 5 μM CFSE solution by diluting 1 μl of 5 mM CFDA, SE stock in 999 μl in PBS. Keep light protected.
4. Add 111 μl of the 5 μM CFDA, SE solution to each 1 ml of cell suspension to give a final concentration of 0.5 μM CFDA, SE.
5. Incubate cell suspension with gentle agitation at 37°C for 10 min in the dark.
6. Transfer cells to a 15 ml conical tube containing 1 ml ice cold PBS supplemented with 20% serum and mix gently.
7. Add a further 8–10 ml cold PBS supplemented with 20% serum.
8. Centrifuge cells at 400 g for 5 min. Decant supernatant.
9. Wash cells once more in 10 ml PBS.
10. Count cells and add 200,000 unlabelled whole bone marrow filler cells per recipient.
11. Centrifuge cells at 400 g for 5 min. Decant supernatant.
12. Resuspend cells in 200 μl PBS per recipient mouse; allow 100 μl extra cell suspension (*see* **Note 24**).

3.5.3. Transplants

1. Inject cell suspension into non-ablated recipient mouse via the lateral tail vein.
2. Allow the cells to engraft for 15 h (*see* **Note 25**).

3.5.4. Transplantation Analysis – Spatial Distribution

1. After 15 h, the femoral bone marrow is fixed by perfusing anaesthetised mice with 2% paraformaldehyde, 0.05% gluteraldehyde through the descending aorta at physiological pressure (20.4 ml/h) (*see* **Note 26**). After inserting the perfusion needle, cut the inferior vena cava to release the blood pressure.
2. Excise the femurs and immerse in fixative for 6–12 h at room temperature, while rotating.
3. Remove the femurs from the fixative and rinse in 10% EDTA.
4. Place the bones in a tube containing 10 ml 10% EDTA.
5. Using a suspension mixer, rotate the tube at 4°C for 10 days. Replace the 10% EDTA daily (*see* **Note 27**).
6. The bones are then either dehydrated in graded ethanol and embedded in paraffin or washed in PBS twice and embedded in OCT.

7. Cut longitudinal sections (3.5 μm) of the femur, de-wax and bring to water (*see* **Note 28**).

8. Wash sections in PBS prior to mounting in Vectashield anti-fade mounting media.

9. Sections are analysed using a fluorescence microscope fitted with a FITC and Texas red dual filter set (*see* **Note 29**).

10. The spatial distribution of the transplanted cells is determined by analysing the location of the CFSE$^+$ cells from at least six longitudinal sections per transplant recipient under the fluorescence microscope. The locations of positive cells are designated as either endosteal (previously arbitrarily defined as within 12 cells of the endosteum) or central (>12 cells from either endosteum) (2) (**Fig. 6.3** and Color Plate 5) (*see* **Note 30**).

3.5.5. Transplantation Analysis – Homing

1. After 15 h, the transplanted mice are euthanised and the central and endosteal marrow harvested from femurs, tibia and iliac crests as described in **Sections 3.1.1.** and **3.1.2.**, respectively. The number of cells in each of the central and endosteal regions is then counted.

2. 3×10^6 bone marrow cells are labelled with CD45-PE, by incubating the cells in 100 μl PBS 2% serum with 0.5 μg/ml antibody on ice for 20 min, prior to washing using centrifugation (5 min at 400 g) with excess PBS 2% serum, to provide a denominator for the total number of cells analysed.

3. 2×10^6 events are run through the flow analyser, and the proportion CD45$^+$ as well as the number of CFSE$^+$ events determined. The denominator (D) is mathematically calculated as the total number of CD45$^+$ cells analysed (D = % CD45$^+$/100 × 2 × 10^6).

4. The proportion donor (% do) of analysed BM is then calculated using the number of CFSE$^+$ events detected (% do = #CFSE$^+$/D × 100).

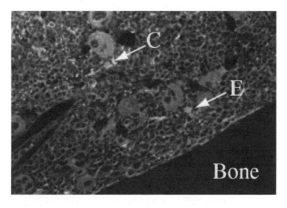

Fig. 6.3. Transplanted LSK cells detected using a spatial distribution assay. Cells are designated as either central (**C**) or endosteal (**E**). (*See* Color Plate 5)

5. The total number of donor cells detected in the bone marrow of each recipient is calculated using the proportion of donor cells, the total number of cells harvested in each region of bone marrow (#BM) and the assumption that this represents 30% of the total number of cells in the mouse (total # do = [%do/100 × #BM] /30 × 100).

6. Finally, the proportion of CFSE$^+$ cells from each transplant (% CFSE$^+$ tplt) detected in each recipient is calculated using the total number of donor cells and the number of CFSE$^+$ cells transplanted (# cells tplt) into each recipient (% CFSE+ tplt = total # do/# cells tplt × 100) (*see* **Note 31**).

4. Notes

1. All the volumes in this method are described for the marrow harvested from 10 mice, which we routinely use for each HSC isolation. These can be adjusted for smaller or larger numbers of animals.

2. This osmolarity is appropriate for murine cells and results in better cell recovery.

3. We use this limited number of lineage antibodies in our cocktail. This results in the removal of approximately 70% of whole bone marrow cells. Additional antibodies can be added for a more complete removal of lineage committed cells.

4. Although this product is packaged as non-sterile, we have not experienced post-sort contamination when using this product for pre-sort filtration.

5. This product is commonly (but incorrectly) referred to as CFSE. When CFDA-SE is incorporated into the cell, it is converted to CFSE by intracellular esterases.

6. Fixative is made in a fume hood. Weigh 400 g of PFA and add to 190 ml of 0.1 M Sorenson's phosphate buffer (dilute from 0.2 M stock with distilled water) in heat proof beaker. Stir on heat (approx. 55°C) until powder dissolves (approx. 1–2 h), then let the solution cool. Filter into measuring cylinder through no.1 Whatman paper and add 400 μl gluteraldehyde (25%). Top up to 200 ml with 0.1 M Sorenson's phosphate buffer. Store frozen in 15 ml aliquots. Sorensens phosphate buffer: 0.2 M is made in two parts. Solution A = 0.2 M of Na$_2$HPO$_4$ (28.3 g/L) and Solution B = 0.2 M of KH$_2$PO$_4$ (27.2 g/L) (5.44 g/200 ml). Both are dissolved in water. Lower pH of solution A to pH7.4 by slowly adding solution B and mixing well.

7. Remove iliac crest, femur and tibia from the spinal cord of the mouse. Next remove the muscle from each bone by carefully scraping the bones with the scalpel blade. Dislocate the femur from the knee. Scrape the area around the head of the femur as clean as possible. To allow a 21-gauge needle to be inserted, shave a small sliver of bone from behind the femoral head. Excise the tibia by pulling the foot and peeling the muscle away. Dislocate the knee. When the muscle is removed from the iliac crest, a flat triangular piece of cartilage is exposed. Remove this cartilage by cutting it through the acetabular notch (Refer to **Fig. 6.1** and Color Plate 4).

8. Flushed bones can be placed into the 50 ml conical tube containing the bone ends in PBS-2% serum for harvesting of the endosteal marrow.

9. To maximise cell recovery, the supernatants should be re-centrifuged.

10. The purpose of this step is to expose the medullary cavity and surrounding cancellous bone to enzymatic digestion without reducing the bone to powder.

11. The bones will be white when properly flushed, ground and washed.

12. Each antibody needs to be individually titred to determine the optimal working concentration for lineage depletion. As a guide we used our antibodies at 1–2.5 μg/ml.

13. It is a good practice to centrifuge the antibody cocktail briefly in a microfuge before use; the supernatant is then used, eliminating non-specific background staining by any protein aggregates formed during storage of the antibodies.

14. If desired, cells may be washed twice.

15. Typically we would resuspend cells as follows: Endosteal cells ($<3 \times 10^8$) in 3 ml buffer in 5 ml polypropylene tube (Falcon #352063), central cells ($>3 \times 10^8$) in 6 ml buffer and place 3 ml in each of two 5 ml polypropylene tubes.

16. The number of Dynabeads used has been determined to maximally remove the number of lineage committed cells labelled by the lineage cocktail.

17. This step is to ensure the removal of any residual Dynabeads from the cell suspension.

18. To maximise cell recovery, Tubes (A) and (B) are held in the magnet and sequentially washed with 1 ml of buffer, which is added to the cell suspensions in the 10 ml collection tube.

19. In order to reduce the incidence of nozzle clogs during sorting, sort as soon as possible after labelling and filter the sample immediately prior to the sort.

20. In addition, the yield and purity of any sort will be optimised by maximising the frequency of the population you wish to sort in your sample.
21. For convenience, we use unfractionated bone marrow for the compensation controls. This does not use our enriched cell population, but requires using a highly expressed marker such as the anti-mouse CD45 for compensation.
22. With our staining method on mouse bone marrow cells, this band of auto-fluorescence can be excluded using this filter combination.
23. A minimum volume of 200 µl is recommended for cell labelling.
24. An aliquot of the cell suspension should be analysed by flow to test the CFDA, SE staining.
25. This time point was chosen to ensure that donor cells found in the marrow will accurately reflect the fate of transplanted cells rather than their progeny *(9)*.
26. To set up the perfusion apparatus, insert a 26-gauge needle into an end of a ~70 cm length of perfusion tubing. This is then attached to the 10 ml syringe filled with fixative and placed in the perfusing apparatus as per manufacturer's specifications. Gently snap the other 26-gauge needle from its base and insert it into the other end of the tubing. This will be inserted into the descending aorta.
27. Decalcification can take up to 3 weeks if 10% EDTA is changed weekly.
28. Sections are de-waxed as per any standard histological protocol.
29. This filter set was specifically chosen because the short emission bandwidths allow CFSE positive cells to be easily distinguished from host bone marrow cells.
30. To ensure that individual cells are only analysed once, every alternate 3.5 µm section is analysed.
31. This data can be represented for each individual region of bone marrow, or for total marrow by adding together the two individual percentage of CFSE[+] cells from each region to give a total of the transplant homed to the bone marrow.

References

1. Schofield, R. (1978) The relationship between the spleen colony-forming cell and the haemopoietic stem cell. *Blood Cells* **4**, 7–25.
2. Nilsson, S. K., Dooner, M. S., Tiarks, C. Y., Weier, H.-U. G., Quesenberry, P. J. (1997) Potential and distribution of transplanted hematopoietic stem cells in a nonablated mouse model. *Blood* **89**, 4013–4020.
3. Nilsson, S. K., Johnston, H. M., Coverdale, J. A. (2001) Spatial localization of transplanted hemopoietic stem cells: inferences for the localization of stem cell niches. *Blood* **97**, 2293–2299.

4. Calvi, L. M., Adams, G. B., Weibrecht, K. W., Weber, J. M., Olson, D. P., Knight, M. C. et al. (2003) Osteoblastic cells regulate the haematopoietic stem cell niche. *Nature* **425**, 841–846.

5. Zhang, J., Niu, C., Ye, L., Huang, H., He, X., Tong, W. G. et al. (2003) Identification of the haematopoietic stem cell niche and control of the niche size. *Nature* **425**, 836–841.

6. Haylock, D. N., Williams, B., Johnston, H. M., Liu, M. C., Rutherford, K. E., Whitty, G. A., Simmons, P. J., Bertoncello, I., Nilsson, S. K. (2007) Hemopoietic stem cells with higher hemopoietic potential reside at the bone marrow endosteum. *Stem Cells* **25**, 1062–1069.

7. Nilsson, S. K., Haylock, D. N., Johnston, H. M., Occhiodoro, T., Brown, T. J., Simmons, P. J. (2003) Hyaluronan is synthesized by primitive hemopoietic cells, participates in their lodgment at the endosteum following transplantation, and is involved in the regulation of their proliferation and differentiation in vitro. *Blood* **101**, 856–862.

8. Nilsson, S. K., Johnston, H. M., Whitty, G. A., Williams, B., Webb, R. J., Denhardt, D. T. et al. (2005) Osteopontin, a key component of the hematopoietic stem cell niche and regulator of primitive hematopoietic progenitor cells. *Blood* **106**, 1232–1239.

9. Nilsson, S. K., Dooner, M. S., Quesenberry, P. J. (1997) Synchronized cell-cycle induction of engrafting long-term repopulating stem cells. *Blood* **90**, 4646–4650.

Chapter 7

Ex Vivo Megakaryocyte Expansion and Platelet Production from Human Cord Blood Stem Cells

Valérie Cortin, Nicolas Pineault, and Alain Garnier

Abstract

The identification and cloning of thrombopoietin was certainly a defining moment for the study of megakaryopoiesis and thrombopoiesis ex vivo. This and other progresses made in the development of culture processes for hematopoietic stem cells have paved the way for ongoing clinical trials and, in the future, for the potential therapeutic use of ex vivo produced blood substitutes such as platelets. This chapter describes a 14-day culture protocol for the production of megakaryocytes (MK) and platelets from human cord blood stem cells. The CD34+ cells are grown in a serum-free medium supplemented with a newly developed cytokine cocktail optimizing MK differentiation, expansion, and maturation. A detailed methodology for flow cytometry analysis of the cells and platelets is also presented together with supporting figures. A brief review on megakaryocytic differentiation and ex vivo MK cultures is first presented.

Key words: Megakaryocytes, platelets, hematopoietic stem cells, cord blood, flow cytometry, ex vivo cell culture, culture medium optimization, stem cells expansion, cytokines, culture medium design, culture medium optimization, statistical design of experiments.

1. Introduction

The capacity to grow and expand hematopoietic stem cells (HSC) ex vivo now makes possible the culture of large quantities of usually rare cells, such as megakaryocytes (MKs), for functional studies or for therapeutic applications that include the co-injection of ex vivo expanded megakaryocyte-progenitors (MK-progenitors) in patient undergoing HSC transplantation. The optimization of culture processes for platelet production ex vivo is also of great value toward a better understanding of the basic mechanisms regulating and orchestrating megakaryopoiesis and thrombopoiesis. Together, these may make possible the ex vivo production of platelets suitable for transfusion, which would represent a major interest for blood banks that still seek a valuable alternative to blood-derived platelets.

Human HSC and MK have for common property to be extremely scarce, representing approximately 0.05% and 0.4% of whole marrow, respectively. Because of their low frequencies, these cells are often studied after enrichment using various cell surface antigens. MKs are commonly enriched by selection of CD41a (GPIIb) or CD61 (GPIIIa) positive cells. Though a large number of techniques exist for the purification of human HSC, the most common technique relies on the isolation of CD34+ cells by fluorescent-activated cell sorting (FACS) or immunoselection (reviewed in *(1)*). In both cases, negative selection is recommended to reduce the impact of the selection procedure on the cells of interest.

Large quantities of MKs and platelets can also be produced ex vivo from various sources of hematopoietic progenitor cells enriched in the CD34+ fraction. These include bone marrow, peripheral blood mobilized CD34+ cells, and cord blood (CB) cells *(2–4)*. CB cells present logistical and safety advantages compared to other CD34+ sources. However, it is important to point out that the maturation of CB-derived MK ex vivo is not as extensive as that seen for bone marrow-derived MKs. The major differences include a greater proliferative potential ex vivo, a reduced terminal cell size and a reduced polyploidization potential for CB-derived MKs *(5, 6)*.

MKs are derived from HSC following a number of differentiation events (recently reviewed in *(7)*) that lead to the emergence of a bi-potent progenitor with MK and erythroid potential *(8)*. Differentiation of this progenitor along the MK lineage is first apparent by the expression of CD41a (GPIIb) and CD61 (GPIIIa/β3), the earliest markers of MK differentiation. CD41a and CD61 are the most abundant glycoproteins on the surface of platelets and form a functional dimeric complex that recognizes and binds to fibrinogen and the von Willebrand factor (vWF). During the course of MK maturation, the MK will express various cell surface molecules also present on platelets. These include a second receptor that recognizes the vWF which is composed of three chains: GPIbα, GPV, and GPIX. Moreover, GP1bα, also referred to as CD42b, is commonly used to discriminate between immature MK and mature MK. The expression of these markers during the course of megakaryopoiesis is schematized in **Fig. 7.1**. As MK maturation progresses, the expression of MK receptors increases as well as the cell size and DNA content. Flow cytometry is perfectly suited to monitor these characteristics of MK differentiation, as shown in **Fig. 7.2**. The release of platelets from proplatelets-bearing MK is normally accompanied by apoptosis (see chapters by Italiano et al. on proplatelet filament formation for more details, *(9)*). Thus, viability of

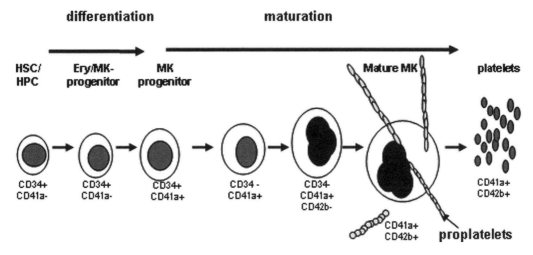

Fig. 7.1. Schematic representation of megakaryocytic differentiation showing the expression of the major cell surface antigen used to characterize various stages of MK differentiation. Note that CD61 (GPIIIa) is not shown but its expression parallels that of CD41a (GPIIb).

MK cell culture is necessarily low once the MKs are fully matured and platelets are being released.

The maturation of MK is a complex process essentially characterized by two phases: endomitosis and cytoplasmic maturation *(10)*. Endomitosis leads to the formation of large polyploid MK with a single nucleus and a DNA content greater than $4n$ ($16n$–$128n$, reviewed in *(11)*). Cytoplasmic maturation is characterized by an increase in size of the cytoplasm and a profound reorganization of the cytoplasmic cytoskeleton that leads to the formation of the invaginated demarcation membrane system from which the proplatelets filaments are derived *(12)*. This intricate process is terminated by the release of newly assembled

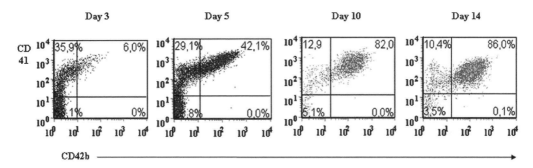

Fig. 7.2. Flow cytometry analysis of CB CD34+ cell culture undergoing MK differentiation. The expression of CD41a (*y*-axis) and CD42b (*x*-axis) is presented in dot-plot analysis as a function of time. Note that the majority of MKs at day 3 are immature since they do not express CD42b. As megakaryocytic maturation progresses, the majority of the MK gain expression of CD42b, and the intensity of both markers increases. These cultures were maintained in the MK medium supplemented with the BS1 cytokine cocktail at 39°C under 10% CO_2 for 14 days, as described in the method.

platelets from the protruding proplatelet filaments, in the marrow-intravascular sinusoidal space.

Because of their low frequencies and the need for large cell numbers for functional and biochemical studies, numerous articles have been published on the expansion of MK cells ex vivo *(13–21)*. The majority of the conditions described in the literature are based on serum-free medium supplemented with various cytokine cocktails, all of which include thrombopoietin (TPO). The use of pleiotropic cytokines, such as IL-3 and IL-6, and early-acting cytokines such as stem cell factor (SCF) and Flt-3 ligand (FL), usually leads to increased MK expansion, but at the expense of purity. In contrast, high MK purity cultures (>90% CD41+) can be easily obtained by culturing CD34+ cells with TPO alone (10–100 ng/mL) in serum-free media, though MK yields are then substantially reduced. Recent studies have also shown that adhesion molecules, allowing MK interactions with bone marrow components, may also be involved in MK expansion *(22)* and maturation *(23)*.

Due to their potential therapeutic value to reduce the extended period of thrombocytopenia in patients undergoing CB transplantation, much effort has also been directed toward the development of cytokine cocktails optimizing the expansion of MK-progenitors ex vivo *(14, 16, 19, 24, 25)*. Cytokines, such as TPO, SCF, FL, IL-3, IL-6, GM-CSF, and IL-11, have all been shown to support the expansion of MK-progenitors to different extents.

A simple and efficient culture protocol developed in our laboratory, together with the related flow cytometry analysis methods, are described in this chapter for the differentiation and maturation of CD34+ cells along the MK lineage, as well as the expansion and maturation of MK. The proposed culture conditions and cytokine cocktail (BS1) were developed specifically for the differentiation and maturation of CB-derived MK cells *(13, 25)*. The cytokine cocktail BS1 was developed by a statistical design of experiment approach which involved the screening of 13 cytokines by two-level factorial designs and the optimization of the concentration of the selected cytokines by a response surface methodology *(13)*. Together, they provide good MK expansion and excellent purity when used with CB CD34+ enriched cells.

2. Material

2.1. CB Mononuclear Cell MNC) and CD34+ Cell Preparation

CB CD34+ enriched cells are used in this method as a starting source of HSC and multipotent progenitors to derive the MK-progenitors, MK and platelets ex vivo. CD34+ cells can be enriched by any means available; the details provided below describe the technique of negative immunomagnetic purification currently used in our laboratory.

1. MNC are isolated from CB (*see* **Note 1**) using Ficoll-Hypaque isotonic solution (1.077 g/mL; Nicomed Pharma, Oslo, Norway).
2. CD34+ cells are enriched using a complete StemSep separation unit (Stem Cell Technologies, Vancouver, BC, Canada). The Human Progenitor Enrichment Cocktail kit (Stem Cell Technologies) was used for lineage depletion with antibody mixtures consisting of anti-CD2, -CD3, -CD14, -CD16, -CD19, -CD24, -CD56, -CD66b, and glycophorin A.
3. Iscove's modified Dulbecco's medium (IMDM; Invitrogen).
4. Cryoprotective medium: 40% fetal bovine serum (FBS, qualified, Invitrogen) and 10% dimethylsulfoxide (DMSO) in IMDM.
5. Centrifuge (Beckman GS-6; Beckman Coulter, Montréal, Qc, Canada).
6. 15 and 50 mL polypropylene centrifuge tubes (Falcon, Beckton Dickinson Labware, Franklin Lakes, NJ, USA).
7. 2 mL propylene cryotubes (VWR International, Mississauga, Ont., Canada).

2.2. CD34+ Cell Proliferation and Differentiation into MK

1. Enriched CD34+ cell suspension, obtained following procedure 3.1.
2. IMDM (Invitrogen).
3. Bovine serum albumin/insulin/transferin (BIT), serum substitute solution (Stem Cell Technologies).
4. 7.5% (w/v) Bovine serum albumin, fraction V, (BSA; Invitrogen).
5. Low density lipoprotein (LDL; Sigma-Aldrich) (*see* **Note 2**).
6. Potassium-buffered saline (PBS) (Gibco BRL).
7. 2-Mercaptoethanol (2-ME), 14.3 M (Sigma-Aldrich).
8. 2-ME, 5×10^{-2} M: 49 µl 14.3 M 2-ME diluted in 14 mL sterile-filtered (0.22 µm) deionized water.
9. PBS-1% BSA: 1.33 mL 7.5% BSA added to 8.67 mL PBS.
10. 10 µg/mL human recombinant cytokines (SCF, TPO, IL-9, and IL-6; PeproTech, Rocky Hill, NJ, USA). Reconstituted by adding aseptically 1 mL PBS–1% BSA to each 10 µg aliquot of lyophilized cytokine. Mix and dissolve well before aliquoting. Store at −35°C (*see* **Note 3**).
11. Basal medium: IMDM aseptically supplemented with 20% BIT, 0.4% LDL (*see* **Note 4**), 0.1% 5×10^{-2} M 2-ME. For instance, to prepare 80 mL of medium: add 16 mL BIT, 320 µl LDL, and 80 µl 5×10^{-2} M 2-ME to 63.6 mL IMDM.

12. MK culture medium: Basal medium supplemented with the required volumes of cytokines to obtain the desired concentrations. We recommend the BS1 cytokine cocktail, which contains the following cytokines at these final concentrations: SCF (1 ng/mL), TPO (30 ng/mL), IL-9 (13.5 ng/mL), and IL-6 (7.5 ng/mL) (*see* **Note 5**).

13. PBS-glucose solution: 2 g/l glucose and 0.5% phenol red in PBS, sterile filtered (0.22 μm).

14. 0.4% Trypan blue solution (Invitrogen).

15. Hemacytometer (Hausser Scientific, Horsham, PA, USA).

16. 24-well cell culture microplates (Costar, Corning Incorporated, Corning, NY, USA, or Beckton Dickinson Labware).

17. 15 and 50 mL polypropylene centrifuge tubes (Falcon, Beckton Dickinson Labware).

2.3. Flow Cytometry Analysis of Megakaryopoiesis

1. BSA (Gibco BRL).

2. 10% (v/v) Triton X-100: dissolve 1 volume of pure Triton X-100 (Bio-Rad Laboratories, Life Science Research, Hercules, CA, USA) in 9 volumes of PBS.

3. 10× MK buffer: 20 mM theophyllin (Sigma-Aldrich) and 272 mM Na-citrate in PBS, dissolved in a water bath at 37°C, sterile filtered (0.22 μm). For instance, add 360 mg theophyllin and 7.643 g Na-citrate in 200 mL PBS. Stable for 3 months at 4°C (*see* **Note 6**).

4. 1× MK buffer: for 200 mL, mix 20 mL 20× MK buffer with 26.8 mL 7.5% BSA and complete volume to 200 mL with PBS (*see* **Note 6**).

5. 4% paraformaldehyde (PFA): 8 g PFA (Sigma-Aldrich) dissolved in 200 mL PBS by heating in a water bath at 55°C inside a chemical hood. Sterile filtered (0.45 μm). Store at 4°C, protected from light.

6. 2 mg/mL RNase A: 10 mg RNase A (DNase free, Sigma-Aldrich) dissolved in 5 mL deionized, sterile water. Aliquoted and stored at −20°C.

7. Fluorochrome coupled human antibodies and their respective isotypic murine controls: phycoerythrin (PE)-coupled anti-CD34 (Immunotech, Beckman Coulter Co., Marseille, France), PE-coupled anti-CD62 (Immunotech), allophycocyanin (APC)-coupled anti-CD41a, and fluorescein isothiocyanate (FITC)-coupled anti-CD42b (Becton Dickinson). Isotypic controls; mouse IgG1 isotype control conjugated to -APC, -PE, and –FITC.

8. 1 mg/mL propidium iodide (PI): dissolve 25 mg PI (Sigma-Aldrich) in 25 mL deionized water. Prepare 1 mL aliquots and store at −20°C.

9. 20 μg/mL DNA marking PI solution: 2% 1 mg/mL PI solution, 1% of 10% Triton X-100 solution and 5% 2 mg/mL RNase A in 1× MK buffer. For instance, to prepare 1 mL, mix 10 μl 10% Triton X-100, 20 μl 1 mg/mL PI, 50 μl 2 mg/mL RNase A, and 9,920 μl MK buffer.
10. Centrifuge (Biofuge Pico, Heraeus Instruments).
11. Flow cytometer (FACS-Calibur, Becton Dickinson Immunocytometry Systems, San Jose, CA, USA).
12. Centrifuge microtubes made of homopolymer, 1.5 mL (Axygen Scientific, Union City, CA, USA).
13. 5-mL polystyrene FACS tubes (Bio-Rad Laboratories).

3. Methods

CB mononuclear cells (MNC) are isolated on a Ficoll-Hypaque density gradient and cryopreserved (*see* **Note 1**). MNC from three to six CB are mixed prior to CD34+ enrichment in order to reduce the inter-donor variability and to allow production of larger cell lots. Following enrichment using negative immunomagnetic selection, CD34+-enriched cells (purity ≥75%.) are aliquoted and cryopreserved. Cultures are initiated in a serum-free MK medium supplemented with cytokines. TPO alone at 10–100 ng/mL is often reported in the literature because it does not support the growth of non-MK cells, resulting in pure MK cell cultures. However, we strongly recommend to use the BS1 cytokine cocktail developed in our laboratory, as it supports both high MK purity and cell expansion (*see* **Section 2.2.14** and **Note 5**).

3.1. Purification of CD34⁺ Cells

3.1.1. MNC Preparation

MNC cells are isolated on a Ficoll-Hypaque density gradient following the manufacturer's instructions. Then, cells are cryopreserved (follow method 3.1.2), or CD34+ cells can be purified immediately (follow method 3.1.4).

3.1.2. MNC Cryopreservation

1. Centrifuge the cells obtained in method 3.1.1. for 10 min at 514 g.
2. Resuspend the cell pellet in cryoprotective medium, maintained at 4°C, at a density not higher than 300×10^6 cells/mL. Prepare 1 mL aliquots of this suspension into cryotubes.
3. Place the cryotubes at −80°C for 24 h, then transfer to liquid nitrogen.

3.1.3. MNC Thawing

1. Thaw cryopreserved MNC in a 37°C water bath, without mixing.
2. Transfer cells into a 15 mL tube. Complete tube volume with IMDM containing 20% FBS at 4°C.
3. Centrifuge for 10 min at 228 g in room temperature.
4. Remove supernatant and resuspend pellet in 10 mL IMDM containing 20% FBS at 4°C.
5. Take a sample to count cells and assess viability (*see* **Note 7**).
6. Centrifuge for 10 min at 228 g in room temperature.
7. Decant supernatant and resuspend cells to obtain a density between 2×10^7 and 8×10^7 cells/mL in PBS-glucose containing 5% FBS for magnetic labeling (*see* Method 3.1.4).

3.1.4. Magnetic Labeling and CD34+ Cell Separation

The procedure for the immunomagnetic removal of non-MK-committed cells follows essentially the instruction of the manufacturer (Human Progenitor Enrichment Cocktail kit, Stem Cell Technologies). It consists in marking cells with a primary antibody mixture conjugated to a hapten and specific to the following human cell surface antigens: CD2, CD3, CD14, CD16, CD19, CD24, CD56, CD66b, and glycophorin A. Cells are then marked with a hapten-specific antibody that is coupled to colloids of magnetic microbeads, and introduced in a magnetic column in which they are retained, except for non-marked CD34+ cells, which are eluted with the appropriate buffer solution. Once CD34+ cells are purified, they are cryopreserved in aliquots ($\sim 5 \times 10^5$/vials), as described in Method 3.1.2.

3.2. Ex vivo Expansion and Maturation of MK Starting from CB CD34+ Cells

Cultures are usually performed in 24-well tissue culture plates with volumes ranging from 0.8 to 1 mL. We recommend to only use the 8-central wells and fill the 16-external wells with PBS, to reduce the risk of overestimation of cell density due to evaporation, especially when culturing cells at 39°C (*see* **Note 8**). The cultures can also be performed in 6- to 96-well plates or in larger flasks, by keeping the same volume to culture surface ratio (*see* **Note 9**). The whole cell culture lasts 2 weeks where the cells mainly proliferate and differentiate into MK progenitors during the first week, while they mature and produce platelets during the second week with a peek of platelet production around day 14. Flow cytometry analysis (**Section 3.3.1**) and cell counts are usually performed on days 0, 7, 10, and 14 to assess cell differentiation and expansion, as well as on day 14 to evaluate platelet production.

1. Thaw the cryopreserved CD34+ cells in a 37°C water bath, without mixing.
2. Transfer the cells in a 15 mL tube. Complete with PBS-glucose solution at 37°C by adding the first 5 mL very slowly.

3. Centrifuge 10 min at 228 g. Discard the supernatant.
4. Resuspend the cells in MK culture medium to a density around 3 to 6 $\times 10^5$ cells/mL.
5. Take a sample to measure the cell density and cell viability (*see* **Note 7**).
6. Dilute the CD34+ cell suspension to 40,000 cells/mL in MK culture medium. Incubate at 39°C, 10% CO_2, in a humidified incubator (*see* **Note 8**). Greater cell density can be used if analyses of the cultures are performed within the first few days of culture.
7. After 4 days, mix gently the wells content, remove half of the cell suspension volume and add an equivalent volume of fresh MK culture medium.
8. Dilute the cells with fresh MK culture medium on day 7 to reach 200,000 cells/mL, as well as on day 10 to reach 200,000–250,000 cells/mL.

3.3. Culture Analysis

MK differentiation can be observed directly by using a phase contrast inversed microscope; mature MKs are discernible as larger cells and can be observed within 7 days, while proplatelet filaments are clearly visible by day 10. Note that the proplatelets are extremely fragile, and great care must be taken when handling the cultures. For a quantitative monitoring of megakaryopoiesis, cells should be characterized by flow cytometry (*see* **Section 3.3.1**), while MK progenitors can be measured using colony assays (*see* **Section 3.3.2**).

3.3.1. Flow Cytometry Analysis of the Culture

Phenotypic analysis by flow cytometry is carried out using fluorochrome-conjugated antibodies against specific MK markers, such as CD41a, CD61, and CD42b. Other markers can also be used, such as CD62 which is used to determine the activation status of platelet (*see* **Note 10**). Note that platelets do not express CD45. MK-progenitors can also be estimated by flow cytometry as the CD34+CD41a+ population, though we recommend using a more robust assay such as MegaCult™. Megakaryocyte polyploidization can also be followed by cytometry (*see* **Section 3.3.1.4**). Two or three color flow cytometry analysis can simultaneously reveal the percentage of platelets and of each cell type based on their specific phenotype as well as the MK ploidy distribution.

3.3.1.1. Flow Cytometry Acquisition Set-Up

The first step prior to the analysis of the samples is to establish a proper cell/platelet analysis acquisition template for the acquisition software used with the cytometer (such as CellQuest™). We recommend using a similar acquisition strategy as that described in **Fig. 7.3D–H**. A hematopoietic cell line can also be used to facilitate the set-up.

1. Cell region R1: In a dot-plot acquisition window, with forward scatter (FSC) and side scatter (SSC) on the *x*- and *y*-axis respectively, draw the R1 region around the nucleated cells, which have FSC and SSC properties significantly greater than platelets and cell debris (**Fig. 7.3D**). This region is used as a gate to specifically select cell events.

2. Propidium iodide (PI) cell region R3: In a dot-plot acquisition window acquiring events from the R1 region only, select FSC and PI (read on FL2 channel) on the *x*- and *y*-axis respectively, draw the R3 region around the PI negative cell events. This region is used as a gate to select viable cells.

3. Platelet region R2: Draw the platelet region R2 in the same dot-plot acquisition window in which the cell region was drawn (step 1 above). Platelet events have reduced FSC and SSC properties. The platelet region is used as a gate to select platelet events and it must be validated using fresh platelets as indicated in **Section 3.3.1.3**.

4. PI platelet region R4: In a dot-plot acquisition window acquiring events from the R2 region only, select FSC and PI (on FL3 channel) on the *x*- and *y*-axis respectively, draw the R4 region around the PI negative platelet events. This region is used as a gate to select events emanating from platelets specifically. Normal platelets do not stain positively with PI (**Fig. 7.3B**). PI+ platelets events seen with culture samples are most likely cellular MK debris and/or MK-derived microvesicles.

5. In dot-plot acquisition windows, select CD41a (if APC antibody, FL4 channel) and CD42b (if FITC antibody, FL1 channel) on the *y*- and *x*-axis respectively to analyze MK (gating R1*R3 events only), or platelets (gating R2*R4 events only) as shown in **Fig. 7.3F and H**.

6. Run the isotopic control stain(s) first to help adjusting the cytometer settings and to measure background levels. Run samples afterwards.

7. Once the flow cytometry acquisition template and flow cytometer settings are validated with culture samples and normal platelets, they should be saved for future use. In the latter case, new samples can be analyzed rapidly without the requirement of preparing fresh normal platelets.

3.3.1.2. Preparation of the Culture Samples for Flow Cytometry Analysis of the Cells

This procedure is designed for the analysis of the cells only, but flow cytometry analysis of the cell and platelet fractions can be performed simultaneously. In this event, it is important to follow the instructions presented in **Section 3.3.1.3** to avoid platelet activation.

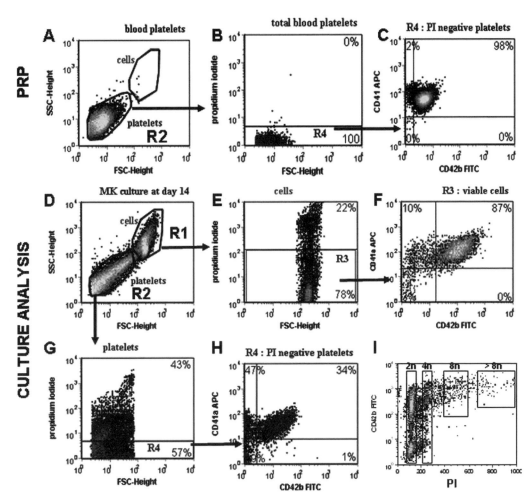

Fig. 7.3. Flow cytometry methodology used to analyze MK and platelets. (A–C) Analysis of fresh platelets from platelet-rich plasma (PRP). (D–I) Analysis of ex vivo produced MK and platelets. In more details; (A) Normal platelets derived from a PRP are first used to set the analytical platelet region 'R2' in a 2D dot-plot acquisition window showing the forward-side scatter (FSC) and side scatter (SSC) properties on the x- and the y-axis respectively. (B) On an independent dot-plot window, platelet events from the R2 region are analyzed for their retention of propidium iodide (PI), which is null for normal platelets. (C) PI-negative platelet events (R2*R4 region) are then analyzed for the expression of CD41a and CD42b, on the y- and x-axis respectively. The use of fresh PRP platelets also ensures proper adjustments of the various settings of the flow cytometer. (D) Cell cultures are analyzed in a similar fashion. Platelet events appear in the previously drawn R2 region, whereas MNC appears as distinctive events of greater size (FSC) and greater granularity (SSC). The cell region 'R1' is drawn around that second population. (E and G) The cell and platelet events from the regions R1 and R2 are then analyzed for their retention of PI. The PI-negative selection regions for the cells (R3) and platelets (R4) events are shown in E and G, respectively. (F and H) The PI-negative cell and platelet events (from R1*R3 and R2*R4 regions, respectively) are analyzed for CD41a and CD42b (or other markers of interest) expression in F and H, respectively. (I) Ploidy analysis of CB-derived MK cells at day 10 of culture. Dot-plot analysis of CD42b and PI intensity on the y- and x-axis, respectively. The regions for MK with DNA content of 2n and greater are shown.

1. Measure cell density and cell viability using a hemacytometer and trypan blue (*see* **Note 7**).

2. Resuspend gently the cell culture and transfer 0.2 mL aliquots of cell suspension (100,000–500,000 cells), into two 5 mL FACS tube.

3. Add antibodies (CD41a, CD42b…) to each tube (*see* **Note 11**) and incubate 15–30 min in the dark on ice. Each specific stain should be accompanied by an isotopic control stain done in parallel at the same antibody dilution.

4. Wash the cell/antibody mixtures by adding 1–2 mL of MK buffer. Centrifuge at 1,000 g for 5 min, discard supernatants and resuspend pellets in 0.3–0.5 mL MK buffer containing 5 µg/mL PI.

5. Proceed to flow cytometry analysis as indicated in **Section 3.3.1.1** as soon as possible since samples are not fixed. Acquire a minimum of 10,000 PI-negative cell events (R1*R3 events).

6. The total number of specific cell populations is calculated by multiplying the viable TNC density by the percentage of the corresponding population revealed by flow cytometry.

3.3.1.3. Preparation of the Culture Samples for Flow Cytometry Analysis of the Platelets

Platelets are routinely enumerated as PI-negative CD41a+ events with scatter properties similar to control platelets prepared from peripheral blood. However, the technique described in this section includes the use of CD42b as an additional platelet marker. We strongly recommend to validate the platelet gating strategy and cytometer settings with a normal platelet sample, which can easily be isolated from a small blood sample. The first step consists of using normal platelets to identify and define the platelet region (R2 on **Fig. 7.3A**). Afterwards, platelets produced in vitro can be analyzed by flow cytometry.

First step: Preparation of control platelet for validation of the platelet region

1. Centrifuge citrated whole adult blood (1–10 mL) at 514 g for 10 min without brake.

2. Collect the platelet-rich plasma (PRP) with a pipette, in the upper phase of the centrifuged sample. Centrifuge this PRP at 1,428 g for 15 min.

3. Discard the supernatant and resuspend platelets in MK buffer.

4. Sample an aliquot for platelet count; this can be done on a hematologic analyzer. Alternatively, counts can be estimated using a hemacytometer. If the latter method is used, dilute sample in 3% acetic acid to lyse red cells; platelets will appear as small dots since they are 10–25 times smaller than nucleated cells (for more details *see* (26).

5. Stain a small fraction of PRP samples: dilute ~1×10^6 platelets in MK buffer pre-warmed to room temperature, add antibodies against platelet antigens, such as CD41a, CD42b, and CD62 and incubate 15–30 min at room temperature instead of on ice (*see* **Note 11**). CD62 is optional (*see* **Note 10**). Each specific stain should be accompanied by an isotopic control stain done in parallel at the same antibody dilution.

6. Add directly to the 0.2 mL stained aliquots, 0.5 mL MK buffer pre-warmed to room temperature containing 7 μg/mL PI in order to reach a final PI concentration of 5 μg/mL (*see* **Note 12**).

7. Analyze by flow cytometry as indicated in **Section 3.3.1.1** as soon as possible since platelets are sensitive to spontaneous activation. Acquire a minimum of 10,000 events to identify the platelet region R2 (**Fig. 7.3A**) with specific FSC and SSC characteristics.

Second step: analyses of platelets produced in vitro

1. Measure cell density and cell viability using a hemacytometer and trypan blue (*see* **Note 7**).

2. Resuspend gently the cell culture and transfer 0.2 mL aliquots of cell suspension (100,000–500,000 cells), into two 5 mL FACS tubes.

3. Add antibodies (CD41a, CD42b,...) to each tube and incubate 15–30 min at room temperature instead of on ice (*see* **Note 11**). Each specific stain should be accompanied by an isotopic control stain done in parallel at the same antibody dilution.

4. Add directly to the 0.2 mL stained aliquots, 0.5 mL MK buffer pre-warmed to room temperature containing 7 μg/mL PI in order to reach a final PI concentration of 5 μg/mL (*see* **Note 12**).

5. Proceed to flow cytometry analysis as indicated in **Section 3.3.1.1** as soon as possible since samples are not fixed and platelets are sensitive to spontaneous activation.

6. Platelet enumeration: The number of platelets produced per culture volume is obtained by multiplying the viable TNC density by the proportion of platelets events that are CD41a+ (or CD41a+CD42b+, as preferred) and the ratio of platelet events to viable TNC events. This latter corresponds to the ratio of the number of events in R4 divided by the number of events in R3 (**Fig. 7.3E** and **G**). Cleavage of CD42b by metalloproteinase is responsible for a significant proportion of the CD41a+CD42b-platelets events.

7. Platelet yield: The number of platelets produced per seeded cells is calculated by dividing the number of platelets produced per culture volume by the concentration of seeded cells and multiplying this ratio by the cumulative dilution factor of the

culture at the time of sampling. The latter is obtained by multiplying all individual dilution factors for each culture dilution performed prior to sampling.

3.3.1.4. MK Ploidy Distribution

This is a short summary of a well-established protocol. More details and troubleshooting on the technique can be found in *(27)*.

1. Resuspend gently cell culture and transfer 0.2 mL aliquots of cell suspension (100,000–500,000 cells), into two 5 mL FACS tube.
2. Add an equivalent volume of PFA 1% solution to each tube. Incubate between 15 and 60 min at 4°C.
3. Resuspend the cell pellet in 1,000 µl MK buffer. Centrifuge at 2,380 g, 5 min. Discard the supernatant. Resuspend the cells in 200 µl MK buffer.
4. Add antibodies, as indicated in **Section 3.3.1.2**, to each tube. Incubate 30 min at 4°C, in the dark.
5. Resuspend in 0.5 mL DNA marking PI solution. Incubate at least 30 min at room temperature in the dark.
6. Transfer the samples in FACS tubes.
7. Analyze by flow cytometry (*see* **Note 13**). The ploidy can be analyzed with dot plots or histograms. The first is usually done plotting CD41a+ (total MK) or CD42b+ (mature MK) cell events versus PI. Polyploid MK cells are relatively rare with CB cells, and usually appear as CD42b bright cells (**Fig. 7.3I**). For histogram analysis, the number of events is plotted as a function of the PI intensity, and similar cell event selection (e.g., CD41+) can be done.

3.3.2. CFU-MK Titration

Culture conditions favoring MK expansion and differentiation, like the one proposed in this chapter, necessarily support the expansion of MK-progenitors. However, the magnitude of the expansion is for the most part dictated by (1) the cytokine cocktail used (nature and concentration) and (2) the length of expansion. Thus, these two parameters must be carefully optimized to obtain the maximal expansion. The BS1 cocktail suggested in this chapter provides good MK-progenitor expansion ranging from 5- to 100-fold for culture maintain for 3–10 days at 39°C (N. Pineault personal results, 2005). There exist a few assays to measure MK-progenitors: the plasma-clot technique, the serum-free fibrin clot assay, and a commercially available collagen-based kit known as MegaCult™. More information on the fibrin clot assay can be found in *(28)*.

We routinely use the MegaCult™ assay (2,250–8,000 cells per plate). In general, the frequency of CFU-MK progenitors in culture decreases as a function of culture time, hence we suggest to plate two different cell doses (1×–2×). For the BS1 cocktail, we recommend to plate 2,250 cells between days 0 and 5 of culture, and 3,500 between days 6 and 10. The MegaCult™ assay uses immunostaining to ensure proper enumeration of MK-progenitors (referred to as CFU-MK; colony-forming unit-megakaryocyte). In contrast to most myeloid colonies, colonies originating from MK-progenitors are small in size and in the number of cells present within. The colonies are scored based on their size, with mature MK-progenitor forming small colony (3–20 cells), and immature progenitor forming larger colony (≥ 50 cells). Background information, material and methods are all supplied in the MegaCult™ kit (StemCell Technologies, Vancouver, Canada). To assess the expansion of MK-progenitors in culture, MK-progenitor frequency (f) should be determined at day 0, and at any other time of interest (day X). MK-progenitor expansion is then calculated as follow: the total cell expansion is multiplied by f on day X and divided by f on day 0.

4. Notes

1. CB cells have been collected after an informed consent from the mother and following an approved protocol from Héma-Québec and Québec City Saint-François d'Assise hospital ethical committee. The samples have all been treated within 6 hours following their collection in aseptic conditions.

2. The type and source of LDL have an important impact on the platelet production. Our MK-culture experience is based on presolubilized LDL solution added to our medium, but lyophilized LDL can also be used. Sonication helps to emulsify lyophilized LDL in aqueous solution.

3. Cytokine aliquots should not be submitted to more than one freeze/thaw cycle.

4. Use 0.4% presolubilized LDL solution in basal medium. Perform dose–response tests to find the optimal concentration for reconstituted lyophilized LDL.

5. The cytokine cocktail BS1 (defined in **Section 2.2.14**) gives the same MK purity (>90% CD41a+ cells on day 14), but increased MK and platelet yields (~5-fold) when compared to TPO (10–100 ng/mL) alone. If final MK purity is not an issue, higher MK expansion and platelet yields can be obtained by increasing the SCF concentration to 40 ng/mL in the BS1 cytokine cocktail during the second week of culture *(13)*.

6. It is important to keep the MK buffer (10× and 1×) sterile even though cytometry analysis does not require aseptic

manipulations. Indeed, bacterial contaminated MK buffer results in numerous false platelet events (R2 region, **Fig. 7.3**) that will perturb flow cytometry analysis. Moreover, MK buffer concentrated more than 10× will crystallize.

7. To make cell counts and viability assessment, a 20 µl cell suspension sample is mixed with 20 µl of 0.4% trypan blue solution. Further dilution is made in order to count no more than 200 events in the hemacytometer.

8. Héma-Québec Laboratory have previously demonstrated that CB cells cultured at 39°C experience an accelerated MK differentiation kinetic and improved MK and platelet yields *(25)*, but the culture protocol described herein can also be followed with culture maintained at 37°C. Recent optimization of the culture period length at 39°C led to the discovery that MK and platelet yields are even greater if the cultures are switch back from 39°C to 37°C on day 4 *(29)*. In any case, if a 39°C incubator is not available, the incubation can be performed at 37°C.

9. The culture size depends principally on the quantity of cells desired or the number of different conditions needed to be tested. Thus, early large cell requirements will necessitate larger culture volume or greater seeding cell-density (cell expansion ratio is around 4-fold at day 4), whereas analysis done after 10 days of culture requires fewer starting cells due to increasing cell expansion (expected final cell expansion ratio of 150–250- and 300–600-fold at days 10 and 14, respectively).

10. Antibodies against other CD42 molecules can also be used instead of CD42b, such as CD42d (GPV). Platelet activation can be conveniently monitored by flow cytometry using an antibody recognizing the P-selectin CD62. This antigen is present in the alpha granules of platelets, and is exposed to their surface following activation of the platelets.

11. Routinely, all antibodies listed in **Section 2.3** are diluted 1:40, with the exception of the anti-CD41a which is used at 1:200. We recommend titrating all antibodies solutions.

12. For platelet analyses, centrifugation is omitted in order to avoid platelet activation. In our experience, no significant difference has been noticed between washed and unwashed stained cells with regard to non-specific staining.

13. Although the cells are fixed with PFA at this point, the flow cytometry analysis should be performed within 4 h after the sample treatment. After this time, a loss of the marker intensity is observed, which leads to an underestimation of the MK population.

Acknowledgments

The authors wish to thank Lucie Boyer for helpful discussions and Jean-François Leblanc for careful review of this manuscript. This work was supported by a Strategic Project grant from the Canadian Natural Science and Engineering Research Council.

References

1. Wognum, A. W., Eaves, A. C., and Thomas, T. E. (2003) Identification and isolation of hematopoietic stem cells. *Arch Med Res* **34**, 461–75.
2. Bartley, T. D., Bogenberger, J., Hunt, P., Li, Y. S., Lu, H. S., Martin, F., Chang, M. S., Samal, B., Nichol, J. L., Swift, S., and et al. (1994) Identification and cloning of a megakaryocyte growth and development factor that is a ligand for the cytokine receptor Mpl. *Cell* **77**, 1117–24.
3. de Sauvage, F. J., Hass, P. E., Spencer, S. D., Malloy, B. E., Gurney, A. L., Spencer, S. A., Darbonne, W. C., Henzel, W. J., Wong, S. C., Kuang, W. J., and et al. (1994) Stimulation of megakaryocytopoiesis and thrombopoiesis by the c-Mpl ligand. *Nature* **369**, 533–8.
4. Norol, F., Vitrat, N., Cramer, E., Guichard, J., Burstein, S. A., Vainchenker, W., and Debili, N. (1998) Effects of cytokines on platelet production from blood and marrow CD34+ cells. *Blood* **91**, 830–43.
5. Schipper, L. F., Brand, A., Reniers, N. C., Melief, C. J., Willemze, R., and Fibbe, W. E. (1998) Effects of thrombopoietin on the proliferation and differentiation of primitive and mature haemopoietic progenitor cells in cord blood. *Br J Haematol* **101**, 425–35.
6. Mattia, G., Vulcano, F., Milazzo, L., Barca, A., Macioce, G., Giampaolo, A., and Hassan, H. J. (2002) Different ploidy levels of megakaryocytes generated from peripheral or cord blood CD34+ cells are correlated with different levels of platelet release. *Blood* **99**, 888–97.
7. Pang, L., Weiss, M. J., and Poncz, M. (2005) Megakaryocyte biology and related disorders. *J Clin Invest* **115**, 3332–8.
8. Debili, N., Coulombel, L., Croisille, L., Katz, A., Guichard, J., Breton-Gorius, J., and Vainchenker, W. (1996) Characterization of a bipotent erythro-megakaryocytic progenitor in human bone marrow. *Blood* **88**, 1284–96.
9. Italiano, J. E., Jr., Lecine, P., Shivdasani, R. A., and Hartwig, J. H. (1999) Blood platelets are assembled principally at the ends of proplatelet processes produced by differentiated megakaryocytes. *J Cell Biol* **147**, 1299–312.
10. Kikuchi, J., Furukawa, Y., Iwase, S., Terui, Y., Nakamura, M., Kitagawa, S., Kitagawa, M., Komatsu, N., and Miura, Y. (1997) Polyploidization and functional maturation are two distinct processes during megakaryocytic differentiation: involvement of cyclin-dependent kinase inhibitor p21 in polyploidization. *Blood* **89**, 3980–90.
11. Zimmet, J., and Ravid, K. (2000) Polyploidy: occurrence in nature, mechanisms, and significance for the megakaryocyte-platelet system. *Exp Hematol* **28**, 3–16.
12. Schulze, H., Korpal, M., Hurov, J., Kim, S. W., Zhang, J., Cantley, L. C., Graf, T., and Shivdasani, R. A. (2006) Characterization of the megakaryocyte demarcation membrane system and its role in thrombopoiesis. *Blood* **107**, 3868–75.
13. Cortin, V., Garnier, A., Pineault, N., Lemieux, R., Boyer, L., and Proulx, C. (2005) Efficient in vitro megakaryocyte maturation using cytokine cocktails optimized by statistical experimental design. *Exp Hematol* **33**, 1182–91.
14. De Bruyn, C., Delforge, A., Martiat, P., and Bron, D. (2005) Ex vivo expansion of megakaryocyte progenitor cells: cord blood versus mobilized peripheral blood. *Stem Cells Dev* **14**, 415–24.
15. Bruno, S., Gunetti, M., Gammaitoni, L., Dane, A., Cavalloni, G., Sanavio, F., Fagioli, F., Aglietta, M., and Piacibello, W. (2003) In vitro and in vivo megakaryocyte differentiation of fresh and ex-vivo expanded cord blood cells: rapid and transient megakaryocyte reconstitution. *Haematologica* **88**, 379–87.
16. Williams, J. L., Pipia, G. G., Datta, N. S., and Long, M. W. (1998) Thrombopoietin

requires additional megakaryocyte-active cytokines for optimal ex vivo expansion of megakaryocyte precursor cells. *Blood* **91**, 4118–26.
17. Ungerer, M., Peluso, M., Gillitzer, A., Massberg, S., Heinzmann, U., Schulz, C., Munch, G., and Gawaz, M. (2004) Generation of Functional Culture-Derived Platelets from CD34+ Progenitor Cells to Study Transgenes in the Platelet Environment. *Circ Res*.
18. Shaw, P. H., Gilligan, D., Wang, X. M., Thall, P. F., and Corey, S. J. (2003) Ex vivo expansion of megakaryocyte precursors from umbilical cord blood CD34 cells in a closed liquid culture system. *Biol Blood Marrow Transplant* **9**, 151–6.
19. Dolzhanskiy, A., Basch, R. S., and Karpatkin, S. (1997) The development of human megakaryocytes: III. Development of mature megakaryocytes from highly purified committed progenitors in synthetic culture media and inhibition of thrombopoietin-induced polyploidization by interleukin-3. *Blood* **89**, 426–34.
20. Tajika, K., Ikebuchi, K., Inokuchi, K., Hasegawa, S., Dan, K., Sekiguchi, S., Nakahata, T., and Asano, S. (1998) IL-6 and SCF exert different effects on megakaryocyte maturation. *Br J Haematol* **100**, 105–11.
21. Proulx, C., Boyer, L., Hurnanen, D. R., and Lemieux, R. (2003) Preferential ex vivo expansion of megakaryocytes from human cord blood CD34+-enriched cells in the presence of thrombopoietin and limiting amounts of stem cell factor and Flt-3 ligand. *J Hematother Stem Cell Res* **12**, 179–88.
22. Fox, N. E., and Kaushansky, K. (2005) Engagement of integrin alpha4beta1 enhances thrombopoietin-induced megakaryopoiesis. *Exp Hematol* **33**, 94–9.
23. Avecilla, S. T., Hattori, K., Heissig, B., Tejada, R., Liao, F., Shido, K., Jin, D. K., Dias, S., Zhang, F., Hartman, T. E., Hackett, N. R., Crystal, R. G., Witte, L., Hicklin, D. J., Bohlen, P., Eaton, D., Lyden, D., de Sauvage, F., and Rafii, S. (2004) Chemokine-mediated interaction of hematopoietic progenitors with the bone marrow vascular niche is required for thrombopoiesis. *Nat Med* **10**, 64–71.
24. Bertolini, F., Battaglia, M., Pedrazzoli, P., Da Prada, G. A., Lanza, A., Soligo, D., Caneva, L., Sarina, B., Murphy, S., Thomas, T., and della Cuna, G. R. (1997) Megakaryocytic progenitors can be generated ex vivo and safely administered to autologous peripheral blood progenitor cell transplant recipients. *Blood* **89**, 2679–88.
25. Proulx, C., Dupuis, N., St-Amour, I., Boyer, L., and Lemieux, R. (2004) Increased megakaryopoiesis in cultures of CD34-enriched cord blood cells maintained at 39 degrees C. *Biotechnol Bioeng* **88**, 675–80.
26. Harrison, P., Briggs, C., and Machin, S. J. (2004) *in* "Platelets and Megakaryocytes" (Jonathan M. Gibbins, M. P. M.-S., Ed.), Vol. 272, pp. 29–46, Human Press, New jersey.
27. Darzynkiewick, Z., and Huang, X. (2004) *in* "Current protocols in immunology" (Collingan, J. E., Kruisbeek, A. M., Margulies, D. H., Shevach, E. M., and Strober, W., Eds.), pp. 5.7.1–5.7.18, John Wiley & Sons, New York.
28. Debili, N., Louache, F., and Vainchenker, W. (2004) *in* "Platelets and Megakaryocytes" (Jonathan M. Gibbins, M. P. M.-S., Ed.), Vol. 272, pp. 293–308, Human Press, New jersey.
29. Pineault, N., Boucher, J.-F., Cayer, M.-P., Palmqvist, L., Boyer, L., Lemieux, R., and Proulx, C. (2008) *Stem Cells Dev* **17**, 483–94.

Chapter 8

Ex Vivo Generation of Human Red Blood Cells: A New Advance in Stem Cell Engineering

Luc Douay and Marie-Catherine Giarratana

Abstract

We describe a technological approach permitting the massive expansion of CD34[+] stem cells and their 100% conversion ex vivo into mature red blood cells (RBC). The protocol comprises three steps: a first step consisting of cell proliferation and induction of erythroid differentiation in a liquid medium without serum in the presence of growth factors (GF), a second based on a model reconstitution of the medullar microenvironment (ME) (human MSC or murine stromal cells) in the presence of GF, and a third in the presence of the ME alone, without any GF. This work highlights the impact of the ex vivo microenvironment on the terminal maturation of erythroid cells. A critical point is that the RBC generated in vitro have all the characteristics of functional native adult RBC. Moreover, this new concept of 'cultured RBC' (cRBC) is important for basic research into terminal erythropoiesis and has major clinical implications, especially in transfusion medicine. The three-step protocol can be adapted to use hematopoietic stem cells (HSC) from diverse sources: peripheral blood, bone marrow or cord blood.

Key words: Hematopoietic stem cells, CD34[+] cells, erythroid culture, terminal differentiation, in vitro red blood cell (RBC) production, stromal cells, three-step protocol, functional cultured RBC.

1. Introduction

The interest of disposing of complementary sources of red blood cells (RBC) for blood transfusion is evident in the present context of the constant difficulty of obtaining supplies, lack of availability of certain phenotypes, occurrence of allo-immunization leading to impossible situations in transfusion, limited applications of oxygen transporters (perfluorocarbons), and disappointing performance of stabilized or recombinant hemoglobins. As a result, an attempt to produce erythroid cells in vitro by amplification of hematopoietic stem cells (HSC), notably of blood or placental origin, makes good sense.

The hematopoietic stem cells, which constitute 0.01–0.05% of the cells of the marrow, differentiate according to a well-defined hierarchy *(1, 2)* and in close contact with the medullar microenvironment. They become myeloid stem cells and then progenitors which will give rise to the different constituents of blood: red blood cells, granulocytes, and platelets. The cells of the medullar environment, known as stromal cells, are very numerous and include notably adipocytes, macrophages, and fibroblasts lying on an extracellular matrix *(3)*. These cells play a key role in the differentiation of the stem cells. They secrete soluble factors regulating activatory and inhibitory cytokines and facilitate interactions between stem cells, determinant in the regulation of hematopoiesis *(4)*.

The difficulty resides in constraining a hematopoietic stem cell to commit itself exclusively to the erythroid line and mature into a red blood cell. Since the stem cells are multipotential, i.e., have the capacity to give rise to any blood cell, one has to determine the environmental characteristics responsible for their differentiation into RBC. Attempts were unsuccessful for several years: experiments in the laboratory produced either mature RBC in very small quantities or large quantities of immature RBC still containing a nucleus. In this context, the objective is indeed triple: (i) to induce the massive proliferation of HSC to generate ex vivo erythroid precursors which after transfusion are capable of maturing in vivo into RBC; (ii) to induce the terminal maturation of HSC into mature and functional RBC so as to reach levels of production of the order of the number of cells in an RBC concentrate (1 to 4×10^{12}); (iii) to establish large-scale production conditions compatible with clinical requirements.

If it is apparently easy to obtain almost complete erythroid differentiation, the data of the literature nevertheless reveal, on completion of the different culture protocols proposed, either an important cell proliferation without terminal maturation or enucleation in close to half of the cells but with a reduced level of amplification *(5, 6)*. The mechanisms of enucleation have still not been formally established *(7–9)* due to lack of establishment of the experimental conditions permitting massive production of RBC. No set of ex vivo conditions has yet been described which allows one to obtain both massive proliferation and enucleation of the totality of the erythroblasts.

Our group has developed a method of producing mature human RBC ex vivo from HSC of diverse origins (bone marrow, cord blood, peripheral blood in the basal state or after mobilization with G-CSF) *(10, 11)*. The protocol comprises three steps. First, cell proliferation and erythroid differentiation are induced by culture for 8 days in a serum-free medium supplemented with stem cell factor, interleukin-3, and erythropoietin (Epo). Second, the cells are co-cultured for 3 days in the presence of additional Epo alone on either the murine MS-5 stromal cell line or human mesenchymal cells. In the third step, all exogenous factors are withdrawn and the cells are incubated on a simple stroma for up to 10 days (12). This protocol permits massive erythroid expansion (up to 10^6-fold for HSC from cord blood) and complete differentiation (100%) into mature RBC perfectly functional in vitro.

2. Materials

2.1. Medium for Erythroid Culture

1. Iscove modified Dulbecco's medium (IMDM) (Biochrom AG, Berlin, Germany) is the basic medium. **Table 8.1** (*see* also **Note 1**) lists the required quantities and final concentrations of all the constituents used to prepare 100 ml of medium. The complete medium can be sterilized by filtering through a 0.22 µm filter, aliquoted, and stored frozen for several months, while respecting the expiry dates of the different constituents including IMDM. The medium is supplemented immediately before use with hydrocortisone and cytokines as required by the protocol.

Table 8.1
Required quantities and final concentrations of all the constituents used to prepare the culture medium

Constituent*	Quantity necessary for 100 ml	Final concentration
IMDM without glutamine (1)	83 ml if BSA 88 ml if HSA	–
l-glutamine 200 mM	2 ml	4 mM
Pen and strep (2)	1 ml	1% vol/vol
Inositol 4 mg/ml in IMDM (3)	1 ml	40 µg/ml
Folic acid 1 mg/ml in IMDM (4)	1 ml	10 µg/ml
Monothioglycerol 0.16 M in IMDM (5)	0.1 ml	1.6×10^{-4} M
Transferrin 15 mg/ml (6)	0.80 ml	120 µg/ml
Insulin 1 mg/ml in 5 mM HCl (7)	1 ml	10 µg/ml
BSA (10%, i.e., 100 mg/ml) (8)	10 ml	10 mg/ml
or HSA (LFB, 20%, i.e., 200 mg/ml) (8)	5 ml	10 mg/ml

*The figures in brackets refer to the list of products in the 'Materials' section.

2. Penicillin and streptomycin (Invitrogen, Cergy-Pontoise, France): the stock solution (a mixture of penicillin at 5,000 U/ml and streptomycin at 5,000 μg/ml) is added at 1% final concentration to the complete medium.

3. Inositol (Sigma, St Quentin Fallavier, France): 20 mg of powdered inositol (kept at room temperature) is dissolved in 5 ml of IMDM and may be stored for up to 1 week at 4°C (*see* **Note 2**).

4. Folic acid (Sigma): 20 mg of powdered folic acid (kept at room temperature) is dissolved in 20 ml of pre-warmed IMDM and may be stored for 1 week at 4°C. The solution should be pre-warmed prior to use (*see* **Note 2**).

5. Monothioglycerol (Sigma): 10 μl of an 11.56 M stock solution (11.56 M, $d=1.25$, 98% purity, MW = 108.16) is dissolved in 712 μl of IMDM. We obtain a 0.16 M solution of monothioglycerol.

6. Holo-transferrin (Sigma), already saturated with 1,200–1,600 μg of iron per gram of transferrin, is dissolved at 15 mg/ml in buffer solution (150 mM NaCl, 0.8 mM Na_2HPO_4, 0.2 mM NaH_2PO_4 [pH 7.5]). The solution is filtered, sterilized (0.2 μm), and stored at 4°C (for up to 1 year, *see* **Note 3**).

7. Insulin (Sigma): powdered insulin (kept at −20°C) is dissolved at 1 mg/ml in 5 mM HCl. The solution is aliquoted to avoid repeated freezing and thawing and may be stored for up to 1 year at −20°C.

8. Albumin: the medium is supplemented with albumin (10 mg/ml, final concentration), which may be of variable origin (*see* **Note 4**).

9. Hydrocortisone (Sigma): 20 mg of hydrocortisone salt (MW = 484.5, kept at −20°C) is dissolved in 4.12 ml of IMDM and then diluted 1/100 in the same medium to give a 10^{-4} M solution, which is filtered through a 0.22 μm filter and may be stored for 1 week at +4°C.

10. Growth factors: stem cell factor (SCF, Amgen, Thousand Oaks, USA), interleukin-3 (IL-3, R&D, Abingdon, United Kingdom Systems), and Epo (Eprex, Janssen-Cilag, Issy les Moulineaux, France) are dissolved in appropriate volumes of phosphate-buffered saline (PBS, Invitrogen) supplemented with 1 μg/ml bovine serum albumin (BSA) and stored at −80°C. Appropriate aliquots avoid freezing and thawing cycles.

2.2. Media for Stromal Cells

1. Trypsin–EDTA 1× (Invitrogen) stored at −20°C.

2. αMEM medium containing ribonucleosides, deoxyribonucleosides, and Glutamax (Invitrogen, ref 32571028)

supplemented with 10% fetal calf serum (FCS). This medium is required for expansion of the murine cell line MS-5.

3. αMEM medium without ribonucleosides or deoxyribonucleosides but with Glutamax (Invitrogen) supplemented with 10% FCS and 1 ng/ml βFGF (Abcys, Paris, France). This medium is required for expansion of mesenchymal stromal cells (MSC).

2.3. RBC Analysis

1. Reticulocyte staining reagent (Sigma), stored at room temperature.
2. May-Grünwald solution (Sigma), stored at room temperature.
3. Giemsa stain, modified solution (Sigma), stored at room temperature.

3. Methods

A three-step protocol permits the massive expansion of CD34+ stem cells and their maturation ex vivo into RBC. First, cell proliferation and erythroid differentiation are induced by culture for 8 days in a serum-free medium supplemented with SCF, IL-3, and Epo. Second, the cells are co-cultured with addition of Epo alone on either the murine MS-5 stromal cell line or human MSC for 3 days. In the third step, all exogenous factors are withdrawn and the cells are incubated on a simple stroma for up to 10 days.

3.1. Stromal Cells

1. The murine MS-5 stromal cell line *(13)* is expanded in αMEM medium containing ribonucleosides, deoxyribonucleosides, Glutamax, and 10% FCS. At confluence, adherent cells are collected after treatment of the cultures for 7–10 min with trypsin–EDTA 1× at 37°C. Cell detachment is controlled under an inverted microscope and the reaction is stopped by adding 0.5 ml FCS/25 cm^2 flask. The cells recovered (usually 10^6/25 cm^2) are washed and replated at 4,000/cm^2 in αMEM medium supplemented with Glutamax and 10% FCS. Cultures are incubated at 37°C under 5% CO$_2$ and adherence is usually reached after 1 week (*see* **Note 5**).

2. Mesenchymal stromal cell cultures are established from whole normal adult bone marrow *(14)*.
The first step involves selection of MSC:
Total cells are plated at 50,000/cm^2 in αMEM medium without ribonucleosides or deoxyribonucleosides but containing Glutamax, 10% FCS, and 1 ng/ml βFGF. After 3–5 days, non-adherent cells are removed and the adherent cells are fed twice a week until confluence (about 10–20 days). At the end of this step, one may consider that the stromal cells are highly selected from total bone marrow cells owing to their adherent properties.

In the second step, MSC are expanded:

Adherent cells are collected after treatment of the cultures for 5–6 min with trypsin–EDTA 1× at room temperature. Cell detachment is controlled under an inverted microscope and the reaction is stopped by adding 0.5 ml FCS/25 cm² flask. The cells recovered are washed, replated at 1,000–3,000/cm² in fresh medium supplemented with 10% FCS and βFGF and fed once or twice a week until confluence. Cultures are incubated at 37°C under 5% CO_2. Subsequently, the adherent MSCs are expanded and purified through at least two successive passages following this second step.

3.2. Sources of Hematopoietic Stem Cells

HSC may be obtained from normal bone marrow (BM), normal peripheral blood mobilized with G-CSF [leukapheresis (LK)] or not (PB) or umbilical cord blood (CB) from normal full-term deliveries. CD34⁺ cells are isolated and may be stored frozen at this stage. The purity of the CD34⁺ cells is determined by flow cytometric analysis (*see* **Note 6**).

3.3. Erythroid Culture

The expansion procedure comprises three steps which may be carried out according to two protocols. See **Note 7** for the variations to this technique.

1. **First step**

   Days 0–8: 10^4/ml CD34⁺ cells are cultured in the presence of 100 ng/ml SCF, 5 ng/ml IL-3, 3 IU/ml Epo, and hydrocortisone (HC). The latter is only added during the first step of culture (10^{-6} M final concentration, add 1 ml of 10^{-4} M HC solution to 100 ml of final culture medium).

   A minimum volume of 4 ml and maximum volume of 10 ml are required when using 25 cm² flasks. Smaller volumes between 2.5 and 3 ml can be used to initiate cultures in six-well plates.

   On day 4, one volume of cell culture is diluted in four volumes of fresh medium containing HC, SCF, IL-3, and Epo. The cell concentration should be monitored between days 7 and 8 (especially when using cord blood cells) and should not exceed 2×10^6 viable cells/ml (*see* **Note 8**).

2. **Second step**

   Day 8–11 (3 days): cells are seeded at 10^4, 2×10^4, 4×10^4 or 6×10^4/cm² (for CB, LK, BM, and PB cells, respectively) (*see* **Note 9**) and co-cultured on the MS-5 stromal cell line or human MSC in fresh medium supplemented with Epo. The cells are usually washed between days 8 and 11 to remove metabolites and other factors and this step is desirable if the dilution into fresh medium is less than 1/40. Routinely the cells are resuspended in 5 ml of medium in 25 cm² flasks.

3. **Third step**

   Up to 10 days: cells are maintained on the adherent stromal layer in fresh medium containing no cytokines. It is preferable to wash the cells to remove residual Epo. All non-adherent cells are removed, centrifuged, counted (*see* **Note 10**), and resuspended in 5 ml of fresh medium without cytokines. Immediately after removing the non-adherent cells from the stroma, 5 ml of fresh medium without cytokines is added to each 25 cm^2 flask to preserve the adherent stromal cells under moist conditions. Note that each 25 cm^2 T-flask finally contains 10 ml of cell suspension. This step maximizes the number of non-adherent cells produced and is most important to obtain highly proliferative cord blood cultures.

If the cultures are maintained for more than 15 days from their initiation, the cells should be washed, resuspended at 5–7×10^6/ml, and co-cultured on a *new* stromal layer (*see* **Note 11**). The culture supernatants are renewed every 3 days to avoid the accumulation of carbonates. During this step, addition of 5–20% heat-inactivated human AB serum to the medium allows better preservation of the cultured red blood cells (cRBC) (*see* **Note 12**).

This last step permits final red cell maturation (notably in terms of the hemoglobin concentration) and elimination of the nuclei, which are engulfed by the stroma. It is important to remove the non-adherent cells gently to obtain a greater percentage of enucleated cells.

Cultures are maintained at 37°C under 5% CO_2.

3.4. Microscopic Observation of the Cultures

We describe below some observations of the cultures under the inverted microscope (*see* **Fig. 8.1**), which give indications as to the expected evolution of the system and are generally to be considered in parallel with the stage of differentiation of the erythroblasts.

1. **First step:** at the end of the first step of culture, the cells are very close to one another but should not form clusters.

2. **Second step:** the erythroid cells seeded on the stromal cells adhere to the microenvironment within 2–3 hours and by the day following initiation of the co-culture are strongly attached to the stromal cells (*see* **Note 13**). At the end of this step, the number of large adherent cells has increased, while one can observe some smaller non-adherent cells in the supernatant of the co-culture.

3. **Third step:** 2–3 days after withdrawal of Epo, the non-adherent cells form numerous hemoglobinized clusters in the supernatant, while the number of cells adhering to the stroma decreases. Around day 15 after initiation of the co-culture, there are almost no more adherent erythroid cells. However, one observes numerous very small circles included in the microenvironment, which correspond to the expelled nuclei. The MS-5 cell line may display at the end of this step, large vesicles of lipids. At this stage of culture, one can transfer the culture medium (including reticulocytes) onto a new stroma in order to complete differentiation of the reticulocytes into mature RBC (*see* **Note 14**).

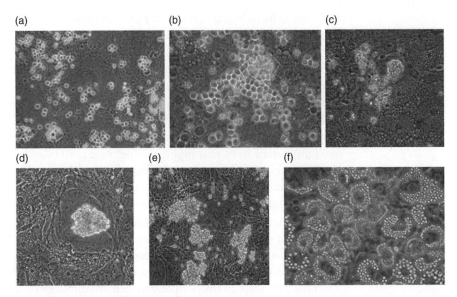

Fig. 8.1. Inverted microscopy images. Morphological characteristics of erythroid cells co-cultured on the murine MS-5 cell line. Co-cultures were initiated with erythroid cells from day 8 of primary culture. (**A**) One day after initiation of co-culture (day 9 of erythroid culture). (**B**) Three days after co-culture (day 11). (**C**) Five days after co-culture (day 13). (**D**) Seven days after co-culture (day 15). (**E**) Nine days after co-culture (day 17). (**F**) Morphological aspect of the MS-5 cell line at the end of the third step of culture.

3.5. In vitro RBC Analyses (see Table 8.2)

1. Morphological analyses

Cells are resuspended in medium (PBS, IMDM, etc.) supplemented with 5% FCS or 2% HSA prior to cyto-centrifugation. The number of cells used for each slide depends on their stage of differentiation, size, and capacity to cross the filter paper (*see* **Table 8.3**).
- Cyto-centrifuged cells are stained with May-Grünwald-Giemsa reagent for morphological analyses.
- Reticulocytes are stained with brilliant cresyl blue colorant (*see* **Note 15** for a description of permanent staining).

2. Standard hematological analyses.

The enucleated cells are monitored for standard hematological variables, including the MCV (fL), MCHC (%), and MCH (pg/cell) (*see* **Note 16**).

3. Progenitor cell assays.

Progenitor cells are assayed in methyl cellulose cultures supplemented with a cocktail of growth factors (SCF, IL-3, Epo, GM-CSF, and G-CSF) *(15)*. **Table 8.4** lists the numbers of normal cells required at each stage of culture to evaluate the progenitor cells. Progenitors are evaluated on day 8 for CFU-E and on day 14 for the other progenitors.

3.6. Deleukocytation

Certain analyses have to be performed with a pure suspension of enucleated cells (ektacytometry, membrane proteins, in vivo

Table 8.2
Numbers of cells required for different analyses

Analysis	Number of cells*
Cytology	See **Note 13**
Standard hematological variables (MCV, MCHC, MCH)	20×10^6
Ektacytometry	$4-5 \times 10^8$
Tonometry	5×10^7
O_2 equilibration curves	1.5×10^8
Enzymatic activity (G6PD, PK)	$2-4 \times 10^7$
Hb analysis/HPLC	10^7
In vivo injection	$4-5 \times 10^9$/mouse

*Cells are washed twice in PBS (pH 7.4) or Bis Tris buffer (2-hydroxyethyl-hydroxymethyl, pH 7.4) before analysis.

injection into animals, etc.). In order to eliminate residual erythroblasts and nuclei from the suspension, the cells have to be passed through a deleukocyting filter, of which we have tested several types as described below (*see* **Note 17**).

1. Filters:
   - **Leucoflex LST2** (Macopharma). Originally designed for total blood separation (450 ml), this system consists of a

Table 8.3
Numbers of cells required for cyto-centrifugation

Day of culture	Degree of differentiation	Number of cells per slide
Population on day 8	Proerythroblasts and basophilic erythroblasts	$5-7 \times 10^4$
Population on day 11	Erythroblasts, basophils, and polychromatophils	$0.8-1.2 \times 10^5$
Population on day 15	Erythroblasts, polychromatophils, Acidophils, and >50% cRBC	$1.5-2 \times 10^5$
Population on day 18	70–100% cRBC	$2-3 \times 10^5$

Table 8.4
Numbers of normal cells required for progenitor assays

Day of culture	Number of cells per plate
Day 0	500
Day 8	1,500
Day 11	10,000
Day 15	30,000

filter with a supple bag underneath. Separation is by gravity and it is necessary to maintain a significant height differential between filter and receptacle.
- **Leucolab LCG2** (Macopharma). Initially designed for the separation of red cell concentrates, this system has a rigid cassette below the filter and offers the advantage of controlling the filtration by pressure.
- **Sepacell**. This is very similar in operation to the LCG2 filter.

2. Preparation for use of an LCG2 filter:

We normally pre-warm 500–800 ml of IMDM supplemented with 1% HSA to 25°C. An empty 20 ml syringe is placed in the downstream opening of the cassette and sequential volumes of about 40–50 ml of IMDM containing HSA are injected to remove the air from the tubing and filter, air bubbles being purged by tapping the tubing. This saturates the system with medium before filtering the cells.

The washed cells to be filtered are resuspended at $50–300 \times 10^6$/ml in IMDM containing 1% albumin (see **Note 18**). One can directly filter fewer cells (less than 50×10^6) in 10 ml of medium.

The contents of the collection syringes are transferred to conical tubes and centrifuged for 10 min at 400 g and the red cells are then washed, pooled, and counted. Filtration takes time but violent pressure detaches the nuclei from the filter and allows the passage of some acidophils, although this is normally less than 1%. Hence the flow rate should not exceed 60 ml/min. Filtration results in a cell loss of 15–50% and the smaller the number of cells filtered, the higher the loss.

4. Notes

1. In our hands, IMDM Biochrom is the most effective for production of the erythroid line. This efficacy may be attributed to the pH and/or osmolarity of the solution. Consequently,

the recommended values for the complete medium are: pH 7.3–7.4 and osmolarity 290–300 mOsm/kg.

2. Supplementation with insulin and folic acid is not critical due to the large amounts of these components present in IMDM.

3. The stability of transferrin in distilled water does not allow it to be stored for more than a few weeks.

4. Albumin constitutes the critical parameter of the culture medium. Whatever their origin, the batches should be tested for efficacy as the outcome of the culture is highly dependent on the quality of this component. *Note:* the tests should be performed at all stages of erythroid culture. In fact, certain albumins strongly support cell proliferation (determined essentially during the first 10 days of culture) but do not allow the maintenance of enucleated cells in the medium and can be responsible for an important hemolysis ... and vice versa. The albumin used to supplement the complete medium can be obtained from various sources: (i) bovine albumin, ready to use (100 mg/ml, i.e., 10%, Stem Cell Technologies) and (ii) human albumin, ready to use (200 mg/ml, i.e., 20%). We do not recommend the use of human albumin at 400(?) mg/ml on account of the higher levels of contaminants.

5. The murine MS-5 stromal cell line can be frozen in aliquots. It is preferable to freeze the cells only 1 week after treatment with trypsin when the stroma is confluent in the flask. For co-culture with erythroid cells, MS-5 cells may be used for up to 2 weeks after trypsin treatment. Trypsin treatment cannot be repeated more than 20 times.

6. Determination of the CD34$^+$ cell purity of the starting product is of interest above all to express the results in terms of the fold amplification. In fact, the cell production during erythroid culture is essentially related to the number of CD34$^+$ cells in the starting product.

7. A modified procedure consists of prolonging the first step of culture by 3 days. In this case, the cells from day 8 are harvested, washed, and resuspended at 5×10^4, 10^5, 2×10^5 or 3×10^5/ml (for CB, LK, BM, and PB cells, respectively) in fresh medium containing SCF and Epo. The duration of the first step is then 11 days. In the second step (usually days 11–15), the cells are seeded at 2×10^4, 4×10^4, 6–8×10^4 or 1.2×10^5/cm^2 (for CB, LK, BM, and PB cells, respectively) and co-cultured on stromal cells in fresh medium supplemented with Epo. The third step (beyond day 18) is carried out as described in section 3 of the 'Methods'.

8. Note that cells from day 7 or 8, cultures can be frozen without damage if one respects the cell concentrations as indicated

above. If the number of necrotic cells is nevertheless elevated after thawing, one can eliminate the non-viable cells by ficoll separation. This is not indispensable because these necrotic cells will be eliminated in the course of co-culture by the stromal cells, which display macrophagic activity. However, we have observed that overloading the stromal cells with necrotic cells diminishes their capacity to eliminate the bare nuclei arising from enucleation during terminal erythroid differentiation.

9. The number of cells seeded on the stromal cells depends on the proliferative capacity of the hematopoietic cells and on their stage of differentiation, which can vary notably with the batch of albumin used in the culture medium. On the other hand, it is useless to increase the number of cells in co-culture as this may saturate the enucleating capacity of the stroma.

10. Cells are preferably counted with a hematimeter. It is nevertheless important to discriminate between large expelled nuclei and enucleated cells, especially when the latter are not fully loaded with hemoglobin. Cell numbers can be overestimated if this precaution is not taken.

11. At this stage of culture, only rare cells still adhere to the microenvironment.

12. It should be noted that mature cRBC can be maintained at 37°C for 3–5 days. However, cRBC can be conserved at 4°C in adenine–glucose–mannitol solution (Sag-Mannitol, Macopharma, Tourcoing, France) for 1 month. Addition of serum to the medium avoids hemolysis of the cells.

13. The adherence of the erythroid cells to the microenvironment depends on their degree of differentiation at the time of co-culture.

14. The 'end point' of culture may differ slightly from the observations described above. The rate of cell expansion and the percentage of enucleated cells can vary (independently of the cell origin) from one batch of complete medium to another (related to albumin, insulin, transferrin, pH, etc.). The behavior of the culture also depends on the quality of the stromal cells, the atmosphere of the incubator, and the pH of the supernatants. Hence, it is necessary to monitor as well as possible the cell concentration and the quantity of medium per cm^2. In general, substituting BSA for HSA slightly modifies the rate of differentiation, enucleated cells being obtained on days 16 and 17 rather than on day 15.

15. Reticulocyte staining: all cells to be stained are washed in PBS to avoid precipitation of the stain on the slide. Staining is

done by placing the washed cells in a glass tube containing 1–2 μl of colorant in 100–140 μl of PBS. The tube is incubated for 10 min at room temperature and the cells are then cyto-centrifuged. For long-term staining, the cells can be fixed for 2 min in 100% methanol, rinsed in water, and stained with Giemsa reagent for 5–6 min. After further rinsing in water, the cells are left to dry.

16. False parameters will be obtained if the sample to be analyzed is rich in excluded nuclei. To ensure that this contamination is minimal (<10%), one may use a deleukocyting filter as described in Section 6 of the 'Methods'.

17. Although sterile filtration is not necessary, the systems employed in transfusion centers are easily adapted and can be used in conjunction with a syringe and Luer lock to control the flow through the filter.

18. WARNING: for higher concentrations, it is preferable to pass the suspension through a 70 μm pre-filter as the expelled nuclei form clumps which block the filter.

Acknowledgments

Thanks are due to the co-authors of this work: Ladan Kobari[1], Hélène Lapillonne[1,2], David Chalmers[1,3], Laurent Kiger[4], Thérèse Cynober[5], Michael C. Marden[4] and Henri Wajcman[6].

References

1. Lemischka, I. R. (1997) Microenvironmental regulation of hematopoietic stem cells. *Stem Cells* **15 Suppl 1**, 63–68.
2. Ogawa, M. (1993) Differentiation and proliferation of hematopoietic stem cells. *Blood* **81**, 2844–2853.
3. Friedenstein, A. J., Deriglasova, U. F., Kulagina, N. N., Panasuk, A. F., Rudakowa, S. F., Luria, E. A., et al. (1974) Precursors for fibroblasts in different populations of hematopoietic cells as detected by the in vitro colony assay method. *Exp Hematol* **2**, 83–92.

---

[1] Laboratoire d'Hématologie EA1638, Université Pierre et Marie Curie, CHU Saint Antoine, 27 rue de Chaligny, 75571 Paris Cedex 12.

[2] Service d'Hématologie Biologique, Hôpital Armand Trousseau, Assistance Publique Hôpitaux de Paris, 26 avenue du Dr Netter, 75571 Paris Cedex 12.

[3] EFS Bourgogne Franche Comté, Boulevard Fleming, BP 1937, Besançon 25290.

[4] INSERM U473, 84 rue du Général Leclerc, 94276 Le Kremlin-Bicêtre Cedex.

[5] Laboratoire d'Hématologie, Centre Hospitalier de Bicêtre, 84 rue du Général Leclerc, 94276 Le Kremlin-Bicêtre Cedex.

[6] INSERM U468, Hôpital Henri Mondor, 94010 Créteil, France

4. Verfaillie, C. M. (1993) Soluble factor(s) produced by human bone marrow stroma increase cytokine-induced proliferation and maturation of primitive hematopoietic progenitors while preventing their terminal differentiation. *Blood* **82**, 2045–2053.

5. Freyssinier, J. M., Lecoq-Lafon, C., Amsellem, S., Picard, F., Ducrocq, R., Mayeux, P., et al. (1999) Purification, amplification and characterization of a population of human erythroid progenitors. *Br J Haematol* **106**, 912–922.

6. Zermati, Y., Fichelson, S., Valensi, F., Freyssinier, J. M., Rouyer-Fessard, P., Cramer, E., et al. (2000) Transforming growth factor inhibits erythropoiesis by blocking proliferation and accelerating differentiation of erythroid progenitors. *Exp Hematol* **28**, 885–894.

7. Bessis, M. (1958) Erythroblastic island, functional unity of bone marrow. *Rev Hematol* **13**, 8–11.

8. Lichtman, M. A. (1981) The ultrastructure of the hemopoietic environment of the marrow: a review. *Exp Hematol* **9**, 391–410.

9. Qiu, L. B., Dickson, H., Hajibagheri, N., Crocker, P. R. (1995) Extruded erythroblast nuclei are bound and phagocytosed by a novel macrophage receptor. *Blood* **85**, 1630–1639.

10. Neildez-Nguyen, T. M., Wajcman, H., Marden, M. C., Bensidhoum, M., Moncollin, V., Giarratana, M. C., et al. (2002) Human erythroid cells produced ex vivo at large scale differentiate into red blood cells in vivo. *Nat Biotechnol* **20**, 467–472.

11. Giarratana, M. C., Kobari, L., Lapillonne, H., Chalmers, D., Kiger, L., Cynober, T., et al. (2005) Ex vivo generation of fully mature human red blood cells from hematopoietic stem cells. *Nat Biotechnol* **23**, 69–74.

12. Dolznig, H., Habermann, B., Stangl, K., Deiner, E. M., Moriggl, R., Beug, H., et al. (2002) Apoptosis protection by the Epo target Bcl-X_L allows factor-independent differentiation of primary erythroblasts. *Curr Biol* **12**, 1076–1085.

13. Suzuki, J., Fujita, J., Taniguchi, S., Sugimoto, K., Mori, K. J. (1992) Characterization of murine hemopoietic-supportive (MS-1 and MS-5) and non-supportive (MS-K) cell lines. *Leukemia* **6**, 452–458.

14. Prockop, D. J. (1997) Marrow stromal cells as stem cells for nonhematopoietic tissues. *Science* **276**, 71–74.

15. Petzer, A. L., Zandstra, P. W., Piret, J. M., Eaves, C. J. (1996) Differential cytokine effects on primitive (CD34$^+$CD38$^-$) human hematopoietic cells: novel responses to Flt3-ligand and thrombopoietin. *J Exp Med* **183**, 2551–2558.

Part III

Regeneration of the Nervous System

Chapter 9

Isolation and Manipulation of Mammalian Neural Stem Cells In Vitro

Claudio Giachino, Onur Basak, and Verdon Taylor

Abstract

Neural stem cells are potentially a source of cells not only for replacement therapy but also as drug vectors, bringing bioactive molecules into the brain. Stem cell-like cells can be isolated readily from the human brain, thus, it is important to find culture systems that enable expansion in a multipotent state to generate cells that are of potential use for therapy. Currently, two systems have been described for the maintenance and expansion of multipotent progenitors, an adhesive substrate bound and the neurosphere culture. Both systems have pros and cons, but the neurosphere may be able to simulate the three-dimensional environment of the niche in which the cells reside in vivo. Thus, the neurosphere, when used and cultured appropriately, can expand and provide important information about the mechanisms that potentially control neural stem cells in vivo.

Key words: Neural stem cells, embryonic neurogenesis, adult neurogenesis, Notch signaling, neurosphere, differentiation.

1. Introduction

The mammalian central nervous system (CNS) is formed by stem cells that reside within the neuroepithelium of the neural tube. After neurulation, putative primitive neural stem cells (NSCs) diminish and give rise to definitive NSCs, which are the progenitors of the neurons and glia of the mature CNS. The formation of the CNS is highly regulated. At mid-embryogenesis (E8.5–E12.5 in the mouse), the NSCs undergo a period of expansion, increasing their numbers in a process of symmetric cell division (1). Subsequently, the neurogenic period commences and evidence suggests that the NSCs switch in their mitotic state and many undergo asymmetric divisions, generating one daughter which remains mitotically active as a stem cell and another that may remain in the cell cycle for a number of divisions but is committed to generate neurons. After the major period of neurogenesis, which ends in most regions of the mouse brain around birth, the NSCs change fate once again to generate glial-restricted progenitors that may undergo a number of cell divisions but will inevitably differentiate. Finally, at least in rodents, NSCs within restricted regions of the neural tube will form adult-type NSCs (aNSCs) that can continue neurogenesis throughout the life of the animal *(2, 3)*. The best-studied regions of neurogenesis in the adult murine brain are the subventricular zone (SVZ) and the dentate gyrus (DG) of the hippocampus. However, cells with stem cells-like properties and neurogenic potential have been isolated from numerous brain regions although their function in vivo is unclear. In the SVZ and the DG, the aNSCs reside within specific niches that likely control proliferation and differentiation. Although in vivo these aNSCs generate interneurons and glia, it is unclear whether they are determined in their fate by factors in the niche and the local environment or whether they are intrinsically restricted to generate specific cell types. Thus, unequivocal identification of the aNSCs and elucidating the factors within the niche that regulate their maintenance and fate is imperative for understanding adult neurogenesis. In addition, although cells with aNSC-like, neurogenic potential in vitro have been shown, and mitotically active cells are found in the adult human brain, recent data suggest that adult cortical neurogenesis is very limited or may not exist in humans. Thus, these findings reinforce the need to address regulatory mechanisms that may control the differentiation of aNSCs.

1.1. What is a Neural Stem Cell?

The term NSC is, and remains, somewhat confusing within the field. Many scientist adopt the term to describe cells that can divide and generate multiple cell lineages. Thus, in order to identify and describe the mechanisms controlling the formation of neural tissue, one needs to define the NSC, and this, unfortunately, may vary from one group to another. In the hematopoietic field, the advent of the adoptive transfer procedure where the potential of cells isolated from a donor animal are assessed for stem cell qualities by injection into irradiated hosts, has enabled a more precise assessment and definition of the term stem cell. For example, through adoptive transfer two classes of hematopoietic stem cell (HSC) have been defined. Those that are able to generate the entire blood system of the host and retain this potential over a long period are classified as long-term HSCs. However, a second class of repopulating cells, the so-called short-term HSC, can regenerate the blood system of the host but only for a limited period of time before they differentiate. Hence, the lack of such a functional assay in vivo for the identification and characterization of NSCs has hampered research in the field and lead to discrepancies in the terminology and definition of NSCs. NSC do have one characteristic however, which has greatly assisted in our understanding of neurogenesis and that is their ability to be cultured and maintained in vitro, in stark contrast to the HSC. Two culture systems have been developed over the last decade or so that enable expansion of multipotent NSC-like cells in vitro. One is the culture of NSCs in a monolayer attached to substrate-coated plates, so-called two-dimensional cultures. However, the most common culture system is the neurosphere *(4, 5)*. Self-replicating multipotent progenitors (NSCs for all intent and purpose) form and can be propagated as neurospheres in suspension culture. The neurosphere provides a three-dimensional environment that can, to some extent, mimic the neurogenic niche from which the NSC is derived. Thus, the neurosphere has been used extensively as a system to address the molecular mechanisms that can regulate NSC-like cell maintenance and differentiation. Here, we describe the procedure for the isolation and culture of multipotent neurosphere-forming cells and the techniques for addressing functional signaling with particular emphasis on the Notch pathway.

1.2. The Notch Pathway Plays a Central Role in the Life of a Stem Cell

Notch signaling is an evolutionary conserved pathway that takes part in cell fate choices in species from nematodes to humans *(6, 7)*. In *Drosophila*, the signaling molecule Notch regulates developmental decisions within the nervous system through a process of lateral signaling between undifferentiated cells (reviewed by *(8)*). Notch controls the number of neuronal precursor cells generated in the fly and modulates their developmental fate by interacting with the transmembrane ligands Delta and Serrate expressed on neighboring cells *(9, 10)*. Mammals have four *Notch* genes (*Notch1-4*) that control developmental processes in a number of tissues through the transcriptional regulation of nuclear factors belonging to the basic helix-loop-helix (bHLH) family (reviewed by *(7, 11)*). Notch receptor signaling can be regulated at different levels, including binding by one of the Delta-like (Dll) or Serrate-related (Jagged) ligands *(12)*, modulation of ligand efficacy to Notch by glucosyltransferases of the Fringe family or by positive and negative intrinsic factors, such as Deltex- and Numb-related proteins, respectively. However, Notch activation requires a complex process of proteolysis (reviewed by *(13, 14)*). Ligand binding induces cleavage of the juxtamembrane region of the ectodomain of Notch which is rapidly followed by proteolysis of the intracellular juxtamembrane via a γ-secretase that contains a member of the Presenilin family (reviewed by *(13, 15)*). The cleaved intracellular domain of Notch activates *Hairy/Enhancer of Split*-related *(E(Spl))* gene transcription (the *Hes* and *Hes-related* families in vertebrates) in a complex including *Suppressor of Hairless*-like *(Su(H))* nuclear factors (the vertebrate CSL family). Subsequently, in the nervous system, Hes proteins suppress expression of the proneural bHLH transcription factors. These proneural genes, such as *Ascl* and *Atoh1*, are pivotal in the induction of neurogenesis and regulation of neuronal differentiation (reviewed by *(16)*). In the murine neural tube, Notch1 is the first member of the family to be expressed and inactivation of the *Notch1* gene results in premature differentiation and loss of NSCs in mouse embryos *(17–19)*.

1.3. The Neural Stem Cell Niche

Stem cells reside within an environment that must supply all of the information that controls maintenance and differentiation. If the niche fails to provide the correct signals, stem cells may either differentiate prematurely and be lost, which would be disastrous for tissue homeostasis, or continue to expand uncontrolled resulting in tumor formation. During embryogenesis, NSCs reside within the neural tube and undergo proliferation and differentiation over an extended and overlapping period of time. Thus, NSCs within the same region of the developing nervous system may, at any one point in time, be proliferating or differentiating. Hence, they must be able to sense a multitude of local signals, while being selective in their response.

In the adult mouse SVZ, four cell-types have been defined based on their morphology and it is likely that interactions between these cells and the local environment form the neurogenic niche. The ventricular surface of the SVZ is lined by a sheet of ependymal cells. Under the ependymal layer resides a population of periventricular astrocytes (B cells), some of which project processes between the ependymal cells (reviewed by *(20)*). Elegant use of transgenic and retroviral approaches suggest that some of these B cells have stem cell properties *(21)*.

The B cells are proposed to undergo asymmetric cell divisions to generate transient amplifying cells (C cells). C cells proliferate rapidly and undergo a limited number of cell cycles before generating neuroblasts (A cells) (reviewed by *(22)*). Although recent data indicate that glial cells both in rodents as well as in other species can have stem cell-like properties *(23, 24)*, the identity of the mammalian SVZ stem cell remains to be determined unequivocally. Recently, we have been able to show that the Notch ligand Jagged1 is a key component of the niche that maintains neural stem cells in an undifferentiated state *(25)*. Jagged1 is expressed by ependymal and B cells within the SVZ, and by endothelial cells of the blood vessels, which have also been suggested to provide niche signals to NSCs. Elucidation of the niche factors will be pivotal to understanding the mechanisms that control neurogenesis and NSC biology.

2. Materials

1. Papain mix: 30 U/ml Papain (Sigma Cat. No: P3125), 0.24 mg/ml cysteine (cell culture grade; Sigma Cat. No: C7477), 40 µg/ml DNAseI Type IV (cell culture grade; Roche Cat. No: 104 159). Always prepare the papain mix fresh, sterile filter and store at 4°C until use (*see* **Note 1**).

2. Ovomucoid mix: 45 mg ovomucoid trypsin inhibitor (Sigma Cat. No: T6522), 21 mg BSA (Sigma Cat. No: A3294), 40 µg/ml DNAse I Type IV (cell culture grade; Roche Cat. No: 104 159), 39 ml L15 medium (Gibco Cat. No: 31415-029). Sterile filtered, the mix can be stored at 4°C for 1 week (*see* **Note 2**).

3. NS Medium: DMEM/F12 + Glutamax (Gibco No: 31331-028) + 10 U/ml penicillin, 10 µg/ml Streptomycin (optional), B27 Supplement (1:50) (Gibco) + Growth factors. Do not sterile filter after addition of B27 (*see* **Note 3**).

4. Growth factors: FGF-2 (20 ng/ml) (R and D Cat. No: 233-FB-025) for 'embryonic-type' NSCs. EGF (10 ng/ml) (R and D Cat. No: 236-EG-01 M) for NSCs.

3. Methods

3.1. Dissociation of Embryonic and Postnatal Brain Structures

Neurosphere-forming NSCs are sensitive to trypsin during the tissue dissociation stage of isolation. Thus, we have obtained the best results using Papain to digest the tissue prior to dissociation. Below, we describe the procedure we use for isolation of NSCs from embryonic and postnatal tissue. NSCs can be isolated from the embryonic neural tube as early as embryonic day 10 in the mouse. The region of the neural tube to be used as a source of NSCs should be isolated from the embryos under sterile conditions. Place the uterus containing the embryos in a dish containing L15 medium (Gibco Cat. No: 31415-029). If necessary, penicillin and streptomycin can be added during the isolation (optional). Under a binocular in a cell culture hood, remove the neural tube by micro-dissection and free of excess tissue. Place the isolated neural tubes in a separate dish of sterile L15 medium.

1. Micro-dissect the region of the neural tube to be used for NSC isolation.

2. Remove the required neural tube region and place in a tube containing digestion mix.

3. For digestion of neural tubes, use 300 µl of a 1:1 mix of Papain and ovomucoid mixes.

4. Mince the tissue in the mixture with tweezers and incubate at 37°C for 30–45 min.

5. After incubation, add an equal volume of ovomucoid mix and incubate at RT for 5–10 min.

6. Dissociate the tissue in the digestion/ovomucoid mix using either a fire-polished glass pipette (preferred) or a 1 ml filter tip (*see* **Note 4**).

7. Wash the suspension by adding and resuspending the dissociated tissue in 9 ml DMEM/F12 (Gibco Cat. No: 31331-028) medium.

8. Separate the cells from the debris by centrifugation 5 min at 80 g (*see* **Note 5**).

9. Remove the supernatant by aspiration.

10. Resuspend the pelletted cells in 1 ml neurosphere (NS) medium containing growth factors (see below) using a fire-polished pipette or 1 ml filter tip (*see* **Note 6**).

11. Plate the cells into flasks or plates. We routinely plate the cells from a single perinatal mouse brain or neural tube in one, maximum two, T25 flasks or in one 10 mm dish (*see* **Note 7**).

12. Add the resuspended cells to culture flasks or dishes containing NS medium (T25: 3–4 ml, 10 mm dishes 9 ml).

13. Culture the cells for 4 days at 37°C in a humid atmosphere of 6% CO_2 (*see* **Note 9**).

14. After 4 days, replace the medium by transferring the culture to 15 ml tubes and collect the neurospheres by centrifugation 5 min at 80 g (*see* **Note 10**).

15. Remove the supernatant leaving a small amount of medium over the pellet.

16. Resuspend the cells in 1 ml of fresh NS medium by gentle pipetting with a 1 ml pipette.

17. Plate into 3 ml of fresh NS medium in a fresh flask or dish.

18. Continue to culture for 2 days or more, exchanging medium every second day (*see* **Notes 10** and **11**).

3.2. Modifications for Preparation of Postnatal Neural Stem Cells

For the preparation of neurospheres from postnatal brains, the procedure is as described but requires a few modifications. The major challenge is to remove the meninges covering the brain and the choroid plexus in the ventricles. The meninges and choroid plexus are difficult to dissociate and thus cause problems during the culture.

1. Dissect mouse brains at postnatal days P1–P9 under sterile conditions and perform the rest of the isolation under a cell culture hood. The removal of the meninges after P5 becomes more difficult and stem cell number reduced rapidly soon after.

2. Place brains into L15 medium (Gibco Cat. No: 31415-029).
3. Make a midline saggital cut to divide the brain into two halves.
4. Remove the hindbrain, midbrain, meninges, choroids plexus, and olfactory bulb (*see* **Note 8**).
5. Transfer cortex (selected micro-dissected brain region) into Papain: Ovomucoid mix (1:1). 300 µl total volume is enough for brains from P1–P5 mice, brains of mice >P5 may need 500 µl (*see* **Note 2**).
6. Mince brains with tweezers until most of the large chunks have been broken up to assist the penetration of the Papain mix into the tissue.

Continue the isolation as described in **Section 3.1**.

3.3. Culture of Embryonic- and Adult-Type Neural Stem Cells

As described above, NSCs are generally characterized based on the region of the nervous system and the age of the animal from which they were isolated and their growth factor requirements. Although this is likely a rudimentary classification, for the current procedure of NSC culture it is sufficient. Two major classes of definitive NSC have been described, those dependent upon FGF2 and those grown in EGF. However, many labs use a combination of both growth factors in their isolation and culture procedures, which results in the isolation of more neurospheres but has the caveat that such conditions may result in mixed cultures of different types of NSC. Therefore, it is highly recommended that when analyzing the NSCs in vitro one tests different NSC populations by at least separating FGF2- and EGF-dependent populations.

The number of neurospheres generated from the postnatal mouse brain falls dramatically with increasing age. This is mainly due to the natural reduction in NSCs in the postnatal brain, but is also potentially the result of increased difficulty in isolating the cells. At later postnatal stages, it is beneficial to restrict the tissue used for the isolation. Thus, isolation of viable neurosphere forming cells increases dramatically if the region of interest is microdissected before digestion of the tissue. Improved NSC isolation from the adult SVZ is obtained by separating the lateral ventricle wall from the underlying striatum and the cerebral cortex. Further, by micro-dissection of the SVZ from 300 µm coronal vibrotome or tissue chopper sections of the live brain also results in improved isolation of neurosphere-forming cells. Careful removal of meningeal membranes and the choroid plexus is imperative to obtain viable cultures from adult mice.

3.4. Passaging of Neurosphere Cultures

After 6 days in culture, the neurosphere should have a smooth shiny appearance (**Fig. 9.1A**) and not have ragged edges, contain dark regions, vacuoles or dead cells. Appearance of the latter, indicates that the culture conditions are not optimal. The neurospheres

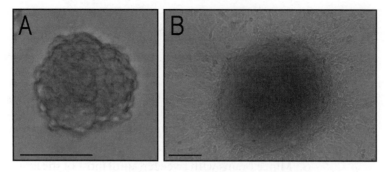

Fig. 9.1. (**A**) A healthy neurosphere is spherical, has a smooth regular surface, and is shiny under transmitted light optics. If the culture conditions are not optimal, or the cells have been 'abused,' the spheres become irregular, contain dark regions, vacuoles, and dead cells. (**B**) When plated clonally onto substrate-coated culture dishes, neurospheres attach rapidly and within 24 hr, cells start to migrate out onto the substrate. Scale bars 50 µm in A 100 µm in B.

should be dissociated and passaged. The procedure is similar to the original isolation, however, we have found that the cells are somewhat more resistant to trypsin once they have been cultured. Therefore, low concentrations of trypsin can be used to dissociate the neurosphere, which we have found is less harsh and produces better results than mechanical dissociation. Furthermore, the use of trypsin enables trituration of the neurospheres to single cells.

1. Transfer the neurospheres to a 15 ml conical tube.
2. Collect the neurospheres by centrifugation 5 min at 80 g.
3. Remove the supernatant and add 0.5 ml 0.05% trypsin diluted in Versene (Gibco).
4. Resuspend the cells by agitating the tube and incubate at 37°C for 5–10 min.
5. Add 0.5 ml ovomucoid mix and mix gently.
6. Incubate at room temperature for 5 min.
7. Triturate the neurospheres with a fire-polished Pasteur pipette (*see* **Note 12**).
8. After trituration, transfer the cell suspension to 9 ml of DMEM/F12 medium in a 15 ml tube and spin the cells through the medium to wash them.
9. Centrifuge the cells for 5 min at 80 g at room temperature.
10. Resuspend the cells in 1 ml of NS medium containing growth factors by gentle trituration with a fire-polished pipette.
11. Count the cells and plate in fresh NS medium (*see* **Notes 13, 14** and **15**).

3.5. Clonal Analysis of NSC Cultures

When plated at low density, 1–10 cells/µl (so-called clonal density), NSCs form clonal neurospheres by proliferation and aggregation of the dissociated cells is low. However, there are reports claiming that fusion of neurospheres can occur, particularly primary neurospheres isolated directly from the brain (26). Thus, to assess NSC potential, it is essential to rigorously control the density of the plated cells for analysis. One key factor that clearly affects the clonality of the neurospheres is the quality of the NSCs and the culture medium. We have observed that NSC that have been treated inappropriately, not passaged early enough (5–6 days) or have been starved, often form neurospheres that are no longer smooth and round (**Fig. 9.1A**) but start to become irregular, extend fine processes, and fuse to other neurospheres. Thus, it is imperative to feed the cells on a regular basis (every second day), and avoid long and frequent periods where the dishes are outside the incubator as extended periods of temperature changes can also lead to neurosphere fusion.

It is often advantageous to perform 'true' clonal cultures, placing single cells in isolated microtiter wells. Culturing NSCs in single cell cultures can prove to be quite challenging, although the neurospheres form by clonal expansion, the efficiency of neurosphere culture at single cell density is markedly reduced. It remains to be shown whether this is due to paracrine factors. We have found that it is possible to sort single cells from dissociated neurospheres by FACS, using forward and side-scatter parameters, and place single cells into 96-well plates. The efficiency of cloning increases if the cells are plated into a 1:1 mixture of fresh and medium conditioned by neurospheres for 2 days.

1. Collect 2-day-old medium from neurosphere culture and centrifuge 10 min at 1000 g to remove cell debris.
2. Collect the neurospheres by centrifugation 5 min 80 g.
3. Remove the supernatant and keep at 4°C.
4. Dilute the neurosphere conditioned medium 1:1 with fresh neurosphere medium.
5. Place 100 µl in each well of a 96 well, flat-bottomed microtiter plate.
6. Sort or pick cells from neurosphere cultures and plate at one cell/well (*see* **Note 16**).

3.6. Manipulation of Stem Cells *In Vitro*

NSCs in the neurosphere culture can be cultured under different conditions and manipulated by growth factor treatment and gene transduction.

3.6.1. Growth Factor Treatment

Definitive mammalian NSCs have been divided into two basic populations reflecting their growth factor requirements. The initial isolation of NSCs from the embryonic mouse nervous system demonstrated a dependence on fibroblast growth factor 2 (FGF-2) as a mitogen in vitro. Subsequently, the aNSCs were shown to required epidermal growth factor (EGF) in vitro. Thus,

FGF2-dependent NSCs are referred to as embryonic-type whereas EGF-dependent NSCs are termed aNSC. Although this classification is convenient for culture analysis, the relationship between embryonic NSCs and aNSCs remains unclear. Indeed, NSCs maintained in the presence of FGF2 and switched into EGF-containing medium are able to give rise to EGF-dependent neurospheres in vitro, suggesting that the aNSC are derived from embryonic NSCs. By contrast, cells cultured as neurospheres in the presence of EGF alone are unable to generate neurospheres that can be expanded in FGF2. Hence, it has been claimed that the aNSC is a product of the embryonic NSC in vivo, although formal proof is lacking. It is clear that both FGF2 and EGF-dependent NSCs reside together within the embryonic neural tube, and both populations of cells can be isolated from late embryo and postnatal mouse brains. It remains a prime goal to functionally discriminate the lineage relationship between these NSC populations in vivo.

3.6.2. Viral Infection of Neurosphere Forming Cells

Due to the difficulties in analyzing the differentiation of NSCs in vivo, many groups, including ours, have turned to the neurosphere assay as a means of addressing questions of gene function in the NSC maintenance and differentiation. The clonal nature and mitotic activity of the cells in the neurosphere enable the use of both retroviral- and adenoviral-based gene transfer to address gene function. We have used adenovirus-mediated conditional gene ablation in the neurosphere to address the role of the Notch1 signaling cascade in regulating self-replication and differentiation *(25)*. The adenovirus and the related adenoassociated virus have proven to be very efficient vectors for gene transfer into neurosphere forming cells in vitro. The recombinant adenovirus results in high-level episomal expression of transgenes, which, due to the replication-deficiency of these recombinant viruses, is lost over time and is not necessarily propagated to the progeny of mitotically active cells. However, for analysis of short-term gene effects or conditional gene rearrangement, for example using the Adeno-Cre system, we have found that the high infection efficiency, broad infection, and episomal expression are suitable for the neurosphere system *(25)*. However, other groups have also successfully used retroviral vectors and Lentivirus-based vector systems to obtain stable viral expression of genes *(27, 28)*. The integration of the retroviral genome and stable expression of the transgene has some clear advantages over the adenovirus for studying long-term effects of genes. We have established the adenoassociated virus system that combines many of the advantages of the adenovirus, including high-level transgene expression in slowly dividing and mitotically inactive cells with a 10–15% stable integration of the genome into defined regions of the host genome.

3.6.3. Notch Signaling in Neural Stem Cells

We have focused on the role of Notch signaling components in the regulation of NSC maintenance and differentiation. Using conditional gene ablation and exposure to soluble forms of the Notch ligands, we have shown that Notch1 plays a key role in NSCs in the neurosphere culture and in vivo.

3.7. Generation of Soluble Forms of Notch Ligands

Fusion of secreted or the ectodomains of transmembrane proteins to the Fc portion of immunoglubulins has been successfully used for purification of ligands and affinity probes. The canonical ligands for Notch are transmembrane molecules, presented to the Notch expressing signal-receiving cell. A number of groups have generated soluble Notch ligands fused to immunoglobulin Fc domains *(25, 29)*. The Fc fusion serves two main purposes with these molecules. The Fc-domain can be used for affinity purification of the fusion protein by passing the medium of transfected cells into which the proteins are secreted over a ProteinA, ProteinG or similar immunoglobulin affinity column. The Fc-portion of the fusion protein binds with high affinity to the matrix, can be washed and eluted in a native form for affinity assays or, in the case of Notch ligands, activation of the signaling pathway. A second advantage of the Fc-domain is that careful selection of the region of the immunoglobulin to be used and inclusion of the so-called hinge region, results in secretion of a dimeric molecule. Frequently ligand induced activation of receptors is potentiated or even requires ligand-induced receptor multimerization. This multimerization or clustering can be enhanced when using ligand-Fc fusions by the addition of an antibody to the immunoglobulin Fc-domain, which results in cross-linking of multiple ligand-Fc proteins and thus enhanced receptor clustering. This antibody-induced multimerization can be titered to obtain optimal receptor activation. In the analysis of Notch signaling, we showed a secreted Jagged1-Fc fusion protein was able to maintain self-replicating aNSCs in neurospheres, in the absence of the EGF. In addition, aNSCs cultured in the presence of exogenous Jagged1 and in the absence of EGF showed a significant increase in neurogenic potential when allowed to differentiate. Thus, these data strongly suggest that the activation of Notch receptors on the aNSCs is pivotal for maintenance of the undifferentiated state and the ability to generate neurons.

1. Jagged1-Fc protein was isolated from the culture medium of transiently transfected Hek293 cells.

2. Adult-type neurospheres were initially isolated and grown in EGF containing NS medium as described above.

3. After dissociation the NSCs were further cultured in NS medium containing Jagged1-Fc lacking EGF.

4. Neurospheres formed in a comparable fashion to EGF-containing medium, unlike daughter cultures plated into NS medium lacking both EGF and Jagged1-Fc.

5. Dissociation of neurospheres cultured in the presence of Jagged1-Fc and expansion in either Jagged1-Fc or EGF supplemented medium, showed that the Notch activation is able to substitute for EGF in vitro and maintains neurosphere-forming NSCs in vitro.

6. Thus, culturing NSCs in the presence of the Notch ligand maintains self-replication.

7. Jagged1-FC treated cells were also plated and under differentiation conditions, they showed multilineage differentiation (*see* **Note 17**).

3.8. Differentiation of NSCs In Vitro

The neurosphere, although a surrogate assay for NSCs analysis, does have some appealing properties. It is a three-dimensional culture where the NSCs seem to generate some form of niche that regulates their development. Naturally, the relationship between this in vitro niche and that in vivo has yet to be fully determined but certain factors seem to be conserved, not least the Notch signaling requirements. However, this in vitro niche also maintains multipotency in the progenitors cells and, as each neurosphere is clonal, if the culture conditions are correct, allows the potential of individual self-replicating stem cells to be addressed. This is exemplified by the differentiation assay for the neurospheres. By plating the neurospheres onto substrate-coated dishes at clonal density and reducing mitogen concentrations, the neurosphere cells will spontaneously differentiate and spread onto the substrate (**Fig. 9.1B**). By the addition of serum or neurotrophic factors, the differentiation of the cells can be enhanced or influenced. Thus, differentiation and fate potential can also be used as readouts for the manipulation of the neurosphere either by changing the culture conditions, growth factor environment or genetic modification. The validity of the neurosphere as a system to study NSCs is supported by the findings that embryonic-derived NSCs in the neurosphere cultures generate neurons prior to and more effectively that they generate astroglia, whereas EGF-dependent postnatal-derived NSC generate predominantly glial cells and few neurons. These traits reflect to a great extend the in vivo situation where during development neurogenesis precedes gliogenesis. Furthermore, if allowed to differentiate in the sphere state, it was shown many years ago that the differentiation can progress to the formation of differentiated neurons that form synapses and axons that are wrapped by a rudimentary myelin sheath.

1. Culture the neurospheres for 5–7 days in suspension culture.

2. Coat glass cover slips or culture dishes with substrate (*see* **Note 18**).

3. Count the neurosphere concentration in the suspension. For clonal analysis, we plate <20 neurospheres/cm^2.

4. Distribute the neurospheres upon plating as they attach rapidly.

5. We find that the neurosphere plate more efficiently if plated initially in the NS medium (*see* **Note 19**).

6. 12–24 hours after plating the distribution of the neurospheres can be controlled and those that have started to plate in close proximity marked on the underside of the dish for future reference and exclusion from a clonal analysis.

7. The medium should be supplemented every 2 days and the mitogen levels can be reduced to facilitate differentiation by adding medium lacking or with reduced FGF-2 or EGF concentration.

8. Differentiation continues with time in culture and cell fate can be assessed after 7–10 days (*see* **Note 20**).

4. Notes

1. Components for the Papain mix for digestion of the cells should be prepared fresh on the day of use. Papain mix should not be stored.

2. Ovomucoid mix should be prepared and sterile filtered and can be stored for 7 days at 4°C.

3. The culture medium should be prepared under sterile conditions; B27 containing medium should not be sterile filtered.

4. Do not make the tip of the glass pipette too small. By applying pressure to the bottom of the tube, the dissociation pressure can be increased. Dissociate until a homogenous suspension is achieved.

5. Do not centrifuge the cells with too high force, the NSCs are sensitive and too much debris will also pellet.

6. Use only moderate pressure to resuspend the cells as they are relatively sensitive in the absence of BSA and in serum-free medium.

7. Many protocols suggest to use bacterial culture dishes for propagation of neurospheres, however, we have found this to be superfluous and have obtained good results using Corning culture dishes. Each group needs to test the plastic that enables neurosphere formation in floating culture and attachment of differentiated cells.

8. During the preparation of the brain for isolation of NSCs, it is important to remove all of the meninges and choroids

plexus as these tissues are difficult to dissociate and will hamper the culture. Maximum neurosphere numbers are achieved from early postnatal mice by using the entire cortex and hippocampus, however, the brain can be dissected into more restricted regions such as the entire wall of the lateral ventricle, SVZ or hippocampus to restrict the NSC origins.

9. Cells should be cultured in a humid atmosphere and 6% CO_2.

10. Replace the medium on day 4 after isolation and then every second day.

11. The best cultures are obtained if the cells are not frequently disturbed, try to use an incubator that is less frequently opened and where vibration is at a minimum. Vibration can result in concentration of the cells to the center of the culture vessel and subsequent neurosphere fusion.

12. The cells are fragile and the fire-polished glass pipette is the most gentle method for dissociation as the diameter of the hole can be modified for optimal results. One can also use a 1 ml pipette tip (using filter-tips can help to avoid contamination).

13. Avoid keeping the cells outside the incubator for long periods of time. Remove individual dishes for medium changes. Long-periods outside the incubator at lower temperature or CO_2 can lead to the cells adhering to the culture plastic.

14. Passage the neurospheres before they become too large (6–7 days maximum) as they can start to die and/or fuse.

15. The cultures can be split 1:10 for expansion and maintenance. This should be carried out every 5–7 days. Check under the microscope that the cells are single before plating. If neurosphere or cell clumps remain continue to dissociate with a fire-polished Pasteur pipette.

16. One can also make classical limited dilution of the cells to obtain clones.

17. Quantitative analysis of differentiation showed that Jagged1-Fc treated cells have an increased neurogenic potential.

18. Commonly used substrates include poly-l-lysine, poly-l-ornithene, Laminin, Fibronectin and collagen, or combinations of these. The NSCs seem to react differently in their plating and differentiation depending upon the substrate.

19. Once attached the medium can be supplemented with 1% fetal calf serum or other trophic factors to be analyzed.

20. The neurons generated in the differentiation cultures are frequently sensitive to medium changes and do not adhere strongly to the substrate or the underlying cells. Thus, when refreshing medium do not remove all of the old medium,

leaving some fluid on the cells will prevent the loss of neurons caused by the tension of the meniscus. In addition, we have obtained the best results when we fix the cells with paraformaldehyde in the medium. Adding fixative directly to the medium reduces the loss of neurons. Antibodies against Nestin (progenitors), glial fibrillary acid protein (GFAP: astroglia), 04 (oligodendrocytes), and β-tubulinIII and neurofillament 160 (neurons), are commonly used for the analysis of differentiation. However, if the culture time is extended, the cells can reach an advanced level of differentiation and can generate synapses and myelinated axons.

References

1. Temple, S. (2001) Stem cell plasticity – building the brain of our dreams, *Nat Rev Neurosci* **2,** 513–520.
2. Alvarez-Buylla, A., and Lim, D. A. (2004) For the long run: maintaining germinal niches in the adult brain, *Neuron* **41,** 683–686.
3. Alvarez-Buylla, A., Garcia-Verdugo, J. M., and Tramontin, A. D. (2001) A unified hypothesis on the lineage of neural stem cells, *Nat Rev Neurosci* **2,** 287–293.
4. Reynolds, B. A., and Weiss, S. (1996) Clonal and population analyses demonstrate that an EGF-responsive mammalian embryonic CNS precursor is a stem cell, *Dev Biol* **175,** 1–13.
5. Reynolds, B. A., and Weiss, S. (1992) Generation of neurons and astrocytes from isolated cells of the adult mammalian central nervous system, *Science* **255,** 1707–1710.
6. Artavanis-Tsakonas, S., Matsuno, K., and Fortini, M. E. (1995) Notch signaling, *Science* **268,** 225–232.
7. Artavanis-Tsakonas, S., Rand, M. D., and Lake, R. J. (1999) Notch signaling: cell fate control and signal integration in development, *Science* **284,** 770–776.
8. Kimble, J., and Simpson, P. (1997) The LIN-12/Notch signaling pathway and its regulation, *Annu Rev Cell Dev Biol* **13,** 333–361.
9. Heitzler, P., and Simpson, P. (1991) The choice of cell fate in the epidermis of Drosophila, *Cell* **64,** 1083–1092.
10. Heitzler, P., Bourouis, M., Ruel, L., Carteret, C., and Simpson, P. (1996) Genes of the Enhancer of split and achaete-scute complexes are required for a regulatory loop between Notch and Delta during lateral signalling in Drosophila, *Development* **122,** 161–171.
11. Robey, E. (1997) Notch in vertebrates, *Curr Opin Genet Dev* **7,** 551–557.
12. Lendahl, U. (1998) A growing family of Notch ligands, *Bioessays* **20,** 103–107.
13. Kopan, R., and Goate, A. (2000) A common enzyme connects notch signaling and Alzheimer's disease, *Genes Dev* **14,** 2799–2806.
14. Chan, Y. M., and Jan, Y. N. (1998) Roles for proteolysis and trafficking in notch maturation and signal transduction, *Cell* **94,** 423–426.
15. Selkoe, D. J. (2001) Presenilin, Notch, and the genesis and treatment of Alzheimer's disease, *Proc Natl Acad Sci U S A* **98,** 11039–11041.
16. Bertrand, N., Castro, D. S., and Guillemot, F. (2002) Proneural genes and the specification of neural cell types, *Nat Rev Neurosci* **3,** 517–530.
17. Lutolf, S., Radtke, F., Aguet, M., Suter, U., and Taylor, V. (2002) Notch1 is required for neuronal and glial differentiation in the cerebellum, *Development* **129,** 373–385.
18. Hitoshi, S., Tropepe, V., Ekker, M., and van der Kooy, D. (2002) Neural stem cell lineages are regionally specified, but not committed, within distinct compartments of the developing brain, *Development* **129,** 233–244.
19. de la Pompa, J. L., Wakeham, A., Correia, K. M., Samper, E., Brown, S., Aguilera, R. J., Nakano, T., Honjo, T., Mak, T. W., Rossant, J., and Conlon, R. A. (1997) Conservation of

the Notch signalling pathway in mammalian neurogenesis, *Development* **124,** 1139–1148.
20. Doetsch, F., Petreanu, L., Caille, I., Garcia-Verdugo, J. M., and Alvarez-Buylla, A. (2002) EGF converts transit-amplifying neurogenic precursors in the adult brain into multipotent stem cells, *Neuron* **36,** 1021–1034.
21. Doetsch, F., Caille, I., Lim, D. A., Garcia-Verdugo, J. M., and Alvarez-Buylla, A. (1999) Subventricular zone astrocytes are neural stem cells in the adult mammalian brain, *Cell* **97,** 703–716.
22. Doetsch, F. (2003) The glial identity of neural stem cells, *Nat Neurosci* **6,** 1127–1134.
23. Noctor, S. C., Flint, A. C., Weissman, T. A., Dammerman, R. S., and Kriegstein, A. R. (2001) Neurons derived from radial glial cells establish radial units in neocortex, *Nature* **409,** 714–720.
24. Malatesta, P., Hartfuss, E., and Gotz, M. (2000) Isolation of radial glial cells by fluorescent-activated cell sorting reveals a neuronal lineage, *Development* **127,** 5253–5263.
25. Nyfeler, Y., Kirch, R. D., Mantei, N., Leone, D. P., Radtke, F., Suter, U., and Taylor, V. (2005) Jagged1 signals in the postnatal subventricular zone are required for neural stem cell self-renewal, *Embo J* **24,** 3504–3515.
26. Singec, I., Knoth, R., Meyer, R. P., Maciaczyk, J., Volk, B., Nikkhah, G., Frotscher, M., and Snyder, E. (2006) Defining the actual sensitivity and specificity of the neurosphere assay in stemcell biology, *Nature Methods* **3,** 801–806.
27. Gaiano, N., Nye, J. S., and Fishell, G. (2000) Radial glial identity is promoted by Notch1 signaling in the murine forebrain, *Neuron* **26,** 395–404.
28. Gaiano, N., Kohtz, J. D., Turnbull, D. H., and Fishell, G. (1999) A method for rapid gain-of-function studies in the mouse embryonic nervous system, *Nat Neurosci* **2,** 812–819.
29. Shimizu, K., Chiba, S., Kumano, K., Hosoya, N., Takahashi, T., Kanda, Y., Hamada, Y., Yazaki, Y., and Hirai, H. (1999) Mouse jagged1 physically interacts with notch2 and other notch receptors. Assessment by quantitative methods, *J Biol Chem* **274,** 32961–32969.

Chapter 10

Isolation, Expansion, and Differentiation of Mouse Skin-Derived Precursors

Karl J.L. Fernandes and Freda D. Miller

Abstract

The isolation and experimental manipulation of multipotent precursors is of increasing therapeutic relevance. We recently reported the generation of cultures of Skin-derived Precursors ('SKPs'), multipotent cells that can be isolated from the dermis of embryonic, neonatal, and adult rodent skin (1), and from adult human skin (2) SKPs have similarities to stem cells of the embryonic neural crest (3), and differentiate into a variety of neural and mesodermal cell phenotypes, including peripheral neurons and glial cells, smooth muscle cells, bone, cartilage, and adipocytes (3–5). Here, we detail the establishment, propagation, neural differentiation, and immunocytochemical analysis of SKP cultures.

Key words: Neural crest, stem cells, SKPs, NCSCs, neurospheres, neurogenic potential.

1. Introduction

The embryonic neural crest is a transient structure that contributes neural and mesodermal cell types to diverse tissues throughout the developing embryo (6, 7). Neural crest-derived cells eventually populate the entire peripheral nervous system and facial skeleton, the skin, the heart, the gut, the adrenal gland, and other tissues. Several recent studies have revealed that, like their CNS counterparts, neural crest-derived stem cells persist postnatally within several of these tissues, such as the gut, skin, heart, and dental pulp.

We have reported that a neurosphere-like culturing approach can be used to isolate neural crest-like stem cells from embryonic and postnatal skin (1, 3, 4), a tissue source that is abundant and highly accessible for potential therapeutic applications. Cultures of skin-derived precursors (SKPs) produce both neural and mesodermal phenotypes, and clonal analysis using multi-passage SKP cultures *(1, 2)* or low-density primary passage cultures *(3)* indicate that skin contains multipotent neural crest-like stem cells. In the present chapter, we provide a detailed protocol for culturing and promoting neuronal differentiation *(4)* of SKPs from mice.

2. Materials

2.1. Cell Culture: Establishment, Expansion, and Differentiation of SKP Cultures

1. Base medium used for culturing SKPs: 3:1 mixture of Dulbecco's modified Eagle's medium (DMEM) containing Glutamax (Invitrogen) and F12 media supplement containing Glutamax (Invitrogen). 1 μg/ml Penicillin/streptomycin (Cambrex) and 1% Fungizone (Invitrogen) are added at a final concentration of 1%.

2. Coated cell culture dishes: If dissociated skin cells will be plated down for differentiation, slides/dishes should be precoated with poly-d-lysine (BD Biosciences) and laminin (BD Biosciences). Poly-d-lysine and laminin are reconstituted by addition of 10 ml tissue culture grade water to each vial, and placing on a rocker at 4°C overnight. Coating mixture consists of 100 μl of poly-d-lysine and 200 μl of laminin per 10 ml of sterile water. Sufficient coating mixture is used to just coat the entire surface. Coated slides/dishes are left covered in the tissue culture hood overnight, then the coating solution is aspirated, the surfaces rinsed 2×5 min with tissue culture water, and the slides allowed to dry.

3. 100 mm Petri dishes with Hank's buffered saline solution (HBSS) for the dissected tissues or with F12 to hold the dissected embryos. Place all on ice.

4. 2×50 ml conical tubes with HBSS for rinsing tissues. Place on ice.

5. Gross dissection tools: large and small forceps, large and small scissors, are washed well with water, then 70% ethanol, and air dried.

6. Fine dissection tools: fine forceps (5/55), spring scissors or fine curved scissors, washed as above.

7. Fibroblast growth factor-2/basic fibroblast growth factor (FGF2/bFGF) (BD Biosciences) and epidermal growth factor (EGF) (BD Biosciences): FGF2 and EGF are reconstituted to stock solutions of 4 and 40 μg/ml, respectively, and aliquots are frozen in tubes at −80°C. We freeze 400 μl aliquots of FGF2 and 40 μl aliquots of EGF. Leave at 4°C until required.

8. SKP growth medium: 20 ng/ml FGF2, 20 ng/ml EGF, 2% B27 media supplement (Invitrogen), 1% N2 media supplement (Invitrogen), 1% CEE Chick Embryo Extract (ICN Biomedical) in base medium.

9. Trypsin (Calbiochem) solution 0.1%, dissolved in HBSS.

10. Collagenase solution (prepared from type XI) (Sigma), 1 mg/ml in F12.

11. Fetal bovine serum (FBS) (Hyclone).

12. NeuroBasal Medium (Invitrogen).

13. Chamber slides (1, 2, 4, and 8 wells) (Nunc).

14. Tissue culture sterile water (Hyclone).

15. Neurotrophins: NGF (Cedarlane), BDNF (Alomone), and NT3 (Cedarlane).

2.2. Immunocytochemistry

1. Paraformaldehyde solution (4%): Add 4 g of paraformaldehyde (Fisher) to 45 ml of MilliQ water that has been preheated in a microwave to 72°C, then add concentrated NaOH dropwise to the stirring solution until it clears. Add as few drops of NaOH as possible. Once the stirring solution has cooled to less than 50°C, filter it into 50 ml of 0.2 M phosphate buffer. Add additional MilliQ water to bring the final volume to 100 ml, then cool the solution to 4°C in a fridge. Filter the cooled solution again prior to use. Note that we use paraformaldehyde crystals, as powdered paraformaldehyde disperses easily in the air. After the paraformaldehyde has been added to water, keep it in the chemical hood or in a sealed container to prevent breathing the fumes.

2. Hoechst 33258 (Sigma) is diluted to a stock concentration of 10 mg/ml in potassium-buffered saline (PBS), and 1 ml aliquots are frozen at −20°C. The working solution is prepared by diluting the stock approximately 1:5,000. Optimize the dilution according to intensity of results.

3. NP40 (EM Science).

4. Normal goat serum (Jackson ImmunoResearch).

5. Geltol (Thermo Shandon).

6. Coverslips (Fisher).

7. Primary antibodies.

8. Secondary antibodies against host species of primary antibodies, and conjugated to fluorophores (from Jackson ImmunoResearch or Molecular Probes).

9. Humidified chambers.

10. PAP pen (RPI).

3. Methods

3.1. Collection of Skin Tissue

For embryonic skin:
1. For embryonic skin cultures, E17–18 pregnant mice are optimal as E17–18 embryos have the highest percentage of skin-derived sphere-forming cells *(3)*. Terminally anaesthetize the pregnant mouse according to approved animal protocols (injection, gas, cervical translocation).
2. With mouse lying on its back, hold the skin just above the genitalia with the large forceps and cut through the skin and underlying muscle towards the sternum.
3. With the abdominal cavity open, use the small forceps and scissors to remove the uterus and transfer it into one of the 100 mm dishes containing cold HBSS on ice.
4. Rinse the uterus of excess blood several times by transferring to fresh HBSS-containing dishes on ice.
5. With fine dissection tools, dissect out the embryos one at a time and transfer them to a fresh Petri dish containing F12 on ice.
6. To dissect the back skin from each embryo, transfer it to a fresh HBSS-filled Petri dish, lay it on its abdomen, and pin down the limbs. Using a dissecting microscope if necessary, gently lift the lower back skin with L-shaped forceps and cut horizontally from hip to hip. Then cut vertically up the side towards the base of the forelegs, and horizontally again to remove a rectangle of skin.
7. Gently and rapidly scrape the underside of the skin to peel away any blood vessels, fat, and muscle that may remain attached. Transfer the skin to a fresh HBSS-filled Petri on ice.

For neonatal/adult skin:
1. Kill neonatal or adult mice by approved method.
2. Lay the dead mouse down on its abdomen in a dry Petri dish, and pin it down through its limbs if necessary.
3. Dissect the back skin and clean it as in 6 and 7 above. Try to avoid the fat underlying the skin of the neck/upper back and sides.

For some purposes, it may be desirable to remove contaminating epidermal cells from neonatal/adult skin cultures. To do so:
4. Prepare a 60 mm Petri dish with the trypsin solution. Place on ice.
5. Place the rectangular piece of back skin in a Petri dish moistened with HBSS and, using a straight razor blade, cut it into squares of 4–5 mm^2.

6. Transfer the skin pieces to the Petri dish containing the trypsin solution, and incubate overnight at 4°C.

7. The next day, peel off the epidermis from each piece of skin, and transfer the remainder of the skin into a fresh Petri dish containing HBSS on ice. Proceed with the dissociation as described below.

3.2. Enzymatic Digestion and Mechanical Dissociation of Skin Tissue

1. Transfer the skin tissue to a fresh 100 mm Petri dish containing a few drops of ice cold F12 medium. For adults, use one back skin, for neonates and late embryos, arrange skins in a rectangle not larger than the length of the rectangular razor blade.

2. Use the razor blade to rapidly chop the skin into as small pieces as possible (*see* **Note 1**).

3. Transfer skin pieces to 35 or 60 mm Petri dish containing F12 on ice (*see* **Note 1**).

4. When all skin has been chopped, transfer skin pieces to a fresh Petri containing the collagenase or trypsin solution and place in 37°C/5% CO_2 tissue culture incubator for required amount of time. Ensure that there is sufficient volume of enzyme solution to completely immerse tissue pieces. Digestion times are typically 15–20 min for embryos, 45–60 min for neonates, and 60–90 min for adults (*see* **Note 2**).

5. Stop the enzymatic digestion by gently transferring the tissue pieces to a 15 ml Falcon tube containing 10% FBS in base medium. Mix gently by inversion several times, and let sit on ice for 5 min. Note that a serum-free alternative for stopping the reaction is to use soybean trypsin inhibitor.

6. The tissue pieces should settle to the bottom of the tube. Remove and discard the overlying serum-containing medium, and fill the tube to the top with fresh tissue culture medium. Mix gently by inversion and allow pieces to settle. Rinse in this manner several times. If pieces fail to sink to the bottom of the tube, a brief (30 s) low speed centrifugation may be necessary.

7. To begin the mechanical dissociation of the digested tissue pieces, transfer them to a 50 ml tube on ice, allow them to settle, and then reduce the volume of the medium as much as possible.

8. Perform the dissociation procedure on ice. For the first dissociation, use a 10 ml pipette to grind the tissue against the sides and bottom of the tube for about 2 min. Add 2 ml of fresh medium and continue dissociating for an additional minute, taking care to avoid splashing from the tube.

9. Add 5 ml of additional medium, mix, and centrifuge briefly to spin down debris.

10. Transfer the overlying medium to a fresh 50 ml Falcon tube (designated #1), where it is further pipetted 20–30 times using a P1000 pipette set to 1 ml, to dissociate small clumps. Then transfer this medium to a fresh 50 ml Falcon tube (#2) and store on ice.

11. Repeat steps 8–10 additional times as required, collecting the dissociated cells in the above 50 ml Falcon tube #2 on ice.

12. After completing the dissociation steps, filter the cells and medium in tube #2 through a 40 μm cell strainer into a fresh 50 ml Falcon tube ('#3') on ice.

13. Collect the dissociated cells by centrifugation at 250 g for 5–7 min.

14. Resuspend the pellet in medium supplemented with 2% B27 (5 ml for embryonic skin, 10 ml for neonates, and 20 ml for adult).

15. Count the cells using a hemocytometer, making appropriate further dilutions if necessary. The bulk cells are now ready to be used for direct differentiation (3.3) or for purification/expansion into skin-derived neurosphere cultures (3.4).

3.3. Direct Differentiation of Skin-Derived Cells

Plating of cells:

1. Calculate the total volume of plating medium necessary for the required chamber slides or dishes. Add 5% to this amount as a pipetting safety factor, becoming the total volume required (V_T) (see **Note 3**).

2. Calculate the volume of cells (V_C) required (see **Note 4**).

3. Prepare the plating medium, which is composed of 20–40 ng/ml FGF2, 20 ng/ml EGF, 2% B27, and 10% FBS, diluted in base medium to $V_T - V_C$.

4. Mix the cells several times by inversion and then transfer the calculated volume of cells to the plating medium.

5. Plate the cells onto slides or Petri dishes as required, mixing the tube by inversion every 30–60 s.

6. Maintain plated cells in a 37°C tissue culture incubator set to 5% CO_2 for 3 days.

Differentiation and maintenance of cells:

7. Differentiation of the SKPs within the plated skin cell population is initiated by withdrawal of FGF2. To prepare FGF2-free differentiation/maintenance medium, prepare the following components calculated for the same V_T as above: 2% B27, 1% N2, 1% FBS, and 50 ng/ml each of NGF, BDNF, and NT-3, diluted to V_T in Neurobasal medium (see **Note 5**).

8. After 3 days in plating medium, carefully remove 75% of the original volume of plating medium (i.e., 450 μl/well in 4-well

chamber slides), adding back 100% the original volume of differentiation/maintenance medium from step 7 (i.e., 600 µl/well in 4-well chamber slides). The excess volume added compensates for evaporation of medium.

9. Every 2–3 days from this point on, remove 25% of the original volume added to the well, re-adding differentiation/maintenance medium prepared as in step 7, at a volume of 40–50% of the original volume (again, the excess compensates for evaporation of medium).

10. Cells are typically differentiated for 2–4 weeks prior to analyses for neuronal differentiation.

3.4. Isolation of SKPs from Rodent Skin

1. To enrich dissociated bulk skin cells (3.2) for SKPs and expand them as non-adherent sphere cultures, first determine the total volume of medium and cells that will be required. Flasks of 75 cm^2 each require 30 ml total volume, 25 cm^2 flasks require 10 ml each. To determine the quantity of cells required, see **Note 6**.

2. Add the required volume of cells in the SKP growth medium, mix well by inversion, and aliquot into the desired flasks.

3. Every 2–3 days, add completely fresh FGF2 and EGF directly to each flask, as these are rapidly depleted at 37°C.

4. After 7–14 days, growing spheres should be visible (**Fig. 10.1**).

3.5. Passaging of Sphere-Forming Skin-Derived Precursors (SKPs)

Method 1: Partial mechanical dissociation approach (*see* **Note 7**)

1. After 10–14 days of growth, floating spheres are pelleted by transferring the entire medium from a flask into a 50 ml Falcon tube and centrifuging at 250 g for 4 min.

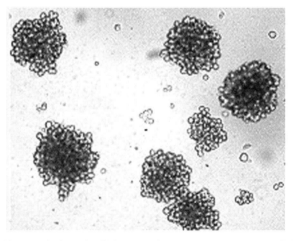

Fig. 10.1. Micrograph of growing SKP spheres from a multi-passage mouse SKP culture.

2. Supernatent is carefully pipetted into a second Falcon tube and saved, as this conditioned medium is vital for expansion of the culture. Leave the pellet in approximately 2 ml of medium.

3. Using a P1000 pipette, the pellet is triturated in the 50 ml Falcon tube with moderate intensity for approximately 30 repetitions, typically yielding a mixture of small spheres and single cells.

4. Flasks are typically passaged one into two. Prepare two fresh Falcon tubes of the appropriate size (15 ml tubes for 10 ml flasks and 50 ml tubes for 30 ml flasks) containing conditioned medium (filtered through a 20 µm syringe filter) to 50% of the final volume. Mix the cells and aliquot half into each of the two new tubes. Add sufficient FGF2, EGF, B27, and N2 for the entire flask (i.e., assuming that the conditioned medium has been entirely depleted), and bring to the final volumes with base medium.

5. Mix the Falcon tubes by inversion, transfer them into flasks, and place in the incubators.

6. Feed cultures every 2–3 days as in 3.4.4.

Method 2: Enzymatic dissociation approach (see **Note 7**)

1. Pellet spheres as in 3.5.1.
2. Remove and save the conditioned medium, leaving the pellet in approximately 1 ml of medium.
3. Add 1 ml of 1 mg/ml collagenase, for a final concentration of 0.5 mg/ml, and incubate cells at 37°C for 10 min.
4. Stop the enzymatic reaction by adding 0.25 ml of FBS.
5. Triturate the spheres with moderate intensity for about 50 repetitions using a P1000 pipette set to 1 ml.
6. Pellet the cells by centrifuging at 250 g for 5–7 min.
7. Remove and discard the supernatant, and resuspend the pellet in 2 ml of base medium.
8. Proceed as in steps 4–6 of Method 1 above.

3.6. Differentiation of SKPs

1. Prepare the required volume of plating medium as in 3.3.1.
2. Proceed in one of the following three ways:
   a. For analysis of single spheres, first aliquot the plating medium into 8-well chamber slides. Then, transfer an aliquot of spheres to a 60 mm Petri dish and, using a P20 pipette and a tissue culture microscope, pick large individual spheres, and transfer them to individual wells of the 8-well slides.
   b. For bulk analysis of spheres, transfer an aliquot of spheres directly into the plating medium, mix well by inversion, and aliquot into chamber slides/Petri dishes. Since spheres rapidly settle, mix the Falcon tube containing spheres by inversion every 30–60 s during aliquoting.

c. For analysis of dissociated spheres, transfer an aliquot of spheres to a fresh 15 ml Falcon tube, add an equal volume of 1 mg/ml collagenase, and incubate 10 min in a 37°C water bath. Place tube of digested spheres on ice, triturate 50 times with moderate intensity using a P1000, and determine the cell density using a hemocytometer. Transfer the desired quantity of cells to the plating medium, mix well, and aliquot into dishes or chamber slides.

3. Incubate plated spheres/cells in plating medium for 3 days in 37°C incubator.

4. Continue differentiation as in 3.3.7.

3.7. Immunocytochemistry

1. Cells are fixed in cold 4% paraformaldehyde for 5–15 min.
2. Rinse slides thrice with 1× PBS, 5 min per wash.
3. Unless extracellular antigens are being targeted, permeabilize cells using 0.2% NP-40 for 5 min.
4. Rinse slides twice with 1× PBS, 5 min per wash.
5. Aspirate majority of PBS, remove chamber, rapidly outline wells with PAP pen, and add sufficient blocking solution to cover entire well.
6. Incubate in blocking solution for 1–2 hr at room temperature, or for convenience, overnight at 4°C in a humidified chamber.
7. Remove blocking solution and add primary antibodies diluted in half-blocking solution.
8. Incubate in primary antibodies for 1–2 days at 4°C in humidified chamber.
9. Wash off primary antibody solution thrice with 1×PBS, 5–10 min per rinse.
10. Prepare secondary antibodies according to manufacturer's specifications.
11. Incubate slides in secondary antibodies for 40–45 min at room temperature in the dark.
12. Rinse twice with 1×PBS, 5–10 min per wash.
13. Incubate for 2 min in Hoechst 33258.
14. Rinse twice with 1×PBS, 5–10 min per wash.
15. Aspirate PBS, add Geltol (one drop per well, minimum four drops per slide), and coverslip (*see* **Note 8**).
16. Let slides dry flat for at least 1 hr in the dark before examination under microscope (**Fig. 10.2**). Slides should be protected from light after addition of fluorescent secondary antibodies.

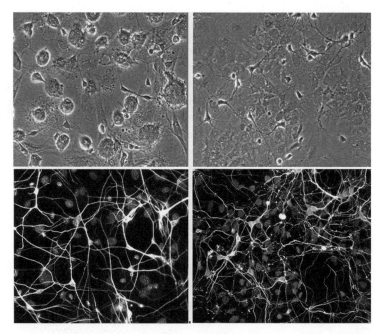

Fig. 10.2. Corresponding phase (*upper row*) and fluorescence (*lower row*) images after immunocytochemistry for the neuron-specific marker βIII-tubulin. Dissociated E18 skin cells (*left column*), or as control, E18 cortical cells (*right column*), were differentiated for 2–3 weeks prior to immunocytochemistry. Cell nuclei in the fluorescent images are labeled by Hoechst 33258.

4. Notes

1. The skin can be diced first in one direction, then in the perpendicular direction, and then at the two 45° angles. The smaller the pieces (i.e., 1 mm^2), the greater the time saved during the dissociation procedure and the better the enzymatic dissociation efficiency.

2. Times should be optimized for individual lab conditions. A convenient rule of thumb is to digest until cellular debris is beginning to float.

3. We use 600 μl/well for 4-well chamber slides, adjusted proportionally for 1-, 2-, and 8-well chamber slides.

4. Overgrowth of smooth muscle results in lifting of cells and can complicate longer differentiation experiments. We therefore typically use relatively low starting concentrations of 10,000–25,000 skin cells/ml of plating medium.

5. The neurotrophins NGF, BDNF, and NT-3 are added to enhance neuronal differentiation, survival, and maturation, but are not required to observe neuronal differentiation.

6. Density is a crucial factor in these selection protocols, as it has been shown that low cell concentrations (i.e., 10 cells/μl) do

not support the survival of differentiated cell types, while single stem cells are able to clonally proliferate into spheres of undifferentiated precursors. In contrast, at higher starting concentrations, cell aggregation results in non-clonally-derived spheres that contain both stem cells and differentiated cell types. Under these latter conditions, overall cell survival is enhanced, but differentiated cell types are not eliminated for several passages. For experiments where clonally-derived spheres are preferable, we seed flasks at concentrations of 10,000 cells/ml or lower and leave the flasks undisturbed during the growth period, whereas for experiments where maximal cell viability over the first few passages is desired, we use a concentration of 50,000 cells/ml.

7. As with CNS neurospheres, multi-passage SKP cultures have been maintained more successfully by mechanical dissociation than by enzymatic dissociation. Mechanical dissociation produces small clusters that can re-grow into new spheres, while enzymatic dissociation produces single cells that have reduced viability.

8. Add four drops of Geltol per slide, in the wells along one of the long edges of the slide. Add the coverslip by making contact along the edge of the slide with the Geltol, and then slowly lowering the opposite end down to avoid airbubbles. Note that for eight chamber slides, one drop of Geltol can be added for each well.

Acknowledgements

We would like to thank all past and present members of the SKP group in the laboratories of Freda Miller and David Kaplan, who have used and contributed to the design and evolution of these protocols over several years. This work has been funded by fellowships and operating grants from the Canadian Institutes of Health Research (CIHR).

References

1. Toma, J.G., Akhavan, M., Fernandes, K.J., Barnabe-Heider, F., Sadikot, A., Kaplan, D. R., and Miller, F.D., *Isolation of multipotent adult stem cells from the dermis of mammalian skin*. Nat Cell Biol, 2001. **3**(9): 778–84.
2. Toma, J.G., McKenzie, I.A., Bagli, D., and Miller, F.D., *Isolation and characterization of multipotent skin-derived precursors from human skin*. Stem Cells, 2005. **23**(6): 727–37.
3. Fernandes, K.J., McKenzie, I.A., Mill, P., Smith, K.M., Akhavan, M., Barnabe-Heider, F., Biernaskie, J., Junek, A., Kobayashi, N.R., Toma, J.G., Kaplan, D.R., Labosky, P.A., Rafuse, V., Hui, C.C., and Miller, F.D., *A dermal niche for multipotent adult skin-*

derived precursor cells. Nat Cell Biol, 2004. **6**(11): 1082–93.

4. Fernandes, K.J., Kobayshi, N.R., Gallagher, C.J., Barnabe-Heider, F., Aumont, A., Kaplan, D.R., and Miller, F.D., *Analysis of the neurogenic potential of multipotent Skin-derived Precursors.* Exp Neurol, 2006. **201**(1): 32–48.

5. McKenzie, I.A., Biernaskie, J., Toma, J.G., Midha, R., and Miller, F.D., *Skin-derived precursors generate myelinating Schwann cells for the injured and dysmyelinated nervous system.* J Neurosci, 2006. **26**(24): 6651–60.

6. Le Douarin, N.M. and Kalcheim, C., *The Neural Crest, edn 2.* 1999, Cambridge, UK: Cambridge University Press.

7. Barembaum, M. and Bronner-Fraser, M., *Early steps in neural crest specification.* Semin Cell Dev Biol, 2005. **16**(6): 642–6.

Chapter 11

Xenotransplantation of Embryonic Stem Cell-Derived Motor Neurons into the Developing Chick Spinal Cord

Hynek Wichterle, Mirza Peljto, and Stephane Nedelec

Abstract

A growing number of specific cell types have been successfully derived from embryonic stem cells (ES cells), including a variety of neural cells. In vitro generated cells need to be extensively characterized to establish functional equivalency with their in vivo counterparts. The ultimate test for the ability of ES cell-derived neurons to functionally integrate into neural networks is transplantation into the developing central nervous system, a challenging technique limited by the poor accessibility of mammalian embryos. Here we describe xenotransplantation of mouse embryonic stem cell-derived motor neurons into the developing chick neural tube as an alternative for testing the ability of in vitro generated neurons to survive, integrate, extend axons, and form appropriate synaptic contacts with functionally relevant targets in vivo. Similar methods can be adapted to study functionality of other mammalian cells, including derivatives of human ES cells.

Key words: Embryonic stem cells, motor neurons, xenotransplantation, chick neural tube, spinal cord.

1. Introduction

Knowledge of the signals that control embryogenesis has reached the point where we can recapitulate large segments of normal development in vitro. In particular, pluripotent ES cells are responsive to extrinsic signals and can be directed to differentiate into a growing number of specific cell types. ES cell-based systems for large-scale production of functional cells in vitro will not only facilitate studies of mammalian embryonic development and provide essential tools for biochemical analysis of rare cell types, but may also lead to novel clinical applications including cell-based drug screening and cell replacement therapy. The utility of ES cell-derivatives for such applications critically depends on our ability to determine that in vitro generated cells are biochemically and functionally equivalent to their in vivo counterparts.

The most rigorous test for quality of in vitro generated cells is their ability to functionally integrate after transplantation in vivo. Nerve cells derived from ES cells are routinely tested by transplantation into the mature central nervous system (CNS) *(1–4)*. Such transplantations rarely lead to physiologically normal integration of ES cell-derived neurons, due to the limited ability of the mature CNS to support axonal growth and due to the loss of many developmentally relevant guidance cues. Therefore, a more appropriate test would be homotopic transplantation of ES-derived neurons into the developing CNS at the time when their counterparts are generated. Such manipulations in mouse embryos are challenging and generally limited to larger embryonic structures that can be successfully targeted with the aid of ultrasound backscatter microscopy *(5, 6)*. Xenografting of mammalian tissues into readily accessible developing chick embryos has been utilized as an alternative to study mammalian development under experimental conditions *(7, 8)*. Here, we adapted these techniques to examine functionality of ES cell-derived spinal motor neurons upon transplantation into the developing chick neural tube.

We present methods for directed differentiation of mouse ES cells into spinal motor neurons (MN) by exposing ES cells aggregates termed embryoid bodies (EBs) to two developmentally relevant signals—retinoic acid (RA), the neuralizing and caudalizing signal and Hedgehog agonist (HhAg1.3), the ventralizing signal *(9)*. The ability of ES cell derived MNs to integrate into the embryonic spinal cord, extend axons along spinal nerves and form neuromuscular junctions in vivo is tested by xenografting of in vitro generated MNs into the developing chick neural tube *(9)*. We conclude that despite the fact that not all developmentally relevant cues and histological structures are conserved between mammalian and avian species, xenografting can serve as an effective surrogate system to examine the functionality of in vitro generated cells that cannot be studied for technical (mouse ES derivatives) or ethical (human ES cell derivatives *(10)*) reasons by allotransplantation.

2. Materials

2.1. General Media

1. Mouse embryonic stem cell medium (ES medium): combine 200 ml DMEM (SLM-220-B, Chemicon), 37.5 ml ES cell tested fetal bovine serum (FBS) (SH30070.03, HyClone), 2.5 ml of 100× non-essential amino acids

(TMS-001-C, Chemicon), 2.5 ml of 100× nucleosides for ES cells (ES-008-D, Chemicon), 2.5 ml of Pen/Strep (15140-122, Invitrogen, 10,000 U/ml Penicillin/10,000 μg/ml streptomycin), 2.5 ml of 200 mM l-glutamine (25030-081, Invitrogen), 180 μl of diluted 2-mercaptoethanol (diluted 1/100 in PBS, M-7522, Sigma), and 25 μl of LIF/ESGRO (ESG1107, Chemicon).

2. Differentiation medium (ADFNK): combine 250 ml of Advanced DMEM/F12 (12634-010, Invitrogen), 250 ml of Neurobasal Medium (21103-049, Invitrogen), 57 ml of knockout-SR (10828-028, Invitrogen), 5.7 ml of Pen/Strep, 5.7 ml of 200 mM l-glutamine, and 400 μl of diluted 2-mercaptoethanol (diluted 1/100 in PBS).

2.2. Reagents Used for Differentiation of ES Cells

1. T25 tissue culture flasks 25 cm^2 (430168, Corning Cell Culture Flask).
2. Suspension culture dishes (35 mm, 430588; 60 mm, 430589; 10 cm, 430591; Corning).
3. 10 cm Nunc tissue culture dishes (150679; Nunc).
4. 0.1% gelatin in ultrapure water (ES-006-B; Chemicon).
5. 0.05% Trypsin–EDTA (05300-054; Invitrogen).
6. 15 ml Falcon tubes (352097; BD Biosciences).
7. Large orifice pipette tips (21-197-2A, Fisher).
8. Swing bucket tabletop centrifuge (Eppendorf, 5702).
9. Tissue culture microscope.
10. 1 mM of all-trans retinoic acid (R2625, Sigma) dissolved in DMSO (1,000× stock solution, aliquot, and store in dark at −20°C).
11. 1 mM of HhAg1.3 (Curis, Inc., Cambridge, MA *(11)*) dissolved in DMSO (1,000× stock solution, aliquot, and store in dark at −20°C).

2.3. Motor Neuron Culture

1. 4-well Nunc plates (12-565-72, Fisher).
2. 24-well Nunc plates (12-565-75, Fisher).
3. 15 mm round glass coverslips (633031, Carolina biological).
4. Poly-l-Ornithine solution (P4957, Sigma).
5. Mouse Laminin (23017-015, Invitrogen) aliquoted and stored at −20°C.
6. DNase I (104159, Roche) 1,000× stock solution is prepared at 25 mg/ml in L-15 medium, aliquoted, and stored at −20°C.
7. Recombinant Rat GDNF (512-GF-050, R&D systems) is resuspended in PBS/0.1% BSA at a concentration of 100 μg/ml, aliquoted and stored at −80°C.

8. Motor neuron culture medium (ADFNB+GDNF): combine 24 ml of Advanced DMEM/F12, 24 ml of Neurobasal medium, 0.5 ml of Pen/Strep, 0.5 ml of l-glutamine, 1 ml of B27 50× supplement (17504-044, Invitrogen), 0.5 μl of GDNF (5 ng/ml final concentration). A pre-mix of Advanced DMEM/F12, Neurobasal Medium, Glutamine, and Pen/Strep can be stored in a refrigerator for up to a month. Reconstituted medium with B27 and GDNF must be used within 1 week.

2.4. Transplantation Experiments

1. Egg incubator (PROFI-I, Lyon Electric Company).
2. Wiretrol 50 μl disposable glass micropipettes (5-000-1050, Drummond).
3. Glass capillary tubing (27-30-0, Frederick Haer & Co.).
4. 21G11/2 needles (305167, BD Biosciences).
5. 3 ml syringe (309585, BD Biosciences) or 5 ml syringe (309603, BD Biosciences).
6. 1 ml of 26G3/8 Intradermal Bevel Needle (309625, BD Biosciences).
7. Microinjector (modified MO-10, Narishige).
8. Eggs (SPAFAS, specific pathogen-free).
9. Leibovitz's L-15 Medium (11415-064, Invitrogen) supplemented with penicillin–streptomycin.
10. Heavy mineral oil (330760, Sigma).
11. Aspirator tube assemblies for mouth pipetting (A5177-5EA, Sigma).
12. Stereomicroscope (Zeiss Stemi2000 with a KL200 fiber optic light source, Carl Zeiss).
13. Black electric tape (Scotch Super33+, 3 M) and vibratome (VT1000, Leica).
14. Glass needles: Two types of glass needles are pulled using a P-30 Vertical Micropipette Puller (Sutter Instruments) with a setting that results in relatively long tapered tips (~15 mm long). Smaller diameter needles (27-30-0; Frederick Haer & Co.) are used to generate suction lesions in chick neural tube and larger needles (Wiretrol 50 μl) are used for transplantation of ES cell derived MNs. Tips of the needles are broken off and observed under the stereomicroscope to ensure that the edges are smooth, oval or round. Needles with rough edges are either discarded or polished using a glass micropipette grinder (EG-44, Narishige).
15. Ink solution: Mix 100 μl Fount India black ink (221143, Pelikan), 100 μl 10× CMF-PBS buffer (BSS-2010-B, Chemicon), and 800 μl of distilled water to generate 1 ml of ink solution that will be used to visualize chick embryos.

16. Microinjector: Hydraulic oil microinjector MO-10 (Narishige) has been modified according to Dr. Arturo Alvarez-Buylla's design, by attaching a sidearm equipped with a glass capillary holder (**Fig. 11.3A**). Transplantation glass needle is backfilled half way with L-15/ Pen-Strep medium followed with a quarter volume of heavy mineral oil. Insert Teflon plunger (provided with glass capillaries) approximately half way into the capillary and mount the needle and plunger into the modified MO-10 microinjector (**Fig. 11.3A**). Make sure that no bubbles are trapped between the medium, oil, and plunger.

3. Methods

Tissue culture work is performed under standard sterile cell culture conditions using laminar flow tissue culture hood and 5% CO_2; 38°C tissue culture incubator.

3.1. Differentiation of ES Cells into Spinal MNs

1. Coat one T25 tissue culture flask with 0.1% gelatin (incubate for >30 min at room temperature (RT)). A vial of expanded Hb9-GFP ES cells (HBG3 ES cell line *(9)*) (*see* **Note 1**) is rapidly thawed in 37°C tissue culture water bath. Dilute cells with warm ES medium (5 ml) and pellet in a clinical centrifuge (200 rcf, 5 min). Aspirate supernatant, disperse ES cells by gently flicking the tube, and resuspend in 5 ml of warm ES medium. Aspirate gelatin solution from T25 flask, count cells using hematocytometer, and plate $\sim 10^6$ ES cells on gelatin-coated T25 flask in 7 ml of ES medium (*see* **Note 2**). Medium is changed daily.

2. ES cell colonies should appear mostly compact, covering 30–50% of the surface area 2 days after plating (**Fig. 11.1A**). Replace ES medium with 7 ml of ADFNK medium 1 hour before trypsinization. Aspirate ADFNK medium and trypsinize ES cells with 3.5 ml of 0.05% trypsin–EDTA (equilibrated to room temperature) in tissue culture incubator (2–5 min). Gently tap on the side of the flask and monitor dissociation under the microscope. Stop trypsin by adding 5 ml of ADFNK medium (*see* **Note 3**) as soon as most colonies are floating, transfer cells to 15 ml tube, and centrifuge cells at 400 rcf for 5 min. Aspirate medium and disperse colonies into single cells by firmly flicking the tube with a finger (~ 10 times). Add 10 ml of ADFNK, mix by repeated pipetting and count ES cells using hematocytometer (5–10 × 10^6 ES cells should be obtained from a single T25 flask). Plate 1–2 × 10^6 ES cells per 10 cm Nunc tissue culture dish in 10 ml of ADFNK medium.

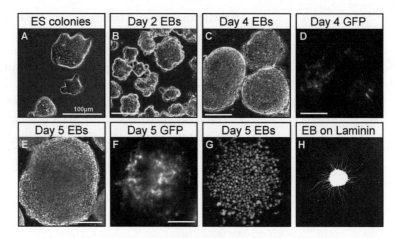

Fig. 11.1. Differentiation of ES cells into motor neurons. Embryonic stem cells grown for 2 days on gelatin-coated dishes should retain relatively compact appearance (**A**). EBs at 2 days of differentiation reach 50–100 μm in diameter (**B**). By 4 days of differentiation, EBs acquire round shape and first postmitotic MNs expressing Hb9-GFP are readily detectable (**C, D**). Most EBs should contain GFP-positive MNs in their cores by day 5 (**E, F**). EBs collected in the center of the dish on day 5 before changing culture medium (**G**). Robust motor neuron axon outgrowth 2 days after plating day 6 EBs on laminin-coated coverslips is a hallmark of healthy MNs (**H**). Scale bar in all panels is 100 μm.

Evenly distribute cells by swirling dishes in a figure eight motion before placing dishes in the tissue culture incubator.

3. Differentiation day 1 (Day 1): Large number of embryoid bodies (EBs)—small floating aggregates of ES cells—should be visible under the microscope (a fraction of the aggregates will be adherent). Swirl dish in a gentle round motion to concentrate floating EBs in the center of the dish and transfer all floating EBs to a 15 ml tube (discard the plate with adherent aggregates). Centrifuge EBs for 3 min at low speed (200 rcf) aspirate supernatant, gently flick the tube to disperse EBs, add 10 ml of ADFNK medium and plate in a new 10 cm Nunc tissue culture dish. Swirl the dish to evenly distribute EBs before placing it in the incubator.

4. Differentiation day 2 (Day 2): EBs should be of a fairly uniform size (50–100 μm in diameter) (**Fig. 11.1B**) and most of them should be floating (*see* **Note 4**). Swirl the dish, collect EBs in a tube, and let them settle by gravity (~15 min). Aspirate medium and resuspend EBs in 400 μl of ADFNK medium. Prepare four 10 cm suspension culture dishes, each containing 10 ml ADFNK medium to split the EBs 1 to 4. Using a 200 μl pipetman and a large orifice yellow tip mix EBs by repeated pipetting and dispense 100 μl of EBs per 10 cm dish. To induce MN differentiation, supplement dishes with retinoic acid (RA, 1 μM final concentration) and Hedgehog agonist (HhAg1.3, 1 μM final concentration) (*see* **Note 5**). To induce differentiation

into other types of nerve cells supplement dishes with 1 μM RA only. Place dishes back in the incubator, swirl, and incubate until day 5. EBs on day 4 reach motor neuron progenitor stage when majority of cells are expressing MN progenitor marker Olig2. Few GFP-positive postmitotic MNs are observed already at this stage (**Fig. 11.1C, D**).

5. Differentiation day 5 (Day 5): Medium in the dishes should turn orange/yellow by day 5. Swirl dishes in a circular motion to collect EBs in the center of the dish (they are by now readily visible to the naked eye) (**Fig. 11.1G**). Aspirate medium from the edges of the dish, taking care not to aspirate EBs, and add fresh ADFNB+GDNF medium. Observe EBs under the fluorescent microscope—most EBs should contain green fluorescent cells in their cores, indicating successful induction of MN differentiation (**Fig. 11.1E, F**). EBs are utilized for MN culture or transplantation on day 6 and for cryosectioning and immunocytochemical analysis (**Fig. 11.2A, B**) on day 7 of differentiation (*see* **Note 6**).

3.2. Motor Neuron Culture

In order to establish the health of in vitro generated motor neurons and the robustness of their axon outgrowth, we recommend culturing MNs in vitro in parallel with transplantation experiments. ES cells differentiated into MNs (day 6 or 7) can be plated either as whole embryoid bodies or as dissociated cells. Whole EB cultures should result in a robust axon outgrowth (**Fig. 11.1H**). Dissociated cultures allow morphometric examination of MNs as well as immunocytochemical analysis.

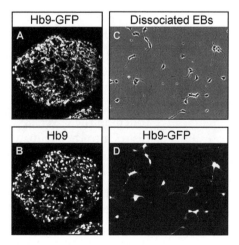

Fig. 11.2. Characterization of in vitro generated MNs. ES cell-derived MNs are characterized by immunostaining of cryosectioned EBs with MN marker (Hb9, Developmental hybridoma bank, 81.5C10). Most GFP-expressing MNs also express Hb9 (**A, B**). Dissociated EBs plated on Laminin substrate and cultured for 16 hr with large number of GFP-expressing MNs extending processes (**C, D**).

3.2.1. Coverslip Preparation

1. Transfer glass coverslips into a beaker and wash with 100% ethanol for 10 min with agitation.
2. Using forceps, remove each coverslip from the ethanol bath and briefly flame with a Bunsen burner (place the burner far from the ethanol containing beaker) before transferring into a sterile beaker under the culture hood. After coverslips cool down transfer them into wells of a 4- or 24-well plate.
3. Dilute poly-l-ornithine in sterile distilled water (1:10, 0.001% final concentration). Add 250 µl of diluted poly-l-ornithine to each coverslip and gently press each coverslip with a pipette tip to ensure that the coverslip is not floating and properly covered with the coating solution.
4. Incubate the plate for more than 1 hr at 37°C (we usually incubate the coverslips overnight). Wash coverslips thrice with 500 µl of sterile distilled water, remove all the water, and allow coverslips to dry.
5. During this period, dilute the laminin stock solution in PBS to a final concentration of 5 µg/ml (for plating dissociated neurons) or 10 µg/ml (for plating whole embryoid bodies).
6. Add 250 µl of the diluted laminin solution to each coverslip and incubate plates for >3 hr at 37°C, wash thrice for 10 min with 500 µl of PBS at RT and keep in PBS until used for MN culture.

3.2.2. Motor Neuron Culture

1. Collect EBs by swirling tissue culture dish in circular motion to concentrate EBs at the center of the dish. Using a yellow 200 µl wide-orifice pipette tip, EBs are gently transferred into a 15 ml tube, allowed to settle by gravity and washed twice with 5 ml of 1×CMF-PBS (Ca^{2+} Mg^{2+} free PBS).
2. For plating whole EBs aspirate PBS from coated coverslips and add 300 µl of ADFNB+GDNF medium. Carefully dispense four to five embryoid bodies in each well with a wide-orifice pipette tip. Distribute EBs on top of the coverslip (avoid touching the coverslip with the pipette tip). Carefully transfer dishes into the incubator and culture undisturbed for >12 hr. EBs will adhere and robust axonal outgrowth should be visible at 12–48 hr after plating (**Fig. 11.1H**).
3. For plating dissociated motor neurons, remove as much PBS as possible from EBs, add 1 ml of cold (4°C) trypsin–EDTA and incubate ~6 min at room temperature. Monitor trypsinization by tapping the bottom of the tube with your finger and stop the reaction with 100 µl of fetal bovine serum as soon as EBs start sticking together.
4. Let the EBs settle, remove trypsin solution, and add 2 ml of L-15 medium supplemented with 25 µg/ml DNase. Triturate

EBs into single cells by repeated pipetting (10–15 times, avoid making any bubbles) using a 1 ml pipette tip.

5. Add 3 ml of L-15 with 25 μg/ml DNase to triturated cells. Pipette up and down twice. Let the non-dissociated cells settle by gravity for 5 min. Collect 3 ml of the supernatant in 15 ml tube. Repeat this step three times.

6. Centrifuge the 15 ml tube containing 9 ml of dissociated cells at 200 rcf for 5 min. Remove the supernatant and resuspend cells in 1 ml of ADFNB+GDNF medium. Count the cells and dilute in ADFNB+GDNF medium to a final concentration of $80–100 \times 10^3$ cells/ml.

7. Aspirate PBS from coated coverslips, and add 300 μl of dissociated MNs per well, mix the plates in a figure eight motion and place in the incubator. Cells should adhere within few hours and processes should be visible in ~6 hr (**Fig. 11.2C, D**).

3.3. Xenotransplantation of Mouse ES Cell Derived MNs into Chick Neural Tube

In order to minimize contamination, the transplantation is performed under a stereomicroscope in an open face laminar flow hood. All egg incubations are carried out in a stationary Profi-I incubator at 39°C and high relative humidity (>70%).

1. Incubate eggs lying on their sides (mark the top of each egg with a permanent marker to identify the site for shell opening) for ~55 hr (Hamburger-Hamilton stage 15–16, (12)) or ~65 hr (HH stage 17–18) (see **Note 7**).

2. Pick EBs (at day 5 or 6 of differentiation) exhibiting bright GFP expression and transfer them into 35 mm dish containing L-15 medium with Pen/Strep (select smaller bright EBs or cut larger EBs into smaller pieces).

3. Remove one egg from the incubator and using a 5 ml syringe equipped with a 21G needle carefully aspirate ~3 ml of albumen in order to sink the embryo (insert the needle through the pointed end of the egg aiming towards the base of the egg to avoid rupturing the yolk sac).

4. Using sharp forceps open up the shell at the top of the egg, creating a window ~2.5 cm in diameter. The embryo should appear beneath the opening. Replace the albumen and if necessary add L-15/Pen-Strep medium to lift the embryo into the window for easier manipulations. Carefully aspirate any bubbles and inject diluted India ink underneath the embryo using a 1 ml syringe with a hypodermal needle. Insert the needle through *area opaca*, position the tip under the embryo, and slowly inject ink until the embryo is clearly visible against the black background (make sure to avoid introduction of air bubbles).

5. Using a 25G–30G needle open the vitelline membrane above the neural tube and make a small nick in the dorsal neural tube

~2–3 mm posterior to the prospective site of transplantation. To generate a lesion in the developing spinal cord, insert the smaller needle (attached to the mouth-pipette assembly) through the opening in the neural tube and thread it tangentially through the neural tube to the site of transplantation (**Fig. 11.3B, C**). Generate a small lesion in the neural tube (adjacent to somites 15–20 for brachial grafts and 20–26 for thoracic grafts) by exerting gentle repetitive suction while retracting the needle (*see* **Note 8**).

6. Load the assembled handheld microinjector with in vitro generated cells. A small portion of an EB is front-loaded under the stereomicroscope by aspirating the EB through the tip of the needle (make sure that no bubbles are introduced into the micropipette assembly) (*see* **Note 9**). Insert and thread the loaded needle through the neural tube to the site of lesion in the same manner as the previously inserted lesion micropipette. Transplant ES cell-derived MNs into the lesion site by slowly releasing ES cell-derived cells while simultaneously retracting the micropipette (**Fig. 11.3D, E**).

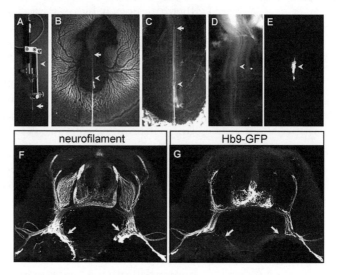

Fig. 11.3. Transplantation of ES cell-derived MNs into the developing chick neural tube. EBs are transplanted using a modified MO-10 microinjector equipped with a sidearm (arrowhead) holding the glass capillary (*arrow*) (**A**). To lesion the spinal cord neural tube is entered ~2–3 mm posterior to the site of transplantation (*arrowhead*) and the tip of the needle is threaded through the spinal cord until the site of lesion (*arrow*) (**B, C**). After lesioning, GFP-expressing MNs are transplanted into the neural tube. Graft fills the lesion site and extends also into the spinal canal of the intact neural tube (**D, E**). Vibratome section of a chick embryo 3 days after transplantation reveals numerous GFP-positive transplanted MN axons extending from the spinal cord into the periphery. Some GFP-positive transplanted cells are observed within the spinal canal. The specificity of axonal projections is demonstrated by the lack of innervation of the sympathetic chain with graft-derived GFP axons (**F, G**, *arrow*).

7. Sink the transplanted embryo by removing ~2 ml of albumen and close the window in the shell using two stripes of electric tape (make sure the opening is completely sealed). Patch up the small hole through which the albumen was aspirated. Place the egg in the incubator and incubate for 2–7 days.

8. Collect surviving embryos and examine grafts under the fluorescent stereomicroscope. GFP-positive axons exiting neural tube should be readily visible in successfully transplanted embryos. Dissect out the graft-containing section of the embryo, including all potential muscle targets.

9. Embryos are formaldehyde fixed, sectioned into 100–200 μm thick sections using vibratome and immunostained (e.g., anti-GFP antibody (A-6455, Invitrogen)) to visualize transplanted MNs and their axons (**Fig. 11.3F, G**).

4. Notes

1. The quality of ES cells is the most important determinant for the efficiency of MN differentiation. We usually expand ES cells on mitomycin treated primary mouse embryonic fibroblast feeders, prepare ~20 cryovials from two T75 flasks and test each lot for its ability to differentiate into motor neurons with efficiency of 40–50%.

2. Some ES cell lines do not form tight colonies when plated on gelatin and tend to rapidly differentiate. To grow such lines, we usually switch to KES medium (same as ES medium but contains knockout SR (Invitrogen) instead of serum). ES cells grow slower in KES medium but colonies remain compact with less differentiated appearance.

3. Make sure that no ES medium is carried over to the differentiation culture. Do not stop trypsin with serum as even a trace amount of serum will result in excessive adhesion of aggregates on day 1.

4. EBs should not stick to each other even at high density. Formation of large aggregates (sometimes appearing as branched structures consisting of tens of EBs) will compromise motor neuron induction (neither too big nor too small EBs differentiate efficiently into MNs).

5. HhAg1.3 agonist provided by Curis, Inc. (under Material transfer agreement) can be substituted with commercially available Hh protein (Shh-N, 461-SH, R&D Systems) or Hh agonist Purmorphamine (540220, Calbiochem). Purmorphamine used at 2 μM concentration induces MN differentiation with significantly lower efficiency than HhAg1.3 or Shh-N protein.

6. The number of GFP-positive MNs peaks on day 7 when 40–60% of cells are GFP positive (based on flow cytometry).

7. The optimal stage for brachial or thoracic grafts is 14–16 and for thoracic or lumbar grafts is 17–18. Once the proper developmental stage is reached, eggs can be taken out of the incubator and stored at room temperature to slow down their development.

8. Prevent embryos from drying during the transplantation by dripping L-15/PenStrep on the embryo whenever necessary. Do not exert excessive suction during lesioning of the neural tube as it will lead to decreased survival rates. Lesion is often visible as a narrowing of the neural tube.

9. The ease and success of transplantations depends on the quality of the transplantation needle. Its edges need to be smooth otherwise it is difficult to load the needle and it clogs easily. In addition, it is important to carefully select EBs—younger and smaller EBs (day 5 or 6) are preferable as they are easier to load into the glass transplantation needle.

Acknowledgements

The authors would like to thank Curis Inc. for providing hedgehog agonist HhAg1.3; Thomas Jessell for his advice, encouragement, and support during the development of described methods; and Esteban Mazzoni and Annette Gaudino for critical reading of the manuscript. This work has been supported by Project A.L.S. foundation.

References

1. Kawasaki, H., Mizuseki, K., Nishikawa, S., Kaneko, S., Kuwana, Y., Nakanishi, S., Nishikawa, S. I., and Sasai, Y. (2000) Induction of midbrain dopaminergic neurons from ES cells by stromal cell-derived inducing activity. *Neuron* **28**, 31–40.

2. Bjorklund, L. M., Sanchez-Pernaute, R., Chung, S., Andersson, T., Chen, I. Y., McNaught, K. S., Brownell, A. L., Jenkins, B. G., Wahlestedt, C., Kim, K. S., and Isacson, O. (2002) Embryonic stem cells develop into functional dopaminergic neurons after transplantation in a Parkinson rat model. *Proc Natl Acad Sci USA* **99**, 2344–9.

3. Harper, J. M., Krishnan, C., Darman, J. S., Deshpande, D. M., Peck, S., Shats, I., Backovic, S., Rothstein, J. D., and Kerr, D. A. (2004) Axonal growth of embryonic stem cell-derived motoneurons *in vitro* and in motoneuron-injured adult rats. *Proc Natl Acad Sci USA* **101**, 7123–8.

4. Tabar, V., Panagiotakos, G., Greenberg, E. D., Chan, B. K., Sadelain, M., Gutin, P. H., and Studer, L. (2005) Migration and differentiation of neural precursors derived from human embryonic stem cells in the rat brain. *Nat Biotechnol* **23**, 601–6.

5. Wichterle, H., Turnbull, D. H., Nery, S., Fishell, G., and Alvarez-Buylla, A. (2001) In utero fate mapping reveals distinct migratory pathways and fates of neurons born in the mammalian basal forebrain. *Development* **128**, 3759–71.

6. Olsson, M., Campbell, K., and Turnbull, D. H. (1997) Specification of mouse telencephalic and mid-hindbrain progenitors following heterotopic ultrasound-guided embryonic transplantation. *Neuron* **19,** 761–72.
7. Fontaine-Perus, J. (2000) Interspecific mouse-chick chimeras. *Methods Mol Biol* **135,** 443–6.
8. Fontaine-Perus, J., Halgand, P., Cheraud, Y., Rouaud, T., Velasco, M. E., Cifuentes Diaz, C., and Rieger, F. (1997) Mouse-chick chimera: a developmental model of murine neurogenic cells. *Development* **124,** 3025–36.
9. Wichterle, H., Lieberam, I., Porter, J. A., and Jessell, T. M. (2002) Directed differentiation of embryonic stem cells into motor neurons. *Cell* **110,** 385–97.
10. Goldstein, R. S. (2006) Transplantation of human embryonic stem cells to the chick embryo. *Methods Mol Biol* **331,** 137–51.
11. Frank-Kamenetsky, M., Zhang, X. M., Bottega, S., Guicherit, O., Wichterle, H., Dudek, H., Bumcrot, D., Wang, F. Y., Jones, S., Shulok, J., Rubin, L. L., and Porter, J. A. (2002) Small-molecule modulators of Hedgehog signalling: identification and characterization of Smoothened agonists and antagonists. *J Biol* **1,** 10.
12. Hamburger, V., and Hamilton, H. L. (1992) A series of normal stages in the development of the chick embryo. 1951. *Dev Dyn* **195,** 231–72.

Chapter 12

Transplantation of Neural Stem/Progenitor Cells into Developing and Adult CNS

Tong Zheng, Gregory P. Marshall II, K. Amy Chen, and Eric D. Laywell

Abstract

Neural transplantation has been a long-standing goal for the treatment of neurological injury and disease. The recent discovery of persistent pools of neural stem cells within the adult mammalian brain has reignited interest in transplant therapeutics. Since neural stem cells are self-renewing, it may be possible to culture and expand neural stem cells and their progenitor cell progeny to sufficient numbers for use in autologous, self-repair strategies. Such approaches will require optimized cultivation protocols, as well as extensive testing of candidate donor cells to assess their capacity for engraftment, survival, and integration. In this chapter, we describe the transplantation of neural stem/progenitor cells—cultivated as either neurospheres or neurogenic astrocyte monolayers—into the persistently neurogenic olfactory bulb system of the adult mouse forebrain, and into the cerebellum of neonatal mutant mice.

Key words: Transplantation, neurosphere, neural stem cell, neurogenic astrocyte, subependymal zone.

1. Introduction

The dream of treating neurological deficits with transplants of donor neurons into areas of central nervous system (CNS) damage or cell loss is perhaps as old as the field of neuroscience itself. Therapeutic grafting into the CNS faces many technical and biological hurdles. In particular, sufficient numbers of appropriate donor cells must be obtained, and this usually requires either pooling of cells from multiple donors, or in vitro expansion of more limited donor sources. Additionally, the receptiveness of the host environment must be such that newly introduced cells can compete for space and metabolic resources. Finally, in order to maintain appropriate spatial cytoarchitectural relationships, cells must be able to extend processes and find their targets which are often located at great distance from their cell bodies. Of these three challenges, the search for an abundant source of transplantable cells has recently received the most scrutiny, particularly with respect to replacing the dopaminergic neurons lost in Parkinson's disease (PD). Both fetal mesencephalon *(1)* and autologous adrenal chromaffin cells *(2)* have been investigated as sources of transplantable dopamine neurons, and each has been used with limited effectiveness in human clinical trials for PD *(3, 4)*. Both of these cell sources, however, have serious shortcomings. In the case of mesencephalic dopamine neurons, many fetuses are required to harvest sufficient numbers of cells for an effective transplant *(5)*. For adrenal chromaffin cells, it has been shown that post-graft survival is quite low *(6)*, limiting their potential in autologous repair approaches.

Since it is now appreciated that the adult mammalian brain harbors—within the periventricular subependymal zone (SEZ)—a pool of multipotent neural stem cells (NSCs) with the capacity for multilineage differentiation, there is a great deal of interest in utilizing these cells in neurotransplantation approaches. NSCs are indigenous and self-renewing, and it is therefore not unreasonable to envision them as an unlimited source of cells for autologous repair strategies. Nevertheless, since NSCs represent only a small fraction of the SEZ cell population, considerable in vitro expansion is required to generate large numbers of transplantable cells. Most frequently, NSCs are cultivated in the form of neurospheres, which are non-adherent clones consisting of both stem and progenitor cells in various states of differentiation *(7, 8)*. Neurogenic SEZ astrocytes with NSC characteristics can also be cultured as adherent monolayers *(9, 10)*, and these cells can be induced to generate neuronal progeny both in vitro and in vivo following re-introduction into the NSC niche of neonatal and adult hosts *(11, 12)*.

Our laboratory has taken a twofold approach to neurotransplantation. First, we view the intrinsic neurogenic niche—the persistent germinal matrix of the SEZ—as an ideal system in which to test the varieties of donor cell sources. NSCs in the SEZ produce migratory neuroblasts which travel via the rostral migratory stream (RMS) to the olfactory bulb (OB), where a percentage of them differentiate and functionally integrate into the existing neural circuitry as granule and periglomerular interneurons (*see* **Figs. 12.1, 12.2 12.3** and Color Plate 6, 7, 8). Transplants into this region, therefore, are designed not to produce a therapeutic effect, but rather to test the ability of engrafted cells to respond to in vivo environmental cues by differentiating as neurons. We have used this system to assess the ability of astrocytes to generate neuronal progeny, as well as to assess the neural transdifferentiation potential of bone marrow-derived mesenchymal stem cells *(13)*. Additionally, we have shown that, as is true for

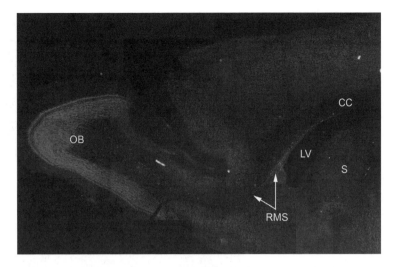

Fig. 12.1. Donor cells migrate normally following engraftment. Parasagittal fluorescence montage through the brain of an adult mouse host following transplantation of GFP+ neurogenic astrocytes into the lateral ventricle. Sagittal sections were immunostained with antibodies against GFP (*green*) and NeuN (*red*) 28 days following engraftment. A large number of GFP+ donor cells can be seen within the rostral migratory stream (RMS) and olfactory bulb (OB). A few donor cells are also present along the roof of the lateral ventricle (LV), and in the septum (S). CC: corpus callosum. (*See* Color Plate 6)

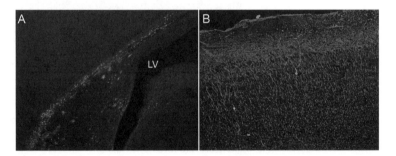

Fig. 12.2. Higher magnification of GFP+ donor cells within the RMS and olfactory bulb. (**A**) Parasagittal section of a host brain immunolabeled for GFP (*green*) and β-III tubulin (*red*). Donor cells are distributed within the SEZ–RMS–OB system, as well as lining the wall of lateral ventricle (LV). (**B**) Parasagittal confocal image of a donor cell that migrated into the olfactory bulb and differentiated into a granule interneuron. (*See* Color Plate 7)

bone marrow transplantation, reduction of the indigenous NSC/progenitor cell population within the SEZ prior to transplantation leads to increased levels of engraftment by donor cells *(14)*. Our second approach to transplantation is directed at more conventional therapeutic interventions. These may be performed in experimental models of disease states, such as neonatal hypoxia/ischemia *(15)*, or in naturally-occurring neurological mutant mice (*see* **Fig. 12.4** and Color Plate 9).

In this chapter, we will describe methods for both generating transplantable neural stem/progenitor cells (i.e., neurospheres and

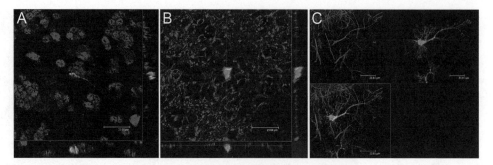

Fig. 12.3. Donor-derived cells are capable of differentiating into neurons and astrocytes. (**A**) and (**B**): Confocal Z-series of GFP+ donor cells (*green*) located in the granular cell layer of olfactory bulb that are double immunolabeled with the neuron markers (*red*) NeuN (A) and β-III tubulin (B). (**C**) A GFP+ (*green*) donor cell that has differentiated into a cortical astrocyte, and is immunolabeled with the astrocyte marker GFAP (*red*). (*See* Color Plate 8)

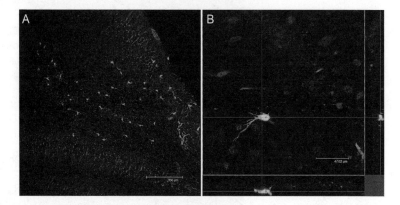

Fig. 12.4. Grafted cells survive, migrate, and differentiate upon transplantation into the cerebellum of both wild type and weaver transgenic mouse pups. (**A**) Donor GFP+ cells (*green*) are seen migrating and differentiating within the cerebellum 3 weeks post-engraftment in the wild-type mouse. (**B**) High magnification confocal z-series of a donor cell transplanted into the cerebellum of a weaver mutant mouse. Pεδ= β-III tubulin in A, and NeuN in B. (*See* Color Plate 9)

neurogenic astrocytes) and transplanting them into the adult neurogenic SEZ niche, and the developing cerebellum of neurological mutant mice. In addition, we will describe our method for improving donor cell engraftment into the SEZ/RMS system by mildly injuring the brain with ionizing radiation prior to transplantation.

2. Materials

2.1. Generation of Neurospheres

1. Base medium: 1× DMEM/F12 medium (Invitrogen cat. no. 12500-062).
2. N2 Culture Supplement: Stock Solution comprised of: Pituitary Extract (Sigma cat. no. P1476, 20 μg/mL final), putrescine (Sigma cat. no. P5780, 100 μM final), insulin (Sigma cat.

no. I5500, 67.5 U), progesterone (Sigma cat. no. P7556, 1.258 μg/mL final), transferrin (Sigma cat. no. T1428, 25 μg/mL final), and sodium selenite (Sigma cat. no. S5261, 30 nM final).

3. Fetal bovine serum (FBS: Atlanta Biologicals, cat. no. S11150).

4. Growth medium: Base medium containing N2 supplements, 5% FBS, EGF (20 ng/mL), and bFGF (10 ng/mL).

5. 0.25% Trypsin/EDTA solution (Atlanta Biologicals cat No. B81310).

6. Ultra-low attachment, anti-adhesive six-well plates (Corning Costar cat. no. 3471).

7. Fire polished Pasteur pipettes: prepare medium and narrow bore sets by briefly exposing the tip of the pipette to the flame of a Bunsen burner to narrow the lumen. Add a cotton plug to the proximal end and autoclave before use.

8. 15 mL Falcon tubes (TPP, cat. no. TP91015).

9. Growth factor stock solution: Cultures require supplementation with 20 ng/mL of EGF (Sigma cat No. E9644) and 10 ng/mL of bFGF (Sigma cat. no. F0291) every 2–3 days; since each culture well will contain approximately 2 mL of medium, we supplement with 50 μL aliquots of 40× stock (8,000 ng of EGF and 4,000 ng of bFGF in 10 mL of DMEM/F12). Stock can be prepared more concentrated if desired, but we do not recommend a less concentrated stock as correspondingly larger aliquots will quickly reduce the viscosity of the neurosphere cloning medium.

10. Dulbecco's PBS (dPBS: Invitrogen cat. no. 14190-144) or base medium containing antibiotic/antimycotic (Sigma cat. no. A9909).

2.2. Generation of Neurogenic Astrocyte Monolayers

1. PBS or base medium containing antibiotic/antimycotic (Sigma cat. no. A9909).

2. Growth medium (as above).

3. 0.25% Trypsin/EDTA solution (Atlanta Biologicals Cat No. B81310).

4. 15 mL Falcon tubes (TPP).

5. T-75 tissue culture flasks (TPP, cat. no. TP90076).

2.3. Transplantation

2.3.1. Transplantation into Adult Mice

1. Avertin anesthetic: 2.5% tribromoethanol (Sigma cat. no. T48402) and 1.5% *tert*-amyl alcohol (Sigma, cat. no. A1685) in H_2O.

2. Betadine solution and 70% ethanol (EtOH).

3. Cotton swabs.
4. Stereotaxic apparatus.
5. Micromanipulator for Hamilton syringes.
6. Surgical microscope and illuminators.
7. One milliliter disposable hypodermic syringe (Becton Dickinson, cat. no. 309602).
8. Autoclaved or disposable surgical scalpel.
9. Dental drill.
10. Hot bead sterilizer (Fine Science Tool).
11. 26 gauge Hamilton syringe (Hamilton, cat. no. 84851).
12. Stainless steel wound clips (Becton Dickinson, cat. no. 427631).

2.3.2. Transplantation into Neonatal Mice

1. Ice-water bath for anesthetizing pups.
2. 70% EtOH.
3. Cotton swabs.
4. Modeling clay mold for holding the pups in place.
5. 26 gauge Hamilton syringe (Hamilton, cat. no. 87900).
6. Hamilton 10 µL repeating dispenser (Hamilton, cat. no. 83700).
7. Incubator or slide warmer (37°C) to warm pups after surgery.

2.4. Irradiation of Adult Mice as a Model for Transient Depletion of the Neural Stem Cell Niche and Enhanced Engraftment of Transplanted NSCs

1. Plexiglass container suitable for holding mice and of the appropriate size for the irradiator.
2. Access to a Gamma Irradiator (our model is a Gamma Cell 40 irradiator, with a Cesium^{137} source).

3. Methods

3.1. Generation of Neurospheres

The following protocol is the standard method our laboratory has developed to produce neurospheres, and has been previously described in detail (16). Any culture dish configuration can be used, but we prefer six-well plates, because they allow for multiple experimental manipulations of the same sample, and they lend themselves to rapid visual screening (without the optical interference common to plates with a smaller well diameter). Moreover, potential infections are contained within single wells, and can be removed without killing the entire sample. This protocol is appropriate for obtaining neurospheres from neonatal or adult animals, but one should expect significantly decreased yield with increasing age of the donor animal. The term 'neurosphere', as commonly used, describes a variety of multipotent clonal and polyclonal aggregates that represent mixtures of self-renewing stem cells and cells with much less capacity for self-renewal. This chapter is not concerned with distinguishing between stem and progenitor cells. For a thorough discussion of this matter, see references *(8)* and *(17)*.

1. Anesthetize the animal with either hypothermia (for pups < P8) or i.p. injection of Avertin (for P8 and >). Neonatal hypothermia is achieved by immersing pups up to the neck in an ice-water bath for 3–4 min. *(18)*.

2. Decapitate and spray the scalp thoroughly with 70% EtOH.

3. Remove the brain, and place it on a clean surface suitable for cutting.

4. With a razor blade, make a coronal block, about 2 mm in thickness, in the area between the rhinal fissure and the hippocampus. Lay the block flat on the cutting surface and use the razor blade to make two parasagittal cuts just lateral to the lateral ventricles, and a horizontal cut to remove the tissue above the corpus callosum. This procedure leaves a small, rectangular chunk of tissue surrounding the lateral ventricles containing a high density of NSCs.

5. Wash the tissue chunk for several minutes in medium or PBS containing antibiotics/antimycotics. All subsequent work should be performed with sterile materials in a laminar flow hood.

6. Remove antibiotics/antimycotics, and incubate tissue in 3–4 mL trypsin/EDTA solution at 37°C for 5 min.

7. Add 1 mL serum and gently triturate the tissue through a series of descending-diameter, fire-polished Pasteur pipettes to make a single-cell suspension. It is important to add the serum *prior* to trituration. The dissociation results in the lysis of many cells, and in the absence of serum this can lead to the formation of a viscous residue that will stick to the inside of the Pasteur pipette and cause a severe reduction in recovered cells.

8. Add several milli-liters of growth medium, and wash twice by centrifugation.

9. Count cells using a hemacytometer.

10. In a 15 mL Falcon tube, combine 6 mL of growth medium, 60,000 cells, and 300 μL of 40× growth stock. Add growth medium to bring the final volume to 12 mL.

11. Mix for several minutes by repeatedly inverting the tube.

12. Distribute 2 mL to each well of an anti-adhesive six-well plate. The final cell density will be about 1,000 cells/cm^2, although the viscosity of the cloning medium makes precise volumetric measurements diffuse.

13. Add 50 µL/well aliquots of 40× growth factor stock every 2–3 days. Neurospheres will be visible under phase optics after 7–10 days.

3.2. Generation of Neurogenic Astrocyte Monolayers

The following protocol describes our method of generating neurogenic astrocyte monolayers that can be induced to produce large numbers of neuronal progeny in vitro *(19)*, and can differentiate into neurons when transplanted into adult brain 12.

1. Follow steps 1–7 above with a P 1–7 mouse pup.
2. It is not necessary to make a single-cell suspension for this initial culture.
3. Centrifuge cells to form a pellet.
4. Aspirate supernatant, and wash by trituration with fresh medium. Pellet and repeat one to two times.
5. Resuspend in growth medium and growth factors, plate in T-75 culture flasks (use one flask for each brain), and place in incubator overnight.
6. Collect the supernatant containing the cells that have not attached during the overnight incubation, and replate them into fresh T-75 flasks (the original flasks may be discarded).
7. The replated cells will attach over the next 1–2 days. Replace medium every 2–3 after that with days with fresh growth medium and growth factors until astrocyte monolayers become confluent.
8. Remove astrocytes from flasks by aspirating culture supernatant and incubating in trypsin/EDTA for 5–10 min.
9. Collect cells in a Falcon tube, add serum to neutralize trypsin, and proceed with step 1 of Section **3.3.1**.

3.3. Transplantation

This section focuses on the transplantation of suspensions of neurogenic astrocytes and neurospheres into the lateral ventricle of both adult and neonatal mice, as well as the neonatal cerebellum.

3.3.1. Transplantation into Adult Mice

1. Trypsinize, dissociate, and suspend donor cells in an appropriate volume of base medium or dPBS. This volume is calculated based on the number of animals and the corresponding number of cells to be transplanted. We typically transplant between 50,000 and 100,000 cells into the lateral ventricle of adult mice, in a volume of no more than 2 µL.
2. Surgeries are performed in a clean and isolated working area, with sterilized (e.g., autoclaved) instruments and surgical supplies. The stereotaxic apparatus and surrounding area are swabbed often with 70% ethanol. Surgery instruments are

autoclaved prior to the surgery and sterilized with a hot bead sterilizer during the surgeries.

3. Animals should be anesthetized according to the local laboratory guidelines. We use Avertin anesthesia in our lab for adult animals because it is an inexpensive, efficient, and reliable way to perform short procedures in mice where their level of anesthesia is deep, but they recover quickly with minimal side effects. Adult mice are anesthetized with an intraperitoneal dose of 0.3–0.6 ml, which can be adjusted to account for variance in animal sensitivity to this drug. *Note:* Avertin should be made within 48 hr of use to avoid chemical breakdown into caustic agents.

4. Place deeply anesthetized mice in a mouse stereotaxic apparatus. The anesthesia should be monitored with pain reflexes (toe pinch, tail pinch, and eye blink reflex) and supplemental doses should be administered if necessary.

5. The hair at the surgical area should be shaved with a single edge razor blade (electric clippers may also be used, but tend to dull and clog frequently).

6. Administer local anesthetic (Xylocaine) under the scalp, and disinfect incision site with Betadine and 70% EtOH.

7. Using a disposable scalpel, make a midline skin incision to expose the underlying muscles and connective tissues. Gently push these away to expose the skull.

8. Using the bregma suture as the starting point, coordinations are calculated and determined according to an atlas of adult mouse brain (Paxinos & Franklin, 2001). For injections into the lateral ventricle of adult mice, we use the following coordinates: AP = –0.5, ML = 1.2, D = 2.5.

9. Mark the location of the target area on the skull and drill a small burr hole. Carefully remove skull fragments using fine forceps, taking care to avoid damage to the underlying dura, and cortex.

10. A Hamilton microsyringe (1–10 µL according to the volume needed) is used for delivering the cell suspension into the target area. Rinse the syringes with distilled water before and after the transplantations. If different types of cells are used in the same experiment, the syringes should be washed in between each injection.

11. As mentioned above, cells are suspended in base medium or dPBS at a concentration of 5×10^4 cells/µL. Keep the cell suspensions on ice during the procedure and gently re-suspend with a pipette before loading into the Hamilton syringes for each injection. Take care to make sure that no air bubbles are present in the syringes. Push the plunger of the syringe very gently to verify that the cell suspension is delivered smoothly through the syringes (*see* **Note 1**).

12. Attach the syringes to the micromanipulator and lower into the lateral ventricles of the anesthetized animal. Slowly pressure inject 1.0 µL of cell suspension. We prefer to inject the cells over small pulses and leave the injection needle in place for 3–5 min before slowly withdrawing it. This helps preventing the reflux of the cell suspension from the needle tracks.
13. Suture the wound site with stainless steel wound clips, and transfer the mice to a warm and quiet area to recover from the anesthesia and surgical procedures. They are returned to their cage after they are fully awake and capable of obtaining food and water.
14. The behavior of the transplanted mice should be closely monitored daily for post-operative complications (i.e., lethargy, lack of grooming, seizures, etc.).

3.3.2. Transplantation into Neonatal Mice

Our procedure for transplantation into neonatal mice has some significant differences compared to the adult transplantation procedure discussed above. First, the anesthesia we use for neonatal mice is hypothermia instead of Avertin. Hypothermia via ice water immersion is an approved method of surgical anesthesia for neonatal rodents *(18)*. Second, to the best of our knowledge, there is no stereotaxic apparatus commercially available for mouse pups. Instead, we use a homemade clay mount to hold the pups and use illuminating lights to help identify landmarks such as sutures and sinuses for visual guidance. Finally, we have found that dams are more likely to kill and eat their pups if an incision is made; therefore, we inject directly through the scalp and skull so no sutures or glues are applied to the pups.

1. As described in adult procedures, prepare the surgical area and the instruments before the pups are removed from the mother to limit the time they are away form the mother.
2. Pups are anesthetized by immersion up to the neck for 3–4 minu in an ice-water bath. Since the injection only takes a short-time period, two to four pups are usually anesthetized at the same time.
3. Place pups are on a pre-chilled homemade clay mold, and use transillumination to identify landmarks, such as sutures and sinuses, through scalp for visual guidance.
4. Cell suspensions are loaded into a Hamilton syringe the same way described for adult surgery except the Hamilton syringes are often attached with a Hamilton repeating dispenser (*see* **Note 4**). For transplantation into neonatal lateral ventricles, 1.0 µL is injected per transplant. However, for transplantation into the neonatal cerebellum, only 0.5 µL can be injected per site without a build-up of pressure on the underlying structures.

5. After the injection, leave the syringe in place for 0.5–1 min before removal. Pups are then cleaned and kept warm in a 37°C incubator until fully awake and active. They are then returned to the mother and closely monitored daily (*see* **Note 3**).

3.4. Irradiation of Adult Mice as a Model for Transient Depletion of Neurogenesis and Enhanced Donor Cell Engraftment

The purpose of this protocol is to transiently diminish the levels of neurogenesis in the brains of adult mice, a condition accomplished by exposing the subject animals to ionizing radiation. We have shown that exposure to 450 Rads (roughly 4.5 Gy) of ionizing radiation (Cs^{137}) results in an immediate depletion of mitotic neuroblasts in the SEZ and RMS, with near full recovery occurring 3 weeks post-irradiation. This model is helpful for increasing the receptiveness of the SEZ to transplanted cells, as we have observed variable but significant increases in levels of engraftment following transplantation of GFP+ neurogenic astrocytes into sublethally (450 Rads) irradiated adult mice at 3 weeks post-irradiation.

1. As each irradiator has a different rate of radioactive exposure, it is important to first determine how much time is required for 450 Rads to be obtained.
2. Transfer adult mice from colony housing to a plexiglass rodent storage container or similar apparatus.
3. Irradiate mice until 450 Rads has been obtained.
4. Return mice to general housing.
5. Three weeks following irradiation, proceed to transplantation as described in **Section 3.3.1**.

4. Notes

1. It is important to use optimal concentration of the cell suspension for transplantations. Too low of a concentration results in insufficient cell engraftment for examination. Conversely, too high of a concentration will cause a dense cellular slurry, which is likely to clog the transplant needle. Additionally, crowding of donor cells as they are forced through the needle results in reduced viability due to the higher pressure required for injection. We have found that a concentration of $5–10 \times 10^4/\mu L$ consistently works best for the procedures discussed here.

2. Hypothermia anesthesia is often used for neonatal surgeries. In our experience, it is very difficult to estimate the amount of Avertin for such young mice. Hypothermia is an approved method for providing sufficient anesthesia for the injection procedure, which usually lasts only a minute or two. During the actual injection procedure, the pups are secured to a

3. Maternal neglect or cannibalism is not uncommon when manipulated pups are returned to their mothers. In an attempt to minimize this, we perform the injections directly through the scalp and skull, without an incision on the skin. This approach causes less bleeding, and does not require suturing or gluing of the skin.

4. Pregnant mothers are separated from the rest of the mice and placed in a clean cage (one mother/cage) before delivery. To avoid introducing new scent, new cages are not provided immediately before or after the transplantation. Additionally, gloves are worn at all times when contacting the pups and during the transplantation procedures. Post-injection, the pups are thoroughly cleaned before returned to the mother. Be sure to handle all pups in the litter, even if not all are serving as graft hosts. Both the mother and the pups are closely monitored. If necessary, another nursing mother can be used as surrogate mother.

5. For transplantation into neonatal mice, the Hamilton syringes are often attached to a Hamilton repeating dispenser. This not only speeds up the injection procedure, but also enables one person to both secure the animal and perform the injection.

6. Most of the neonates we injected are allowed to survive until adulthood. However, occasionally, we need to kill the neonates. The transcardial perfusion in neonates is very similar to the procedure performed on adult mice, except smaller gauge needles and less perfusate are used.

7. Occasionally, exposure to 450 Rads of ionizing radiation can result in immunocompromised mice which show increased morbidity. This can be ameliorated by the addition of an antibiotic (i.e., Bactrim) to the drinking water for 2 weeks following irradiation, allowing time for the animal's hematopoietic system to fully recover.

References

1. Bjorklund, A., Stenevi, U. (1979) Reconstruction of the nigrostriatal dopamine pathway by intracerebral nigral transplants. *Brain Res.* **177**:555–60.
2. Madrazo, I., Drucker-Colin, R., Diaz, V., Martinez-Mata J., Torres, C., Becerri, J.J. (1987) Open microsurgical autograft of adrenal medulla to the right caudate nucleus in two patients with intractable Parkinson's disease. *N Engl J Med.* **316**:831–4.
3. Fernandez-Espejo, E., Armengol, J.A., Flores, J.A., Galan-Rodriguez, B., Ramiro, S. (2005) Cells of the sympathoadrenal lineage: biological properties as donor tissue for cell-replacement therapies for Parkinson's disease. *Brain Res Brain Res Rev.* **49**: 343–54.

4. Winkler, C., Kirik, D., Bjorklund, A. (2005) Cell transplantation in Parkinson's disease: how can we make it work? *Trends Neurosci.* **28**:86–92.
5. Correia, A.S., Anisimov, S.V., Li, J-Y., Brundin, P. (2005) Stem cell-based therapy for Parkinson's disease. *Ann Med.* **37**:487–98.
6. Freed, W.J., Cannon-Spoor, H.E., Krauthammer, E. (1987) Intrastriatal adrenal medulla grafts in rats. Long-term survival and behavioral effects. *J Neurosurg.* **65**:664–70.
7. Reynolds, B.A., Weiss, S. (1992) Generation of neurons and astrocytes from isolated cells of the adult mammalian central nervous system. *Science* **255**:1707–10.
8. Reynolds, B.A., Rietze, R.L. (2005) Neural stem cells and neurospheres—re-evaluating the relationship. *Nat Meth.* **2**:333–6.
9. Laywell, E.D., Rakic, P., Kukekov, V.G., Holland, E.C., Steindler, D.A. (2000) Identification of a multipotent astrocytic stem cell in the immature and adult mouse brain. *Proc Natl Acad Sci USA.* **97**:13883–8.
10. Scheffler, B., Walton, N.M., Lin, D.D., Goetz, A.K., Enikolopov, G., Roper, S.N., Steindler, D.A. (2005). Phenotypic and functional characterization of adult brain neuropoiesis. *Proc Natl Acad Sci USA.* **102**:9353–8.
11. Zheng, T., Steindler, D.A., Laywell, E.D. (2002) Transplantation of an indigenous neural stem cell population leading to hyperplasia and atypical integration. *Clon Stem Cells* **4**:3–8.
12. Zheng, T., Marshall, G.P., II, Laywell, E.D., Steindler, D.A. (2006a). Neurogenic astrocytes transplanted into the adult mouse lateral ventricle contribute to olfactory neurogenesis, and reveal a novel intrinsic subependymal neuron. *Neurosci.* **142**:175–85.
13. Deng, J., Steindler, D.A., Petersen, B.E., Laywell, E.D. (2003). Neural trans-differentiation potential of hepatic oval cells in the neonatal mouse brain. *Exp Neurol.* **182**:373–382.
14. Marshall, G.P., II, Scott, E.W., Zheng, T., Laywell, E.D., Steindler, D.A. (2005). Ionizing irradiation enhances the engraftment of transplanted in vitro-derived neural multipotent astrocytic stem cells. *Stem Cells* **23**:1276–85.
15. Zheng, T., Rossignol, C., Leibovici, A., Anderson, K.J., Steindler, D.A., Weiss, M.D. (2006) Transplantation of multipotent astrocytic stem cells into a rat model of neonatal hypoxic-ischemic encephalopathy. *Brain Res.* **1112**:99–105.
16. Laywell, E.D., Kukekov, V.G., Suslov, O., Zheng, T., Steindler, D.A. (2002) Production and analysis of neurospheres from acutely dissociated and postmortem CNS specimens. *Meth Mol Biol.* **198**:15–27.
17. Seaberg, R.M., van der Kooy, D. (2003) Stem and progenitor cells: the premature desertion of rigorous definitions. *Trends Neurosci.* **26**:125–31.
18. Phifer, C.B., Terry, L.M. (1986). Use of hypothermia for general anesthesia in pre-weanling rodents. *Physiol Behav.* **38**:887–90.
19. Walton, N.M., Sutter, B.M., Laywell, E.D., Levkoff, L.H., Kearns, S.M., Marshall, G.P., II, Scheffler, B., Steindler, D.A. (2006). Microglia instruct subventricular zone neurogenesis. *GLIA* **54**:815–25.

Chapter 13

Transplantation of Embryonic Stem Cell-Derived Dopaminergic Neurons in MPTP-Treated Monkeys

Jun Takahashi, Yasushi Takagi, and Hidemoto Saiki

Abstract

One of the target diseases of cell-replacement therapy is Parkinson's disease. Clinical experiences with fetal dopaminergic cell graft have shown that the therapy is effective, but limited and accompanied by side effects, such as dyskinesia. So, the therapy needs to be further improved and sophisticated. Embryonic stem (ES) cells are expected to be another donor cell for the treatment, because of its proliferative and differentiation capacities. For clinical application, experiments using non-human primates are important, because size, anatomy, and biological characteristics of the brain are different between rodents and primates. Here, we would like to discuss induction of dopaminergic neurons from monkey ES cells and cell transplantation into the brain of monkey Parkinson's disease model.

Key words: Embryonic stem cell, Parkinson's disease, transplantation, dopaminergic neuron.

1. Introduction

Parkinson's disease is characterized by the loss of midbrain dopaminergic (DA) neurons in the substantia nigra, so being considered as one of the target diseases of cell-replacement therapy. Transplantation of fetal DA neurons has been shown to be effective at producing symptomatic relief in both animal models and patients of PD (1, 2). This therapy is limited, however, due to both the technical and the ethical difficulties in obtaining sufficient and appropriate donor fetal brain tissue. Embryonic stem (ES) cells are self-renewing, pluripotent cells derived from inner cell mass of the preimplantation blastocyst, that have both proliferative and differentiation capacities, thus fulfilling many of the characteristics required of a cell source for cell-replacement therapy (3).

DA neurons are induced by stromal cell-derived inducing activity (SDIA), which is present on the surface of mouse stromal feeder cells, PA6 cells *(4, 5)*. Our group generated neurospheres composed of DA neuron progenitors from monkey ES cells, by the combination of a SDIA method and a sphere culture with fibroblast growth factors (FGFs)-2 and -20. Using the primate model for PD, the 1-methyl-4-phenyl-1,2,3,6-tetrahydropyridine-treated (MPTP-treated) cynomolgus monkey, we also discovered that transplantation of these neurons could diminish Parkinsonian symptoms and found that the transplanted cells were able to function as DA neurons *(6)*. In vitro differentiation of DA neurons was evaluated by immunofluorescence study, RT-PCR, and HPLC. In vivo function of the grafted cells was evaluated by neurological scoring, PET study, and immunofluorescence study.

In this chapter, we would like to focus on cell culture and transplantation, because there is no special technique for in vitro studies. While, dealing with monkeys needs some experience and techniques. We would be glad if this chapter helps readers to learn these techniques.

2. Materials

2.1. Tissue Culture

2.1.1. Cell Lines

1. Cynomolgus monkey ES cells (Asahi Techno Glass, Chiba, Japan) *(7)*.
2. Mouse embryonic fibroblasts (STO) (Riken BioResource Center, Tsukuba, Japan).
3. MC3T3-G2/PA6 cells (Riken BioResource Center).

2.1.2. Media for STO Cells

1. Dulbecco's modified Eagle's medium (DMEM; Gibco/Invitrogen, Carlsbad, CA, USA) supplemented with 10% fetal bovine serum (FBS; HyClone, Logan, UT, USA).
2. Solution of trypsin (0.25%, Invitrogen, Carlsbad, CA, USA) and ethylenediamine tetraacetic acid (EDTA; 1 mM, Gibco) in PBS (Gibco).
3. Mitomycin C (Wako Co., Osaka, Japan): Prepare in the tissue culture hood by adding 5 mL of PBS to a 10 mg vial (2 mg/mL), and sterilize through 0.2 μm pore filter. Protect from light and store at −20°C. Stock solutions are stable for 2 weeks at 4°C (*see* **Note 1**).

2.1.3. Media for PA6 Cells

1. Minimum essential medium α-(MEM) (Gibco) supplemented with 10% FBS (HyClone).

2.1.4. Media for Monkey ES Cells

1. DMEM/F12 (Sigma-Aldrich, St. Louis, MO, USA) supplemented with 0.1 mM 2-mercaptoethanol (ME; Sigma), 1,000 units/ml leukemia inhibitory factor (LIF; Chemicon, Temecula, CA, USA), 20% knockout serum replacement (KSR; Gibco), 0.1 mM of non-essential amino acids (NEAA; Gibco), 200 mM of L-glutamine (Sigma), and 4 ng/ml of fibroblast growth factor-2 (FGF-2) (Invitrogen).
2. FGF-2 is dissolved at 20 μg/ml in 10 mM of Tris (pH 7.6) containing 0.1% BSA, and stored in single use aliquots at −80°C.
3. Solution of 0.25% trypsin in PBS with 20% KSR and 1 mM $CaCl_2$ (Wako). Stored at −20°C.

2.1.5. Induction of Neural Differentiation from ES Cells

1. SDIA medium: Glasgow minimal essential medium (GMEM; Gibco) supplemented with 5% KSR, 1 mM of pyruvate (Sigma), 0.1 mM of NEAA, and 0.1 mM of 2-ME. Combine 500 mL of GMEM with 27 mL of KSR, 5 mL of pyruvate, 5 mL of NEAA, and 0.5 mL of 2-mercaptoethanol.
2. Sphere culture medium: neurobasal medium (Invitrogen) supplemented with B27 supplement (Invitrogen), 20 ng/ml of FGF2, and 20 ng/ml of epidermal growth factor (EGF; R&D Systems).
3. Differentiation medium: neurobasal medium supplemented with B27 supplement, 20 ng/ml of brain-derived neurotrophic factor (BDNF; Sigma), and 20 ng/ml of glial cell line-derived neurotrophic factor (GDNF; Sigma).
4. Papain dissociation system: A kit for cell dissociation, including Papain, DNase, Eagle's balanced salt solution, and Ovomucoid protease inhibitor (Worthington Biochemical Corporation, Lakewood, NJ, USA).
5. 5-bromo-2′-deoxy-uridine (BrdU; Sigma), dissolved in PBS at 1 mg/ml and stored at 4°C in a dark container (foil-wrapped).

2.1.6. Gelatin-Coated Dishes

1. Gelatin (Sigma) solution: for coating, prepare a 0.1% gelatin solution in distilled water and sterilize by autoclave.
2. Add 5 mL of the gelatin solution to a 10 cm dish (2 mL to a 6 cm dish), and incubate at 37°C for more than 30 min. Remove the gelatin solution before use.

2.2. Cell Transplantation

1. Adult male cynomolgus monkeys (*Macaca fascicularis*) weighing 2.5–3.5 kg, purchased from Shin Nippon Biomedical Laboratories Ltd (Kagoshima, Japan). (1) Tuberculin test: negative. (2) Bacillus dysenteriae and Salmonella: negative. (3) Parasites including Entamoeba histolytica: negative. Cercopithecine herpesvirus (CHV-1): negative (*see* **Note 2**).

2. 1-Methyl-4-phenyl-1,2,5,6-tetrahydropyridine (MPTP) HCl (0.4 mg/kg as free base, Sigma) (*see* **Note 3**).

3. 3-Tesla Signa Horizon Lx VH3 MRI scanner (General Electric Healthcare, Milwaukee, WI, USA).

4. Matlab: The MathWorks 3 Apple Hill Drive Natick, MA 01760-2098; Phone: 508-647-7000; Fax: 508-647-7001; URL:http://www.mathworks.com/

5. SPM: free software by members & collaborators of the Wellcome Department of Imaging Neuroscience, The Institute of Neurology, University College London; URL: http://www.fil.ion.ucl.ac.uk/spm/

6. MRIcro: free software by Chris Rorden; URL: http://www.sph.sc.edu/comd/rorden/mricro.html

7. Pentobarbital (Dainippon Sumitomo Pharma Co., Osaka, Japan), and ketamine (Sankyo Co., Tokyo, Japan).

8. Surgical frame system (SN-1 N, SM-15, EB-2C; Narishige Co., Tokyo, Japan).

9. Hamilton syringe (1702RN, PT-3; GL Sciences Inc., Tokyo, Japan).

10. KDS 100 Infusion pump (Muromachi Kikai Co., Tokyo, Japan).

11. Immunosuppressant: cyclosporine A (Calbiochem, San Diego, CA, USA) is dissolved in saline at 10 mg/mL before use.

2.3. Behavioral Assessment

1. A rating scale previously proposed by Akai et al. *(8)*, with modifications (**Table 13.1**)

Table 13.1
Neurological scores of MPTP-treated monkeys

Alertness
0. Alert
1. Cloudy
2. Drowsy

(continued)

Table 13.1 (continued)

Head checking movement
0. Always
1. Sometimes
2. No checking movement
Eyes
0. Normal blinking and eye movement
1. Reduced blinking and eye movement
2. Eyes closed
Posture
0. Standing or hanging
1. Stand or hang for more than 5 s in response to stimuli
2. Stand or hang in response to stimuli, but less than 5 s
3. Lying
Balance
0. Normal
1. Sometimes stagger
2. Stagger, or fall down
3. Unable to stand or walk
Motility, at rest
0. Move extremities well
1. Less movement of extremities
2. Move extremities only sometimes
3. Almost no movement of extremities
Motility, reaction to stimuli (When an examiner tries to touch,)
0. Retract extremities quickly without being touched
1. Touched, but retract extremities simultaneously
2. Touched, and then retract extremities
3. Touched, but do not retract extremities

(continued)

Table 13.1 (continued)

Walking
0. Walk always
1. Walk sometimes
2. Walk rarely
3. Unable to walk
Tremor
0. No tremor
1. Sometimes
2. Always
3. Always and severe

2.4. Positron Emission Tomography (PET) Study

1. Propofol (AstraZeneca, Osaka, Japan) and vecuronium-bromide (Boehringer Ingelheim, Ingelheim, Germany).
2. (S)-(-)-carbidopa monohydrate (Wako).

2.5. Immunofluorescence Study of Grafted Cells

1. Paraformaldehyde (PFA; Wako): Prepare a 4% (w/v) solution in PBS freshly for each experiment. To dissolve, heat the solution to 40–50°C (use a stirring hot-plate in a fume hood) and add about 1/2,000 volume of 10 N NaOH, adjust the pH to 7.4, and then cool to room temperature for use. It can be stored at 4°C for a couple of days (*see* **Note 4**).
2. Sucrose solution: 30% sucrose in 0.2 M phosphate buffer.
3. Blocking solution: 0.3% Triton X (Sigma) with 5% skim milk (Dainihon-Seiyaku, Osaka, Japan) in PBS.
4. Antibody solution: 0.3% Triton X and 2% skim milk in PBS.
5. Primary antibodies: mouse anti-BrdU (1:200; Becton Dickinson, Franklin Lakes, NJ, USA), rabbit anti-TH (1:60; Chemicon), and rat anti-DAT (1:100; Chemicon).
6. Secondary antibodies: FITC-labeled anti-mouse Ig (1:200; Jackson Immunoresearch, West Grove, PA, USA), and Cy3-labeled anti-rabbit Ig (1:200; Jackson Immunoresearch).
7. Mounting solution: Permafluor (Beckman Coulter K.K., Tokyo, Japan).

Color Plates

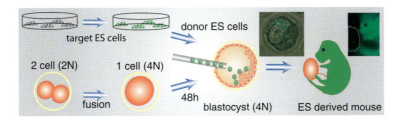

Color Plate 1. Tetraploid blastocyst injection. This procedure can be used in order to decrease the time required to generate mice from targeted ES cells. Two-cell embryos are fused as described in **Section 3.11** and injected with the diploid-targeted ES cells (here shown as GFP-positive ES cells). The tetraploid cells of the blastocysts will only give rise to extra-embryonic tissues such as the placenta, whereas injected diploid ES cells will give rise to the embryo *(17)*. This procedure takes only a few weeks to generate completely ES-derived mice, in contrast to first generating chimeras and then breeding to obtain germline transmission. (*See* discussion on p. 8)

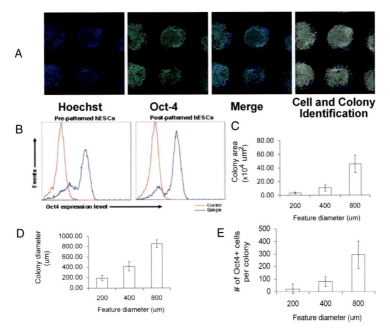

Color Plate 2. Analysis of patterned H9 hESC colonies (D = 400 μm, P = 500 μm). (**A**) Immunohistochemistry to identifying Oct4+ cells (*green*), all nuclei by Hoechst 33342 (*blue*). Individual Oct4+ cells (*light green*) and Oct4- cells (*dark blue*) are identified by the Target Activation algorithm of the Cellomics™ ArrayscanV™ software by drawing overlay masks around the nuclei. (**B**) Representative flow cytometry plots of H9 hESCs before and after patterning by immunocytochemistry. Data demonstrating how the Morphology Explorer algorithm of the Cellomics™ Arrayscan V™ software can be used to identify cells and colonies, (**C**) calculate their area, (**D**) diameter, and (**E**) the number of Oct4+ cells in the colony. Error bars represent the standard deviation over 150 colonies. (*See* discussion on p. 31)

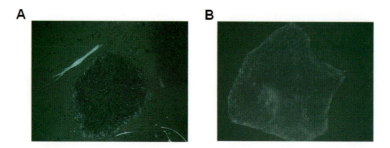

Color Plate 3. Infection of hESCs with GFP marked lentiviral RNAi vectors. (**A**) hESC colony infected with a GFP marked lentivirus displays a patchwork of infected cells. (**B**) hESC colony after enrichment of GFP+ cells by FACS. (*See* discussion on p. 41)

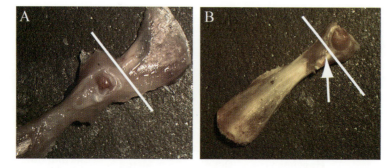

Color Plate 4. The iliac crest with a line through where to remove the flat triangular piece of cartilage (**A**) and where the acetabular notch is removed ready to be flushed (**B**). (*See* discussion on p. 97)

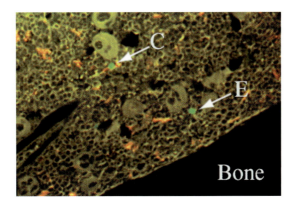

Color Plate 5. Transplanted LSK cells detected using a spatial distribution assay. Cells are designated as either central (**C**) or endosteal (**E**). (*See* discussion on p. 103)

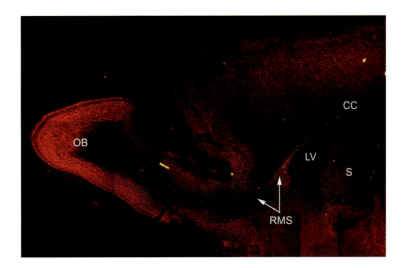

Color Plate 6. Donor cells migrate normally following engraftment. Parasagittal fluorescence montage through the brain of an adult mouse host following transplantation of GFP+ neurogenic astrocytes into the lateral ventricle. Sagittal sections were immunostained with antibodies against GFP (*green*) and NeuN (*red*) 28 days following engraftment. A large number of GFP+ donor cells can be seen within the rostral migratory stream (RMS) and olfactory bulb (OB). A few donor cells are also present along the roof of the lateral ventricle (LV), and in the septum (S). CC: corpus callosum. (*See* discussion on p. 187)

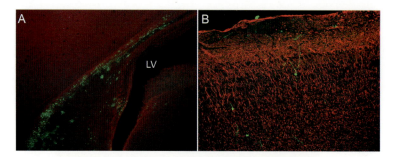

Color Plate 7. Higher magnification of GFP+ donor cells within the RMS and olfactory bulb. (**A**) Parasagittal section of a host brain immunolabeled for GFP (*green*) and β-III tubulin (*red*). Donor cells are distributed with in the SEZ–RMS–OB system, as well as lining the wall of lateral ventricle (LV). (**B**) Parasagittal confocal image of a donor cell that migrated into the olfactory bulb and differentiated into a granule interneuron. (*See* discussion on p. 187)

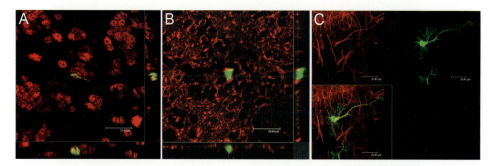

Color Plate 8. Donor-derived cells are capable of differentiating into neurons and astrocytes. (**A**) and (**B**): Confocal Z-series of GFP+ donor cells (*green*) located in the granular cell layer of olfactory bulb that are double immunolabeled with the neuron markers (*red*) NeuN (A) and β-III tubulin (B). (**C**) A GFP+ (*green*) donor cell that has differentiated into a cortical astrocyte, and is immunolabeled with the astrocyte marker GFAP (*red*). (*See* discussion on p. 188)

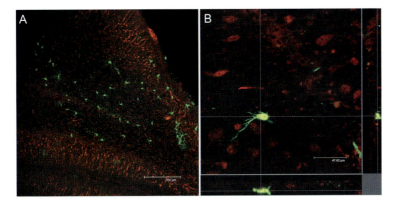

Color Plate 9. Grafted cells survive, migrate, and differentiate upon transplantation into the cerebellum of both wild type and weaver transgenic mouse pups. (**A**) Donor GFP+ cells (*green*) are seen migrating and differentiating within the cerebellum 3 weeks post-engraftment in the wild-type mouse. (**B**) High magnification confocal z-series of a donor cell transplanted into the cerebellum of a weaver mutant mouse. P$\epsilon\delta$= β-III tubulin in A, and NeuN in B. (*See* discussion on p. 188)

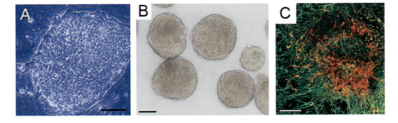

Color Plate 10. Monkey ES cells. Undifferentiated ES cells on STO feeder (**A**), and neural progenitors induced from monkey ES cells. Detached ES cell colonies formed spheres similar to those of neural progenitor cells (**B**). These spheres give rise to Tuj1- (green, neuronal marker) and TH-positive (red, DA neuron marker) cells in vitro. Bar=100 μm (A, B), 100 μm (**C**). (B, C: Reproduced from ref. *(6)* with permission). (*See* discussion on p. 205)

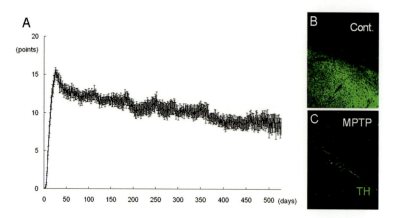

Color Plate 11. Parkinson's disease model monkey induced by MPTP. Neurological scores of MPTP-treated monkeys (**A**, $n = 34$). Immunofluorescence study revealed that neuro-terminals of the nigro-striatal tract, which is composed of DA neuron fibers and immunoreactive for TH, are severely reduced (**B, C**). (*See* discussion on p. 210)

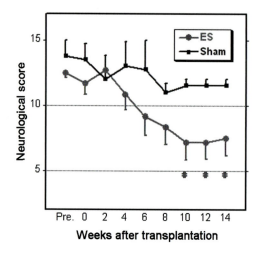

Color Plate 12. Function of ES cell-derived neurospheres in MPTP-treated monkeys. Neurological scores of ES cell-transplanted ($n=6$) and sham-operated animals ($n=4$) are plotted. All values are the mean±SD. *$p<0.05$. (*See* discussion on p. 211)

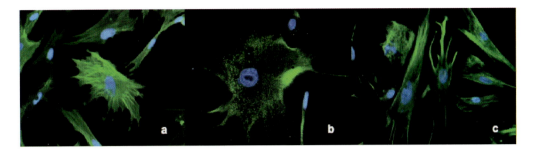

Color Plate 13. HUCPVCs expressing α-actin (**a**), vimentin (**b**), and desmin (**c**). (*See* discussion on p. 275)

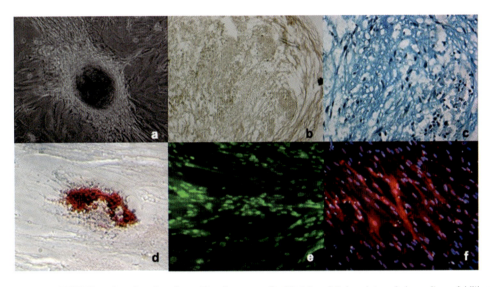

Color Plate 14. HUCPVCs reduce lymphocyte proliferation, even if added 3 and 5 days into a 6 day culture. Addition of HUCPVCs showed a significant decrease in lymphocyte cell number compared to control (no HUCPVCs) over 6 days in a two-way MLC. There is no significant difference among HUCPVCs added on day 0, 3, or 5 ($n = 6$). This figure shows the average cell numbers, + standard deviations. (*See* discussion on p. 276)

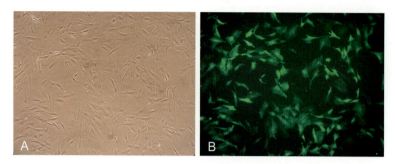

Color Plate 15. Transduction efficiency by fluorescent microscopy. (**A**) Phase contrast microphotograph of P5 bone marrow-derived MSCs. (**B**) The same microscopic field observed under fluorescent light shows that the majority of the cells express GFP, meaning that they were successfully transduced. In our hands, this protocol allows a transduction efficiency higher than 80%, as confirmed also by FACS analysis. (*See* discussion on p. 288)

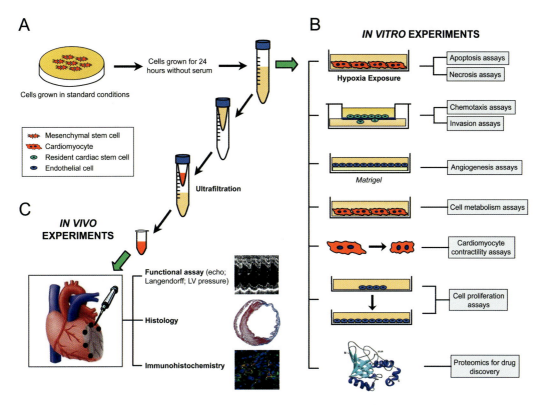

Color Plate 16. Conditioned media production and different in vitro and in vivo assays. (**A**) Stem cells are expanded in normal conditions until they are 90% confluent. The growth medium is then exchanged with medium not containing serum and the cells are left for 24 h in a CO_2 incubator. The medium is then collected and tested either in vitro or in vivo. (**B**) For the in vitro experiments, the conditioned medium is transferred into culture dishes containing different kind of cells according to the goal of the specific experiment. Several different properties of the medium can be tested in vitro. For example, the cytoprotective effects exerted by the conditioned medium can be tested on murine cardiomyocytes. After exposing the cardiomyocytes to hypoxia in the presence of control medium or conditioned medium, apoptosis and necrosis assays are performed and the results compared. To verify if the conditioned medium contains chemotactic factors, a specific cell type (i.e., endothelial cell or cardiac stem cell) is seeded on the membrane of the upper chamber of a dual chamber dish. The number of cells migrating into the lower chamber, containing either conditioned medium or control medium, is then counted.

To test the pro-angiogenetic properties of conditioned medium, endothelial cells are seeded on a matrigel and the number of capillaries is quantified after exposure to conditioned medium or control medium. To verify if cell metabolism is influenced by factors present in the medium, the cell type of interest (i.e., murine cardiomyocytes) is exposed to the conditioned and to the control medium and then collected to perform metabolic assays. Another example: cardiomyocyte contractility may be assessed in the presence of control or conditioned medium; if inotropic factors are present, then cell contractility will be increased in the presence of conditioned medium. Proliferation assay may also be performed. Finally, proteomic analysis of conditioned medium may allow the discovery of new therapeutic molecules and targets. (**C**) Conditioned medium may be tested also in vivo using different experimental disease models. For example, the effects of conditioned medium on ischemic myocardium may be assessed in a murine model of myocardial infarction. Small volumes of concentrated conditioned medium obtained by ultrafiltration are injected at the infarct border zone after left coronary ligation. At established time points, heart function is analyzed by echo (an M-mode image of the left ventricle is depicted in the figure) or other methods. The heart may also be collected for histology to determine, for example, the infarct size (a cross-section of mouse heart stained with Masson Trichrome is depicted in the figure). Furthermore, immunohistochemistry staining may performed to determine different parameters, such as cardiac regeneration, neoangiogenesis, apoptosis. (*See* discussion on p. 289)

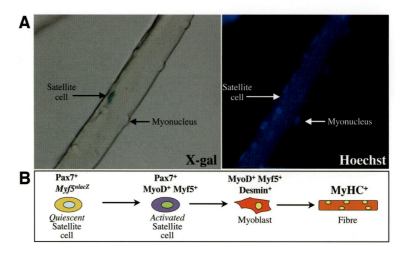

Color Plate 17. Lineage progression of an adult skeletal muscle satellite cell to a differentiated cell. (**A**) Isolated fibre from the Tibialis Anterior muscle of a *Myf5$^{nlacZ/+}$* mouse showing nuclear β-galactosidase activity by X-gal staining. Hoechst staining reveals myonuclei inside the fibre as well as the nucleus of the satellite cell on the fibre. (**B**) Lineage progression of a quiescent to an activated satellite cell, hallmarked by MyoD expression, to a myoblast. Myoblasts fuse homotypically or to pre-existing differentiated fibres or myotubes after leaving the cell cycle. Differentiated fibres are characterised by the expression of myosin heavy chain (MyHC). The expression of different commonly used markers is indicated. (*See* discussion on p. 296)

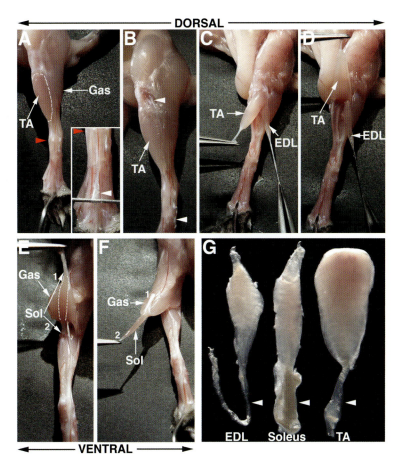

Color Plate 18. Dissections of TA, EDL and Soleus muscles. The skin was removed from a 4-week-old mouse to reveal the underlying muscles. (**A–D**) Removal of TA and EDL muscles. (**E-F**) Removal of Soleus muscle. Lower tendon (1) is sectioned first, the muscles are lifted, then the upper Soleus tendon (2) is sectioned. Note in older mice, the soleus appears more red and distinguishable. G) Actual sizes of dissected muscles with associated tendons (white arrowheads). Muscle attachment proximal to the knee is at the top of the photo. See main text for details. Red arrowhead, cartilage bridge in foot under which tendons are transit. White arrowheads, tendons. (*See* discussion on p. 301)

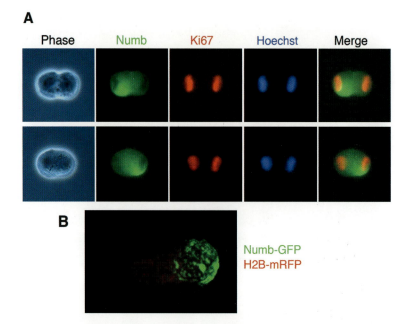

Color Plate 19. Asymmetric segregation of Numb in non-differentiating satellite cell-derived myoblasts. (**A**) Two examples of asymmetric distribution of endogenous Numb protein in mitotic satellite cell-derived myoblasts after 4 days in culture; immunocytochemistry with anti-Numb and anti-Ki67 antibodies; (**B**) p66Numb-EGFP and H2B-mRFP fusion proteins were overexpressed in satellite cell-derived myoblasts using pCAG-p66Nb-EGFP and pCAG-H2B- mRFP plasmids (transfected at 96 h after plating of myofibres). At 4 days after plating, cells were harvested by mitotic shake-off, re-plated at low density and grown for several hours on poly-D-lysine-coated dishes. The cells were fixed and stained with anti-Numb antibody (green), and visualised using EGFP and mRFP epifluorescence. Phase contrast on left; reconstituted confocal stack on right was rendered with Imaris software. Note asymmetric distribution of the majority of the Numb-GFP protein to one pole of this cell. (*See* discussion on p. 306)

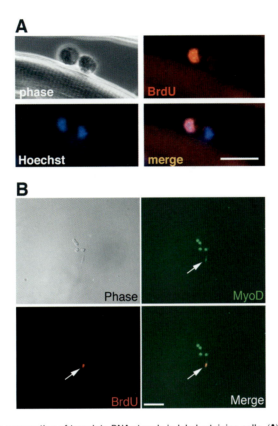

Color Plate 20. Asymmetric segregation of template DNA strands in label retaining cells. (**A**) Asymmetric segregation of BrdU label to only one daughter satellite cell after mitosis, on a freshly isolated EDL fibre. $Myf5^{nlacZ/+}$ mice were pulsed with BrdU from P3-P7; chase for 7 days. (**B**) Clonal analysis of LRCs after 4 weeks chase in vivo, 72 h in culture. Antibody stainings for detecting MyoD protein and BrdU in the nuclei of satellite cell derived myoblasts. The LRC in this case has a lower expression of MyoD. Scale bars: A, 15 μm; B, 40 μm. (*See* discussion on p. 311)

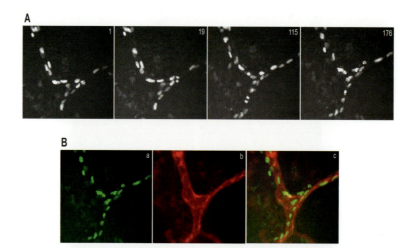

Color Plate 21. Time-lapse imaging of ES cell differentiation cultures. An ES cell line expressing endothelial cell-specific H2B-GFP was differentiated to day 8. It was imaged on a Perkin Elmer spinning disk confocal microscope. Culture is fixed right after imaging and stained for PECAM-1. (A) Frames of a time-lapse movie. Time is in minutes at the top right corner of each panel. (B) Last frame of the movie in green (**a**), the same field with PECAM-1 stain in red (**b**), and the overlay of the two images (**c**). (*See* discussion on p. 341)

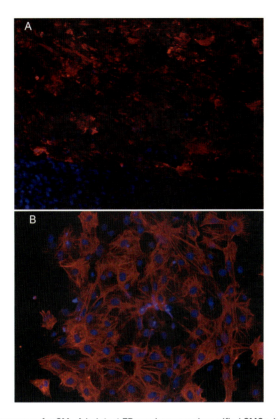

Color Plate 22. Immunofluorescence for SMαA in intact EBs and puromycin purified SMCs. Immunostaining was carried out as described in 3.5 on intact day 28 EBs (**A**) and on cells following enzyme dispersal and overnight selection (**B**). A "sheet" of SMCs may be seen in an intact day 28 EB with adjacent non-staining non-SMC at the bottom of the image (**A**). Such groups of SMC may be seen to contract spontaneously under the microscope. Following puromycin selection, relatively pure populations of cells with a typical SMC morphology were seen (**B**). Red staining: SMαA immunofluorescence, Blue: DAPI nuclear staining. (*See* discussion on p. 356)

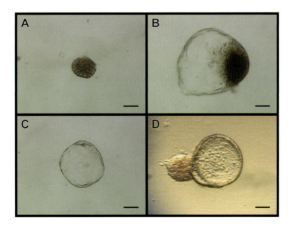

Color Plate 23. Transformation of embedded islets into DLS, and subsequent regeneration of ILS. (**A**) Immediately after embedding, islets are characterized by a solid-spheroid shape. (**B**) DLS formation appears to initiate in specific foci, (**C**) until a DLS replaces the islet. (**D**) Treatment with INGAP induces the budding of regenerating ILS from DLS (bar = 100 μm). (*See* discussion on p. 377)

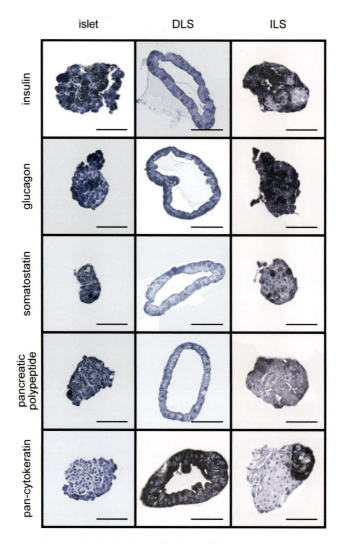

Color Plate 24. Immunocytochemistry. Analysis of islets, DLS, and ILS demonstrates that while islets and ILS express endocrine hormones (insulin, glucagon, somatostatin, and pancreatic polypeptide) in approximately the same proportions, these hormone+ cells are absent in DLS. Conversely, staining for a ductal cell marker (pan-cytokeratin) is observed primarily in DLS. Furthermore, while few or no ductal cells are observed in islets, ILS can be observed to be "budding" from ductal structures (bar = 100 μm). (*See* discussion on p. 380)

8. Micro slide glass (76×52 mm^2) (Matsunami Glass Ind., Ltd, Osaka, Japan).
9. Fluoview FV300 laser scanning confocal microscope (Olympus Optical Co., Tokyo, Japan).

3. Methods

3.1. Tissue Culture

3.1.1. Preparation of Feeder Layers for ES Cells

1. Culture STO cells in DMEM with 10% FBS until they reach confluency in a 10 cm gelatin-coated dish. Add 50 µL of mitomycin C stock solution into the dish per 10 mL medium, and incubate it at 37°C for 1.5–2 hr.
2. Replace the medium with DMEM with 10% FBS, and incubate at 37°C for at least 3 hr (or overnight).
3. Rinse twice with PBS, and dissociate the cells with 0.25% trypsin–EDTA, suspend them in DMEM with 10% FBS to make $0.8–1.0 \times 10^5$ cells/mL.
4. Add 12 mL of cell suspension to a 10 cm gelatin-coated dish (4 mL to a 6 cm dish), to make 5×10^3 cells/cm^2. (*see* **Note 5**).

3.1.2. Maintenance of Monkey ES Cells

1. Maintain undifferentiated monkey ES cells on a feeder layer of mitomycin C-treated STO cells in supplemented DMEM/F12 (**Fig. 13.1A** and Color Plate 10). Change the medium every day.
2. For subcloning, add 1 mL of trypsin solution to a 10 cm dish (0.5 mL to a 6 cm dish) and incubate for 5–10 min until colonies are ready to detach. Then, add 4 mL of DMEM/F12 (2 mL to a 6 cm dish) and suspend the cells by gentle pipetting (*see* **Note 6**).
3. Centrifuge the cell suspension and passage them into new dishes by 1:2–3.

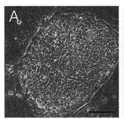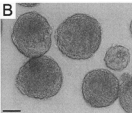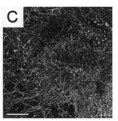

Fig. 13.1. Monkey ES cells. Undifferentiated ES cells on STO feeder (**A**), and neural progenitors induced from monkey ES cells. Detached ES cell colonies formed spheres similar to those of neural progenitor cells (**B**). These spheres give rise to Tuj1- (green, neuronal marker) and TH-positive (red, DA neuron marker) cells in vitro. Bar=100 µm (A, B), 100 µm (**C**). (B, C: Reproduced from ref. *(6)* with permission). (*See* Color Plate 10)

3.1.3. Induction of Neural Precursor Cells by SDIA Method

1. Prepare confluent PA6 cells in a 10 cm non-coated dish, which is just on the day when the cells reach confluence.
2. Dissociate ES cells using trypsin solution and make cell suspension as described in **Section 3.1.2** (*see* **Note 7**). Collect cell pellets by centrifugation, and re-suspend them in SDIA medium. Replate the cells onto a gelatin-coated dish, and incubate at 37°C for 1 hr. ES cells are floating or easily detach from the bottom, while STO cells attach to the bottom.
3. Collect floating ES cells and leave STO cells on the bottom, and replate the ES cells on PA6 cells at a density of 500–1,000 colonies per 10 cm dish. Culture the cells in SDIA medium, and change the medium every 3 days (*see* **Note 8**).

3.1.4. Induction of DA Neurons

1. After 2 weeks, detach differentiated colonies from PA6 cells using a papain dissociation system. The colonies and PA6 cells come off as a sheet, but gentle pipetting with a P-1000 pipet separates the colonies from the cell sheet.
2. Collect differentiated colonies by centrifugation, and suspend them in sphere culture medium. Culture the cells as spheres in a non-coated culture dish for 1 week (**Fig. 13.1B**), changing the medium every 2–3 days (*see* **Note 9**).

3.2. Cell Transplantation

3.2.1. Animal Model

1. Inject MPTP HCl (0.4 mg/kg as free base) intravenously twice a week to adult male cynomolgus monkeys (*Macaca fascicularis*) weighing 2.5–3.5 kg until persistent Parkinsonian behavioral disturbances, such as tremor, bradykinesia, and impaired balance, became evident (*see* **Note 10**).

3.2.1.1. Targeting

1. For accurate orientation of the putamen, each animal was subjected to MRI examination using a 3.0 Tesla SIGMA system (**Fig. 13.2A, B, D**).
2. Re-construct MRI images to adjust each plane parallel to that passes the lower edge of the orbit and the external meatus. Overlay the real-sized grid to the image, set the midpoint of both meatus as the zero point (**Fig. 13.2A**), and fix the coordinates of the targets by the aid of the brain atlas *(9, 10)* (**Fig. 2C**).

3.2.2. Cell Preparation

1. For labeling, add BrdU (5 μg/ml) to the sphere culture medium for 7 days before cell transplantation.
2. Take 1/10 volume of the sphere culture, dissociate the cells by 0.25% trypsin, and count total cell numbers.

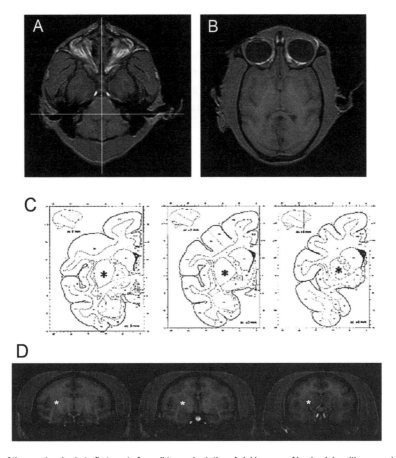

Fig. 13.2. MRI of the monkey brain to fix targets for cell transplantation. Axial images of basic plain with zero point (cross-point of two lines, (**A**) and the plain which passes striatum (**B**). Stars in the atlas (**C**) and coronal MRI images (**D**) indicate targets.

3. Collect rest of the neural progenitor cells by centrifugation, re-suspend in the sphere culture medium to make cell suspension of 50,000 cells per μL for use (*see* **Note 11**).

3.2.3. Transplantation

1. Following anesthesia with pentobarbital (7.5 mg/kg, i.m.) and ketamine (10 mg/kg, i.m.), fix the head of an anesthetized monkey firmly to a surgical frame. Make sure to put the ear bars exactly into both meatus, which correspond to the zero line (*see* **Note 12**).

2. Set a micromanipulator (SM-15) to a surgical frame (SN-1 N). Aspirate cell suspension into an injection syringe, set it to a manipulator, and connect with an injector. Move the manipulator with a syringe to the point where the tip of the needle comes to a center of the ear bar, and read the each scale (Ant-Post, depth). For a midline, adjust the needle to Bregma or Nasion and read the scale. Then, transfer the coordinate of the target points for transplantation to a surgical frame system, and estimate the needle tract on the skull.

3. Drill the skull to expose dura mater. Then, expose brain surface for injection. *Note:* We make a small hole on the skull which covers three tracts. Then, make a small dural incision for each tract.

4. Insert the needle slowly into the brain until it passes 1 mm over the target point, and retract to the target. Inject the cells at the speed of 1 µL/min, stay another minute, and retract 1 mm. Inject the cells again, and repeat this four times. Pull the need slowly out of the brain. Repeat this procedure for three tracts on both sides.

5. Close periosteum and skin tightly.

6. After surgery, all animals were given antibiotics for 1 week and a daily immunosuppressant (cyclosporin A, 10 mg/kg, i.m.) until killing.

3.3. Behavioral Assessment

1. Evaluate Parkinsonian behavior daily using a rating scale previously proposed by Akai et al. *(8)*, with modifications (**Table 13.1**).

2. Compare the scores of control and grafted groups statistically by using a Student's *t*-test. (*see* **Note 13**).

3.4. Positron Emission Tomography (PET) Study

1. Anesthetize the monkey with a continuous infusion of propofol (4 mg/kg/hr) and vecuronium-bromide (0.25 mg/kg/hr).

2. After a 15 min transmission scan using a 68Ge–68 Ga rod source for correction of attenuation of radioactivity, administrate 185 MBq of ^{18}F-fluorodopa by intravenous injection. Simultaneously, start a 90 min emission scan with a 2D dynamic mode, consisting of 37 frames of gradually increasing individual durations (12×15 sec, 4×30 sec, 5×1 min, and 16×5 min). Assess radioactivity using an ECAT EXACT HR PET scanner for 90 min in animals that had received carbidopa (10 mg/kg) 30 min prior to the PET scan.

3. Generate parametric images of the dopamine-irreversible metabolic rate of Ki (min^{-1}), considered to be a measure of presynaptic dopaminergic function, using the time course of radioactivity in each voxel by multiple-time-graphical-analysis using the bilateral occipital lobes as a reference region (**Fig. 13.3**).

3.5. Immunofluorescence Study for Grafted Cells

3.5.1. Fixation of the Brain

1. After deep anesthesia with pentobarbital and ketamine, perfuse the monkey transcardially with 0.1 M PBS (4 L/animal, start at 200 mL/min and then 300–400 mL/min) first, and then fix the brain with 4% paraformaldehyde (8 L/animal, 300–500 mL/min) (*see* **Note 14**).

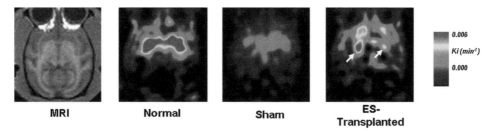

Fig. 13.3. PET study of the monkey brain. Uptake of 18F-fluorodopa in the putamen of ES cell-transplanted animals is increased, which means that grafted cells function as DA neurons in the brain. (Reproduced from ref. *(6)* with permission.)

2. Excise the fixed brain out, paying attention not to damage. Cut the brain into pieces and dip them into 4% paraformaldehyde overnight (*see* **Note 15**). On the next day, transfer the brain tissues into 15% sucrose solution and leave at 37°C until they sink. Then, transfer the tissues into 30% sucrose solution. When the tissues sink, they are ready to be cut.

3.5.2. Staining

1. Cut the brain with a microtome at a 50 μm thickness and stain the sections by the free-floating method.
2. Incubate the brain sections in blocking solution for 30 min.
3. Incubate with the primary antibody in diluent overnight at 4°C.
4. Rinse 3×10 min in PBS.
5. Incubate with the secondary antibody for 1 hr at room temperature.
6. Rinse 3×10 min in PBS and mount the sections onto a slide glass. Then, analyze them using a Fluoview FV300 laser confocal microscope (**Fig. 13.4**).

Fig. 13.4. Survival of ES cell-derived DA neurons in the striatum. Grafted cells (BrdU-labeled, *green* in **A**) survived and differentiated into DA neurons (TH-positive, *red* in **B**). Colocalization (arrows in a merged image: **C**) of BrdU (*green*) and TH (*red*) shows that graft-derived cells have dopaminergic character. (Reproduced from ref. *(6)* with permission.)

4. Notes

1. For use, thaw the aliquots in foil to protect from light.

2. All animals were fed with commercial pellets and fresh fruits and had free access to clean water. Monkeys were cared for and handled according to Guidelines for Animal Experiments of Kyoto University and the National Cardio-Vascular Center and the Guide for the Care and Use of Laboratory Animals from the Institute of Laboratory Animal Resources (ILAR) in Washington, USA.

3. MPTP can induce Parkinsonian syndrome in human, so it should be handled with special care *(11)*.

4. PFA is toxic, so be careful not to inhale it. Work in a hood, use a mask and gloves.

5. STO cells should not be confluent. Use this feeder cells within 4 days.

6. Avoid pipetting too strongly. Unlike mouse ES cells, monkey ES cells do not form colonies from single cells on PA6 cells. Colonies with about 100 cells result in a good efficiency.

7. This time smaller colonies with 10–50 cells will be better. But, single cell density results in almost no cell survival.

8. Just aspirate αMEM of PA6 cell culture and replace it with SDIA medium. No need for rinse with PBS in between.

9. Keep the cell density high like 500–1,000 colonies per milliliter.

10. Animals were given an average of 14.7 MPTP shots, and exhibited stable Parkinsonism within approximately 30 days. To prevent any possibility of spontaneous recovery, only those monkeys that presented stable deterioration for a

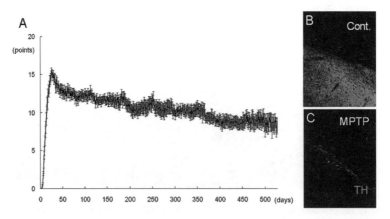

Fig. 13.5. Parkinson's disease model monkey induced by MPTP. Neurological scores of MPTP-treated monkeys (**A**, *n* = 34). Immunofluorescence study revealed that neuro-terminals of the nigro-striatal tract, which is composed of DA neuron fibers and immunoreactive for TH, are severely reduced (**B, C**). (*See* Color Plate 11)

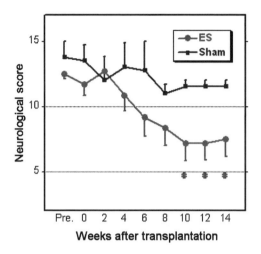

Fig. 13.6. Function of ES cell-derived neurospheres in MPTP-treated monkeys. Neurological scores of ES cell-transplanted ($n = 6$) and sham-operated animals ($n = 4$) are plotted. All values are the mean ± SD. *$p < 0.05$. (*See* Color Plate 12)

period greater than 12 weeks were used for transplantation experiments. The change of neurological scores of all the monkeys tested is shown in **Fig. 13.5A** and Color Plate 11. In the striatum of the MPTP-treated monkey, the neuro-terminals of DA neuron fibers are severely reduced (**Fig. 13.5B, C**).

11. If these spheres are subjected to differentiation in vitro, they give rise to a substantial number of DA neurons which express tyrosine hydroxylase (TH, **Fig. 13.1C**), and secrete dopamine into the culture medium *(6)*.

12. Add pentobarbital and ketamine properly to induce appropriate anesthesia. We use special ear bars with sharpened tip (EB-2C).

13. To prevent subjective biases, the evaluation was performed by a well-trained examiner, who was not involved in the culturing and transplantation of the cells and not informed of the specific procedure to which each monkey was subjected. The change of the scores of both control- and MPTP-treated monkeys is shown in **Fig. 13.6** and Color Plate 12.

14. Use a 14G needle. Clump thoracic aorta to fix the brain more effectively.

15. We cut the brain into six pieces, namely anterior, middle, posterior part of both sides. Grafts are located in the middle part.

Acknowledgments

The authors thank Drs T. Yamamoto (Shin Nippon Biomedical Laboratories Ltd.) and N. Teramoto (National Cardiovascular

Center) for technical help with monkeys; Dr T. Hayashi (National Cardiovascular Center) for PET study; Drs H. Suemori and N. Nakatsuji (Institute for Frontier Medical Sciences, Kyoto University), Y. Sasai (Center for Developmental Biology, RIKEN) for cell culture and differentiation. This study was supported by the following grants: Grants-in-Aid for Scientific Research, Grants in Kobe Cluster, and Establishment of International COE for Integration of Transplantation Therapy and Regenerative Medicine from the MEXT; Health Sciences Research Grants in Research on Human Genome, Tissue Engineering and Food Biotechnology from the MHLW; Grants in Organization for Pharmaceutical Safety and Research.

References

1. Freed, C.R., Greene, P.E., Breeze, R.E., Tsai, W.Y., DuMouchel, W., Kao, R., Dillon, S., Winfield, H., Culver, S., Trojanowski, J. Q., Eidelberg, D., Fahn, S. (2001) Transplantation of embryonic dopamine neurons for severe Parkinson's disease. *N. Engl. J. Med.* **344**, 710–719.
2. Olanow, C.W., Goetz, C.G., Kordower, J.H., Stoessl, A.J., Sossi, V., Brin, M.F., Shannon, K.M., Nauert, G.M., Perl, D.P., Godbold, J., Freeman, T.B. (2003) A double-blind controlled trial of bilateral fetal nigral transplantation in Parkinson's disease. *Ann. Neurol.* **54**, 403–414.
3. Goldman, S. (2005) Stem and progenitor cell-based therapy of the human central nervous system. *Nat. Biotechnol.* **23**, 862–871.
4. Kawasaki, H., Mizuseki, K., Nishikawa, S., Kaneko, S., Kuwana, Y., Nakanishi, S., Nishikawa, S-I., Sasai, Y. (2000) Induction of midbrain dopaminergic neurons from ES cells by stromal cell-derived inducing activity. *Neuron.* **28**, 31–40.
5. Kawasaki, H., Suemori, H., Mizuseki, K., Watanabe, K., Urano, F., Ichinose, H., Haruta, M., Takahashi, M., Yoshikawa, K., Nishikawa, S., Nakatusji, N., Sasai, Y. (2002) Generation of dopaminergic neurons and pigmented epithelia from primate ES cells by stromal cell-derived inducing activity. *Proc. Natl. Acad. Sci. USA.* **99**, 1580–1585.
6. Takagi, Y., Takahashi, J., Saiki, H., Morizane, A., Hayashi, T., Kishi, Y., Fukuda, H., Okamoto, Y., Koyanagi, M., Ideguchi, M., Hayashi, H., Imazato, T., Kawasaki, H., Suemori, H., Omachi, S., Iida, H., Itoh, N., Nakatsuji, N., Sasai, Y., Hashimoto, N. (2005) Dopaminergic neurons generated from monkey embryonic stem cells function in a Parkinson primate model. *J. Clin. Invest.* **115**, 102–109.
7. Suemori, H., Tada, T., Torii, R., Hosoi, Y., Kobayashi, K., Imahie, H., Kondo, Y., Iritani, A., Nakatsuji, N. (2001) Establishment of embryonic stem cell lines from cynomolgus monkey blastocysts produced by IVF or ICSI. *Dev. Dyn.* **222**, 273–279.
8. Akai, T., Ozawa, M., Yamaguchi, M., Mizuta, E., Kuno, S. (1995) Combination treatment of the partial D2 agonist terguride with the D1 agonist SKF 82958 in 1-methyl-4-phenyl-1,2,3,6-tetrahydropyridine-lesioned parkinsonian cynomolgus monkeys. *J. Pharmacol. Exp. Ther.* **273**, 309–314.
9. Sabo, J., Cowan, W.M. (1984) A stereotaxic atlas of the brain of the cynomolgus monkey (Macaca fascicularis). *J. Comp. Neurol.* **222**, 265–300.
10. Martin, R.F., Bowden, D.M. (2000) *Primate Brain Maps: Structure of the Macaque Brain*. Volume 8. Elsevier. New York, USA. 160 pp.
11. Przedborski, S., Jackson-Lewis, V., Naini, A. B., Jakowec, M., Petzinger, G., Miller, R., Akram, M. (2001) The parkinsonian toxin 1-methyl-4-phenyl-1,2,3,6-tetrahydropyridine (MPTP): a technical review of its utility and safety. *J. Neurochem.* **76**, 1265–1274.

Part IV

Regeneration of the Epidermis

Chapter 14

Isolation and Culture of Epithelial Stem Cells

Jonathan A. Nowak and Elaine Fuchs

Abstract

In the skin, epithelial stem cells in the hair follicle contribute not only to the generation of a new hair follicle with each hair cycle, but also to the repair of the epidermis during wound healing. When these stem cells are isolated and expanded in culture, they can give rise to hair follicles, sebaceous glands, and epidermis when combined with dermis and grafted back onto *Nude* mice. In this chapter, we provide a method for isolating hair follicle epithelial stem cells from the skin of adult mice using immunofluorescent labeling to allow for the specific purification of epithelial stem cells by fluorescence-activated cell sorting (FACS). Notably, this method relies exclusively on cell surface markers, making it suitable for use with any strain of mouse and at various stages of the hair cycle. We also provide a detailed protocol for culturing epithelial stem cells isolated by FACS, allowing for analysis using a wide variety of culture assays. Additionally, we provide notes on using cultured cells for specific applications, such as viral manipulation and grafting. These techniques should be useful for directly evaluating stem cell function in normal mice and in mice with skin defects.

Key words: Epithelial stem cells, skin stem cells, hair follicles, bulge stem cells.

1. Introduction

The skin is a tissue which undergoes continuous self-renewal throughout the lifetime of an organism and also has an extensive ability to repair wounds. For these reasons, a stem cell population has long been suspected to exist within the epithelial compartment of the skin. Numerous lines of evidence, including grafting, lineage tracing and pulse-chase experiments to measure cell cycle times, have suggested that the bulge region of the hair follicle acts as a reservoir for skin epithelial stem cells *(1–3)*.

The bulge is the lowest permanent epithelial portion of the hair follicle. At the start of each hair cycle, bulge stem cells migrate downward to regenerate the bulk of the hair follicle and produce a new hair (anagen phase). In mouse backskin, at the end of each hair cycle, the bulge progeny degenerate (catagen phase) to form a thin epithelial strand. When the strand regresses, it drags the dermal papilla (DP) cells of the hair follicle upward to rest beneath the bulge (telogen phase). After a variable period of time in telogen, bulge cells become activated and start the hair cycle anew. In addition to fueling the normal hair cycle, bulge stem cells are also able to move upwards to repair the epidermis and sebaceous glands during wound healing, and thus these cells are referred to as multipotent stem cells *(4, 5)*.

Functional evaluation of bulge stem cell character has been greatly accelerated by recent advances in isolating pure populations of viable bulge cells. Several techniques have been developed for the isolation of bulge cells by fluorescence-activated cell sorting (FACS). Typically, these methods have relied upon transgenic mice expressing fluorescent proteins that preferentially label bulge cells based on their slow-cycling nature or specific promoter activity *(4, 5)*. Such methods are often best suited for isolating bulge cells at the telogen stage, a time at which most of the hair follicle is missing. However, a combination of antibodies against the cell surface proteins CD34 and α6-integrin can also be used to identify distinct populations of cells by FACS, which are greatly enriched for bulge stem cells *(6, 7)*.

When these FACS-isolated bulge cell populations are placed in culture, they give rise to colonies that display stem cell properties, including long-term self-renewal, and when the progeny of an individual FACS-purified bulge cells are grafted onto *Nude* mice, the cells are able to regenerate hair, interfollicular epidermis, and sebaceous glands, reflecting their multipotency. The α6/CD34 bulge cell isolation technique is particularly advantageous because it can be performed in the absence of fluorescent transgenic reporters, making it suitable to use with any mouse strain (*see* **Note 1**). Additionally, this technique can be used to isolate bulge cells from both anagen and telogen stage hair follicles, and, furthermore, it allows for the division of the bulge cell population into two distinct subpopulations based on differential α6-integrin expression. Culture conditions that have been optimized for the propagation of sorted bulge cells allow for the functional assessment of bulge stem cell properties using a variety of assays *(7, 8)*.

2. Materials

2.1. Cell Culture

1. Dulbecco's modified Eagle's medium/Ham's F-12 Nutrient Medium (3:1 Mix) without Calcium (Gibco Invitrogen, special order custom powdered media, Cat. #90-5010EA)
2. Dulbecco's modified Eagle's medium (Gibco Invitrogen)
3. F-12 Nutrient Mixture (Ham's) (Gibco Invitrogen)
4. Fetal bovine serum (US characterized or defined) (Hyclone)
5. Bovine calf serum (defined, supplemented) (Hyclone)
6. Penicillin–streptomycin (Gibco Invitrogen)
7. Sodium bicarbonate (Invitrogen)
8. L-glutamine (Invitrogen)
9. Hydrocortisone (Calbiochem)
10. Cholera toxin (MP Biomedicals)
11. Insulin (Sigma)
12. T_3 (3,3′,5-triiodo-L-thyronine) (Sigma)
13. Chelex (100 resin, sodium, 200–400 dry mesh, 75–150 μm wet bead) (Bio-Rad)
14. Trypsin–EDTA 0.25% (1×) (Gibco Invitrogen)
15. Antibiotic–antimycotic (100×) (Gibco Invitrogen)
16. Calcium chloride 0.06 M (Cascade Biologics)
17. Mitomycin C (Roche)
18. Phosphate-buffered saline
19. Concentrated HCl and NaOH solutions for pH adjustment and preparing stock solutions
20. 95% Ethanol
21. Glass distilled H_2O; Milli-Q tissue culture grade H_2O if glass distilled is not available
22. Compressed CO_2 source.

2.2. FACS

1. Phycoerythrin-conjugated rat anti-human CD49f [integrin α_6 chain] (clone GoH3) (BD Pharmingen)
2. Biotin-conjugated rat anti-mouse CD34 (clone RAM34) (eBioscience)
3. Streptavidin–allophycocyanin conjugate (BD Pharmingen)
4. Propidium iodide (Sigma).

2.3. Instruments and Supplies

1. 4 inch dissecting forceps (Biomedical Research Instruments)
2. 4 inch Iris Scissors (Miltex)
3. #21 blade steel scalpel (Fisher Scientific)
4. Dissection pad and pan (Fisher Scientific)
5. 20 g Syringe needles, or equivalent (Becton Dickinson)
6. Electric Shaver (Oster Professional Products)
7. Reichert Brightline Hemocytometer (Hausser Scientific)
8. Trypan blue solution (0.4%) (Sigma)
9. 50 mL conical tubes (BD Falcon)
10. 40 µM nylon cell strainers (BD Falcon)
11. 70 µM nylon cell strainers (BD Falcon)
12. 5 mL round-bottom cap tubes (BD Falcon)
13. 5 mL cell strainer round-bottom cap tubes (BD Falcon)
14. Tissue culture dish, 100×20 mM (BD Falcon)
15. Tissue culture plate, 24-well flat bottom (BD Falcon)
16. Millipore Millex 0.22 µM PVDF syringe filter (Fisher)
17. Nalgene MF75 Disposable Sterile Filtration Unit, or equivalent (Fisher Scientific)
18. 4-color Flow Cytometer; BD FACSCalibur or equivalent (BD Biosciences).

2.4. Mice

1. Any standard laboratory mouse strain, d20 or older.

2.5. Feeder Cells

1. Swiss 3T3 cells or equivalent (ATCC).

2.6. Preparation of Epithelial Media Components

2.6.1. Preparation of Stock Media Additives (see Note 2)

1. *5 mg/ml insulin*: Dissolve in 0.1 N HCl, store in 10 ml aliquots at 4°C.
2. *5 mg/ml transferrin*: Dissolve in sterile PBS, store in 10 ml aliquots at −20°C.
3. *2×10^{-8} M T_3 (3,3',5-Triiodo-L-thyronine)*: Dissolve in 0.02 N NaOH to produce initial stock solution of 2×10^{-4} M of T_3. Perform serial solutions in PBS to produce 10 ml of 2×10^{-8} M T_3 aliquots. Store stocks from each step at −20°C.
4. *100× Additive Cocktail:* Mix 20 ml of 5 mg/ml insulin, 20 ml of 5 mg/ml transferrin, 20 ml of 2×10^{-8} M of T_3, and 140 ml sterile PBS. Sterilize with Nalgene filter and store in 37.5 ml aliquots at −20°C.

5. *4 mg/ml Hydrocortisone*: Dissolve one vial in 95% ethanol to produce desired concentration. Filter sterilize and store in 1 ml aliquots at −20°C.

6. *10^{-6} M Cholera toxin*: Dissolve one vial in glass distilled water to produce desired concentration. Filter sterilize and store in 1 ml aliquots at 4°C.

2.6.2. Preparation of Chelated Fetal Bovine Serum

Epithelial cells are exquisitely sensitive to calcium, and it is essential to carefully control the calcium levels to which cells are exposed at all times. In order to prepare media with a defined calcium concentration, it is necessary first to remove residual calcium normally present in fetal bovine serum via chelation. The protocol listed below generates approximately 1 L of calcium-free fetal bovine serum, which is sufficient for preparing the epidermal media in the next section as well as for use in preparing FACS staining buffer.

Day 1

1. Add 400 g of dry Chelex to a 4 L beaker with a stir bar. Add glass-distilled H_2O to a total volume of 4 L. Cover and stir continuously.

2. Adjust pH to 7.4 using 10 N HCl. Stir for 20 min, readjust pH with 10 N HCl, and repeat as needed until pH remains stable for more than 20 min.

3. Place the beaker at 4°C overnight to allow the Chelex to form a compact pellet.

Day 2

1. Carefully decant H_2O. Add fresh H_2O to 4 L.

2. Adjust the pH to 7.4 as in Day 1.

3. Place the beaker at 4°C for 1 hour to allow the Chelex to form a compact pellet.

4. Carefully decant H_2O.

5. Slowly add two 500 ml bottles of characterized or defined fetal bovine serum to the Chelex.

6. Stir slowly at 4°C for 1 hr, with a speed set to minimize bubbles.

7. Place the beaker at 4°C for 1 hour to allow the Chelex to form a compact pellet.

8. Decant the serum into a 1 L glass bottle and filter the serum through a Nalgene bottle top filter under sterile conditions.

9. Store unused FBS at −20°C or use immediately to make E media without calcium.

2.6.3. Preparation of E Media Without Calcium for Epidermal Cells (see Note 3).

1. In a 6 L Erlenmeyer flask, combine six packets of Gibco Invitrogen customized DMEM:F12 (3:1) without calcium with glass distilled H_2O to reach a final volume of 5.5 L (*see* **Note 4**).

2. Add 18.42 g of sodium bicarbonate, 2.85 g of L-glutamine, and 60 mL of 100× penicillin–streptomycin solution.

3. Adjust pH to 7.2 using 10 N HCl and adjust the volume to 6 L with H_2O.

4. Apply compressed CO_2 to the media for 15 min. The media should reach an amber color.

5. Combine stock additives with chelated FBS prepared in 3.1.2 by adding 75 ml of 100× cocktail, 750 μl cholera toxin, and 750 μl hydrocortisone to 1 L of chelated FBS.

6. Produce final 15% FBS media in 1 L batches by combining 850 ml of the DMEM:F12 media base from step 4 with 150 ml of the supplemented chelated FBS from step 5 and sterilize using a Nalgene bottle top filter.

7. Media can be stored in 250 or 500 mL bottles at −20°C.

8. Note that calcium must be added to media used for cell culture, but must be omitted from media used in the cell preparation for FACS.

2.7. Preparation of F Media for 3T3 Feeder Cells

1. In a 4 L Erlenmeyer flask, combine three 1 L packets of Dulbecco's modified Eagle's medium and one 1 L packet of F-12 Nutrient Mixture (Ham's) with glass distilled H_2O to reach a final volume of 3.5 L.

2. Add 12.28 g of sodium bicarbonate, 1.9 g of L-glutamine, and 40 mL of 100× penicillin–streptomycin solution.

3. Adjust pH of medium using 1 N NaOH or 1 N HCl to 7.20 and adjust volume to 4 L with H_2O (*see* **Note 6**).

4. Filter sterilize using a Nalgene bottle top filter.

5. Media can be stored in 500 mL bottles at −20°C.

6. Add bovine calf serum to a final concentration of 10% to make complete media.

3. Methods

3.1. Propagation of 3T3 Feeder Cells

Healthy feeder cells are a critical determinant of success in culturing epithelial cells *(8, 9)*. Feeder cells can be produced by treating 3T3 cells with mitomycin C to permanently arrest cell division without killing cells. Treated cells are then pre-plated as a monolayer before co-culturing with epithelial stem cells *(10)*. Swiss 3T3 is not a cloned cell line, and certain precautions should be taken when growing them to ensure that they maintain optimal properties for supporting epithelial stem cell growth (*see* **Note 5**).

1. 3T3 cell lines are conveniently propagated in 100 mm tissue culture dishes with 10 mL of F media per plate. Do not allow actively growing cells to become confluent before passaging.
2. Always feed 3T3 cells the day before transferring.
3. When plates reach 80% confluence, trypsinize cells, and split at a 1:10–1:15 ratio.

3.2. Mitomycin C Treatment of 3T3 Cells

1. Dissolve mitomycin C in sterile PBS to produce a 0.4 mg/mL stock. Mix well and sterile filter using 0.22 µM syringe filter. Store protected from light at 4°C.
2. Remove media from 100 mm plates of confluent 3T3 cells. Add 5 mL of fresh F media per 100 mm plate.
3. Add 100 µL of mitomycin C stock solution to each 100 mm plate.
4. Incubate plates for 2 hr.
5. Wash cells two times with 5 mL of PBS per plate.
6. Add fresh F media to plates and return to incubator. Cells treated with mitomycin C will remain viable for several weeks, provided they are given fresh media one to two times per week.

3.3. Preparation of Single Cell Suspension from Adult Epidermis

The preparation of a high quality, single cell suspension is the most essential component of successful FACS. Careful shaving and dissection to minimize the amount of non-epidermal tissue subject to enzymatic dissociation helps to ensure a clean, high yield preparation of epidermal cells. Handling cells gently and keeping suspensions on ice at all times helps to maintain cell viability. Practicing sterile technique will help to minimize the chance of contamination if cells are to be used for culture experiments.

1. Kill mice following standard laboratory protocol and use the electric shaver to remove all hair from body of an adult mouse, excluding the head and limbs. Shave as close to the skin as possible while being careful to avoid nicks.
2. Wash skin with 70% ethanol to remove hair residue and reduce the chance of microbial contamination. Blot dry.
3. Use forceps and scissors to make a full length dorsal midline incision from the head to the tail. Starting from the dorsal incision, make two circumferential incisions—one just behind the forepaws and the other just before the hindpaws. Use the forceps to carefully peel off the skin in one piece from the mouse.

4. Place the skin hair side down on the dissecting pad and use the needles to pin down two adjacent sides of the skin.

5. Using the scalpel, gently scrape away the fat and blood vessels covering the dermis until the dermis is clearly and uniformly exposed. Note that the dermis has a much duller appearance than fat, which is shiny and reflective (*see* **Note 7**).

6. Fill a 100 mm TC dish with 10 mL of PBS and carefully float the skin, dermis side down, on the PBS. Ensure that the skin is flat and unfolded and that the epidermal side remains dry and unsubmerged.

7. Once all skins have been processed, aspirate the PBS from each dish and add 10 mL of 0.25% trypsin per dish, making sure that the skin is freely floating with an unsubmerged epidermis.

8. Incubate skins overnight at 4°C.

9. The next day, use the scalpel and forceps to separate the epidermis from the dermis. Hold the skin in one corner with forceps while gently scraping along the top to remove the epidermis, which will easily peel off in small chunks. Note that the dermis is very durable and cannot be easily disrupted by scraping.

10. Remove the dermis from the dish and use the scalpel and forceps to briefly mince the largest pieces of tissue.

11. Pour the trypsin and epidermis mixture into a 50 ml Falcon tube.

12. Repeatedly pipette up and down with a 10 ml pipette to triturate the tissue. It may be necessary to tap the end of the pipette on the bottom of the tube to pick up all of the tissue. Continue until the mixture can be easily pipetted without tapping.

13. Place a 70 μM cell strainer on top of a new 50 mL Falcon tube and pass the mixture over the strainer. Wash the strainer with 5 mL of cold E no calcium media to collect residual cells, and inactivate the trypsin.

14. Pass the resulting cell suspension over a 40 μm cell strainer and collect in a new Falcon tube. Add cold E no calcium media to bring the final volume up to 25 mL. Place the suspension on ice until all skins have been processed.

15. Spin tubes for 10 min at 250 g to pellet cells. After the spin, it is normal for the supernatants to be cloudy because they contain dead cells and other debris which does not pellet efficiently. Carefully aspirate supernatants without disturbing the cell pellet.

16. Make fresh staining buffer for use in subsequent steps by adding 1 mL of chelated fetal bovine serum prepared in 3.1.2 to 49 mL of sterile PBS. Keep on ice.

17. Resuspend each pellet in 5 mL of staining buffer. Spin again for 5 min at 250 g and pour off supernatant (*see* **Note 8**).

18. Resuspend cells from both pellets in 1 mL total of staining buffer.

19. Strain the cells once more by passing through the cap of a 5 mL cell strainer cap FACS tube (*see* **Note 9**).

3.4. Staining and FACS Isolation of Epithelial Stem Cells

1. From the 1 mL cell suspension prepared in 3.4, remove 4_50 µL aliquots and place each in a separate FACS tube. Add 250 µL of staining buffer to each 50 µL aliquot to match the volumes listed in **Table 14.1** and label samples as indicated (*see* **Note 10**).

2. Add the amount of primary antibody indicated in **Table 14.1** to the appropriate tubes. Gently mix cells by flicking with the index finger, place tubes on ice, and cover with aluminum foil to protect from light for all subsequent steps (*see* **Note 11**).

3. Incubate cells on ice for 30 min, with gentle mixing every 10 min to prevent cells from settling at the bottom of the tube.

Table 14.1
Summary of control and analyte samples necessary for FACS analysis. Small volumes of cells are used to create an unstained control (#1) and singly stained controls (#2-4) for each fluorescence channel. These samples are used for calibration of the flow cytometer. Cells for analysis (#5) are stained with both antibodies and resuspended in a large volume of staining buffer with propidium iodide to permit identification of viable bulge stem cells by FACS.

Sample	Sample type	Volume (µL)	Primary antibody (µL)	Secondary antibody (µL)	Final resuspension
1. Unstained	Control	300	–	–	Staining buffer
2. A6-PE	Control	300	10 α6-PE (1:30)	–	Staining buffer
3. CD34-APC	Control	300	6 CD34-biotin (1:50)	1 Streptavidin-APC (1:300)	Staining buffer
4. PI	Control	300	–	–	Staining buffer + PI
5. Analysis	Analyte	800	26.7 α 6-PE (1:30) 16 CD34-biotin (1:50)	2.7 Streptavidin-APC (1:300)	Staining buffer + PI

4. Wash cells by adding 1 mL of ice cold PBS to each tube and spinning for 5 min at 250 g at 4°C. Carefully decant supernatant. Cells will remain as a pellet on the bottom of the tube.

5. Resuspend cells in the same initial volume of staining buffer and add secondary antibody as indicated in **Table 14.1**.

6. Incubate cells on ice for 30 min, with gentle mixing as before.

7. Wash cells by adding 1 mL of ice cold PBS and centrifuging as before.

8. Prepare staining buffer containing propidium iodide (PI) by adding 1 μL of the 10,000× PI stock to 10 mL of staining buffer and mixing well. The PI stock is made by dissolving 10 mg of PI in 1 mL of PBS. Store for up to 1 month protected from light at 4°C.

9. Resuspend cells in the buffer indicated in **Table 14.1** at their initial volumes.

10. Set up the flow cytometer and ensure that excitation is available from both the 488 nm (for measurement of FSC, SSC, PE, and PI) and 633 nm (for measurement of APC) lasers and that the instrument is recording from detectors for all three fluorescent channels, typically FL2, FL3, and FL4 (*see* **Note 12**).

11. Run the unstained control cells and adjust the voltage on each detector so that the cells are widely distributed but still all visible in a plot of FSC versus SSC, and so that cells are visible, but negative (i.e., $< 10^1$ units of fluorescence), in the three fluorescence channels. The FSC and SSC detectors should be set to a linear scale, while all fluorescence channels should be set to logarithmic scales. A detailed flowchart for instrument setup is provided in **Fig. 14.1A**. Plots for unstained cells should appear similar to those in **Fig. 14.1B, C**. Create region R1 on the FSC versus SSC plot to use for excluding non-cell debris from further analysis and gate all subsequent plots on R1.

12. Beginning with the PE single color control, run each single stained sample, and adjust the compensation between channels to eliminate overlap in the fluorescence signal. Plots for single-stained controls should appear similar to those in **Fig. 14.1D–F**.

13. After adjusting the voltage and compensation for the unstained and single stained controls, create region R2 on the FSC versus PI plot to exclude PI positive, dead cells. Finally, define a gate which includes cells only in regions R1 and R2, and display these cells on a plot of PE (α6) versus APC (CD34). Depending on the age of the mouse from which the cells were isolated, one or two populations of bulge cells expressing high levels of CD34 should be visible (**Figs. 14.1G and 14.2**).

14. Create regions around the bulge cell populations for subsequent analysis and/or sorting.

Isolation and Culture of Epithelial Stem Cells 225

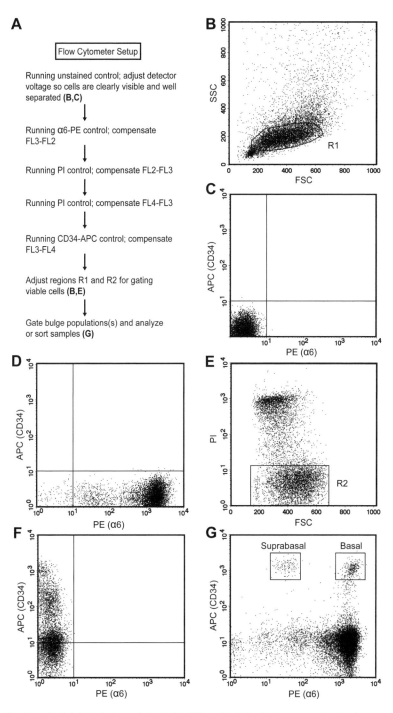

Fig. 14.1. Identification of hair follicle bulge cells by FACS. Skin epithelial cells from 7-week-old mice were enzymatically isolated to produce a single cell suspension which was stained for α6-integrin and CD34. (**A**) Protocol used for setting detector voltage and compensation on the flow cytometer. (**B**) Forward (FSC) versus side scatter (SSC) plot of unstained cells determines the location of the R1 gate, used for excluding debris from all subsequent plots. (**C**) Unstained cells are visible on the plot of FL2 (PE) and FL4 (APC), but remain within the double negative quadrant. (**D,F**) Singly stained α6-PE and CD34-APC control cells distribute along the expected axes. (**E**) Propidium iodide (PI) staining is used to identify viable cells, shown here in region R2. (**G**) Double staining for CD34-APC and α6-PE, along with gating on regions R1 and R2, demonstrates two distinct bulge cell populations—the suprabasal $\alpha 6^{Low} CD34^{High}$ and basal $\alpha 6^{High} CD34^{High}$ population.

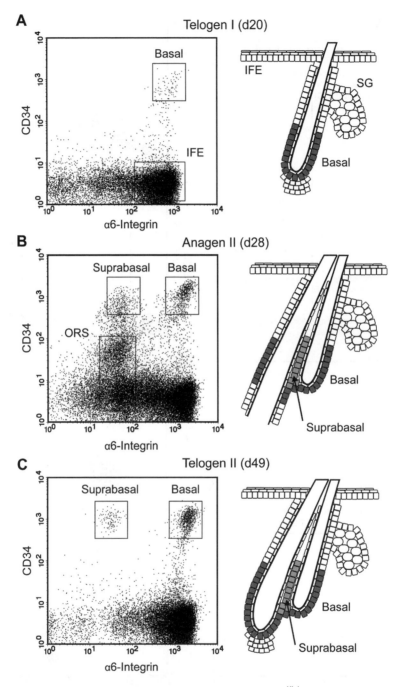

Fig. 14.2. Expression levels of α6-integrin distinguish two populations of CD34[High] bulge cells. (**A**) CD34 is specifically enriched in the bulge region of the hair follicle beginning at the first telogen stage of the hair cycle, approximately day 20. A single layer of uniformly α6[High] basal bulge cells (*dark grey*) surround the lower portion of the club hair, below the sebaceous gland (SG). Basal cells of the interfollicular epidermis (IFE) and upper hair follicle are present as a α6[High]CD34[−] population. (**B**) During the next anagen stage of the hair cycle, the growth of a second hair shaft creates a new population of α6[Low]CD34[High] suprabasal bulge cells (*light grey*). During this stage, cells from the lower, cycling portion of the outer root sheath (ORS) are also visible as a population of α6[Low]CD34[Low/−] cells. (**C**) During the second telogen stage of the hair cycle, both the α6[High]CD34[High] and α6[Low]CD34[High] populations are distinct, and the cycling portion of the outer root sheath is absent.

15. Cells collected for culture can be sorted directly into microfuge tubes containing antibiotic–antimycotic supplemented E medium with no added calcium. To prevent cells from sticking to the sides of the collection tube, pre-coat the tube by filling it with 50% chelated FBS in PBS and incubating at room temperature for 30 min. Discard the coating solution before adding collection media. Once sorted cells have been collected, keep them on ice at all times, and place them in culture as soon as possible.

3.5. Culture of FACS-Isolated Skin Epithelial Stem Cells

1. At least 1 hour before plating sorted cells, plate mitomycin C treated feeder cells growing in F medium in the desired culture vessel, typically a 24-well plate, and allow them to adhere. One confluent 10 cm plate of treated feeder cells typically provides enough cells to prepare an entire 24-well plate for co-culture.

2. Prepare E medium for cell culture by adding stock calcium to a final concentration of 0.3 mM (500 µL of 0.06 M $CaCl_2$ stock per 100 mL of E medium) and adding antibiotic–antimycotic solution to a final 1× concentration (1 mL of 100× stock solution per 100 mL of E medium).

3. Aspirate F medium from feeder cells and replace with pre-warmed calcium and antibiotic–antimycotic supplemented E medium.

4. Gently vortex tubes contained sorted bulge cells and spin for 5 min at 250 g at 4°C to pellet cells out of collection media. Carefully remove supernatant and resuspend cells in a small volume of E no calcium media, typically 100–200 µL per 1.5 mL microfuge tube. Note that the pellet is often invisible.

5. Use the hemocytometer and trypan blue to count the number of viable cells in the suspension (*see* **Note 13**).

6. Plate 10^3–10^4 viable bulge cells per single well in a 24-well plate.

7. Feed cells with fresh E medium supplemented with 0.3 mM calcium every 3 days. Addition of antibiotic–antimycotic is not necessary if there is no evidence of contamination after 3 days. Colonies will begin to be visible by 7 days (**Fig. 14.3**) and will typically be ready for passaging by 14 days (*see* **Note 14**).

3.6. Use of FACS Isolated Bulge Cells for Specific Applications

1. *BrdU Quantification*: Measuring BrdU incorporation in the bulge cell populations by FACS is valuable for quantitative assessment of cell cycle kinetics in BrdU pulse or pulse-chase experiments. BrdU detection can easily be added to the α6-integrin/CD34 staining protocol using monoclonal antibodies directed against BrdU (available from BD Pharmingen and other sources) and established staining protocols *(11, 12)*.

2. *RNA Isolation*: Bulge cells can be collected for subsequent RNA isolation by sorting cells directly into lysis buffer from

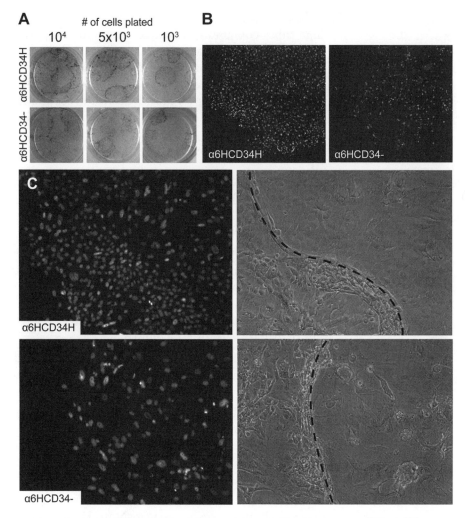

Fig. 14.3. Bulge cells display distinct growth properties when placed in culture. The $\alpha 6^{High}CD34^{High}$ bulge cells and $\alpha 6^{High}CD34^-$ epidermal cells were isolated by FACS from mice expressing a GFP-tagged histone H2B under the control of a keratin 14 promoter. (**A**) Rhodamine B staining of cultured cells shows that although both cell populations have similar colony forming efficiencies, colonies derived from bulge cells tend to be larger and more regularly shaped. (**B**) GFP fluorescence demonstrates that $\alpha 6^{High}CD34^{High}$ cells give rise to densely packed colonies with regular borders, while $\alpha 6^{High}CD34^-$ cells generate sparser colonies with a more irregular appearance. (**C**) GFP fluorescence (*left*) and phase contrast images (*right*) show that colonies derived from $\alpha 6^{High}CD34^{High}$ cells consist largely of small, undifferentiated cells with a uniform morphology, while $\alpha 6^{High}CD34^-$ colonies contain a mix of larger cells with more variable morphology. Note that colonies begin growing underneath the feeder cell layer, and feeder cells are gradually pushed aside as the colony expands in size. Dotted lines indicate the approximate boundary between feeder cells (*left*) and epidermal cells (*right*).

any method appropriate for purifying RNA from small populations of cells. The lysis buffer should be chilled during collection to minimize RNA degradation, and care should be taken to collect only the recommended number of cells per volume of lysis buffer used.

3. *Grafting*: A wide variety of techniques exist for reconstituting skin by grafting mixtures of epidermal and dermal cells onto

immunocompromised *Nude* mice. Although the number of bulge cells which can be isolated by FACS is often too small for direct grafting experiments, their numbers can easily be expanded by culturing for several passages *(7, 13)*.

4. *Viral Manipulation*: Cultured epithelial cells can be successfully infected with several types of retroviruses as well as VSV-G pseudotyped HIV-derived lentiviruses *(14, 15)*. Regardless of the system used, it is important to note that the high calcium concentration of the media used to collect the virus can cause epithelial cells to differentiate and cease proliferating. To avoid this, cells must be exposed to viral supernatant for only a limited period of time, or the virus must be either purified or concentrated from the supernatant.

4. Notes

1. Although the use of just α6-integrin and CD34 is sufficient to isolate distinct and substantially enriched populations of bulge cells by FACS, and may be sufficient for most applications, the addition of a transgenic fluorescent marker expressed exclusively in epithelial cells (such as K14-GFP) is needed to guarantee that α6-integrin$^+$/CD34^{High} cells are derived from the bulge and do not represent contaminating cells from other non-epithelial tissues. GFP expressing cells are particularly useful for quick visualization in culture and for lineage marking in grafting experiments. If a flow cytometer with sufficient fluorescence channels is available, an antibody directed against β4-integrin, which is specific for epithelial cells in the skin, can also be used in conjunction with α6-integrin and CD34 to increase the purity of bulge cells.

2. Use plastic containers since all of these compounds adhere to glass.

3. α6^{High}CD34^{High} bulge stem cells have also been grown successfully in culture media based on William's E medium (Gibco Invitrogen) and non-chelated FBS *(6, 16)*. Media based on commercial base preparations specifically optimized for keratinocytes, such as KBM (Cambrex Bio Science) and DK-SFM (Gibco Invitrogen) can support the growth of bulk adult mouse keratinocytes and may also be suitable for bulge stem cells *(17, 18)*.

4. If desired, this protocol can be scaled down to produce a smaller amount of media. However, media can be conveniently stored at 20°C until needed and producing a larger batch of media allows for improved consistency across experiments.

5. When grown in DMEM with 10% FBS, 3T3 cells grow with a doubling time of approximately 18 h in sparse culture and reach a saturation density of about 4 million cells per 100 mm plate. Cells can be maintained at this saturation density for several weeks, provided the media is changed twice weekly. In order to maintain this property of high contact inhibition, it is necessary to passage cells only at high dilutions, otherwise variants tend to be selected having reduced contact inhibition. It is also simple to make clonal isolates from 1:100,000 dilutions of confluent plates, selecting for clones with a high degree of contact inhibition. Once a good line of line of 3T3 cells has been derived, glycerol stocks should be made to allow for consistent propagation from a similar passage number over time.

6. If the medium is not red to red/orange in color, bubble compressed CO_2 through the medium before attempting any pH adjustments. Usually no adjustment will be necessary.

7. It is essential to remove all of the fat attached to the dermis in order for trypsin to easily penetrate the tissue and dissociate cells. If the fat has been insufficiently removed, separation of the epidermal cells from the dermis will be difficult, and the yield of single cells will be very low. In the event that separation of the epidermis is difficult after overnight incubation at 4°C, the skins can be additionally incubated at 37°C for 30 min prior to epidermal separation.

8. Use of chelated fetal bovine serum in staining buffer is essential because even trace amounts of calcium will cause cells to adhere to each other and form clumps. It is often easier to resuspend the cell pellet using a P1000 micropipette than a 10 mL pipette.

9. As an optional control for culture experiments, cells from this step may be placed directly in culture to assess initial cell viability before subjecting cells to staining and sorting.

10. The volumes listed are appropriate for staining cells isolated from the entire epidermis of two adult mice. However, the preparation can be easily scaled up or down in increments of 500 μL per total yield of cells from one epidermis.

11. Multiple fluorophore combinations can be used as labels for α6-integrin and CD34, and this choice may be dictated by reagents already at hand or the capabilities of a particular flow cytometer. We routinely use the combination of PE and APC fluorophores because they are relatively photostable, provide a broad dynamic range and because this combination leaves the FL1 channel available for detecting GFP in cells isolated from transgenic mice or measuring BrdU incorporation via anti-BrdU antibodies directly coupled to FITC.

12. The proper setup and operation of flow cytometers are complex and beyond the scope of this chapter. For general background and theory of flow cytometry, a comprehensive reference should be consulted *(11)*.

13. Although the flow cytometer will provide a count of sorted cells, this number often substantially exceeds the cell count determined by trypan blue exclusion because not all cells survive the collection process intact.

14. Cultured bulge cells are most easily passaged by first removing feeder cells before trypsinization to collect bulge cells. After aspirating the media, wash the well twice with PBS. Use a P1000 pipettor to repeatedly direct a gentle stream of PBS perpendicular to the surface of the dish. The feeder cells, which are less adherent than the epithelial cells, will be dislodged from the surface of the tissue culture dish. Colonies of bulge cells will now be clearly visible. Wash away floating feeders before trypsinizing epithelial cells and passaging onto fresh feeder cells.

References

1. Cotsarelis, G., T.T. Sun, and R.M. Lavker, (1990) *Label-retaining cells reside in the bulge area of pilosebaceous unit: implications for follicular stem cells, hair cycle, and skin carcinogenesis.* Cell, **61**(7): 1329–37.
2. Morris, R.J. and C.S. Potten, (1999) *Highly persistent label-retaining cells in the hair follicles of mice and their fate following induction of anagen.* J Invest Dermatol, **112**(4): 470–5.
3. Taylor, G., et al., (2000) *Involvement of follicular stem cells in forming not only the follicle but also the epidermis.* Cell, **102**(4): 451–61.
4. Tumbar, T., et al., (2004) *Defining the epithelial stem cell niche in skin.* Science, **303**(5656): 359–63.
5. Morris, R.J., et al., (2004) *Capturing and profiling adult hair follicle stem cells.* Nat Biotechnol, **22**(4): 411–7.
6. Trempus, C.S., et al., (2003) *Enrichment for living murine keratinocytes from the hair follicle bulge with the cell surface marker CD34.* J Invest Dermatol, **120**(4): 501–11.
7. Blanpain, C., et al., (2004) *Self-renewal, multipotency, and the existence of two cell populations within an epithelial stem cell niche.* Cell, **118**(5): 635–48.
8. Rheinwald, J.G. and H. Green, (1977) *Epidermal growth factor and the multiplication of cultured human epidermal keratinocytes.* Nature, **265**(5593): 421–4.
9. Rheinwald, J.G. and H. Green, (1975) *Serial cultivation of strains of human epidermal keratinocytes: the formation of keratinizing colonies from single cells.* Cell, **6**(3): 331–43.
10. Todaro, G.J. and H. Green, (1963) *Quantitative studies of the growth of mouse embryo cells in culture and their development into established lines.* J Cell Biol, **17**: 299–313.
11. Shapiro, H.M., (2003) *Practical flow cytometry.* 4 ed.: Wiley-Liss. 736.
12. Gratzner, H.G. and R.C. Leif, (1981) *An immunofluorescence method for monitoring DNA synthesis by flow cytometry.* Cytometry, **1**(6): 385–93.
13. Weinberg, W.C., et al., (1993) *Reconstitution of hair follicle development in vivo: determination of follicle formation, hair growth, and hair quality by dermal cells.* J Invest Dermatol, **100**(3): 229–36.
14. Kuhn, U., et al., (2002) *In vivo assessment of gene delivery to keratinocytes by lentiviral vectors.* J Virol, **76**(3): 1496–504.
15. Garlick, J.A., et al., (1991) *Retrovirus-mediated transduction of cultured epidermal*

keratinocytes. J Invest Dermatol, **97**(5): 824–9.
16. Wu, W.Y. and R.J. Morris, (2005) *Method for the harvest and assay of in vitro clonogenic keratinocytes stem cells from mice.* Methods Mol Biol, **289**: 79–86.
17. Redvers, R.P. and P. Kaur, (2005) *Serial cultivation of primary adult murine keratinocytes.* Methods Mol Biol, **289**: 15–22.
18. Yano, S. and H. Okochi, (2005) *Long-term culture of adult murine epidermal keratinocytes.* Br J Dermatol, **153**(6): 1101–4.

Chapter 15

Regeneration of Skin and Cornea by Tissue Engineering

Danielle Larouche, Claudie Paquet, Julie Fradette, Patrick Carrier, François A. Auger, and Lucie Germain

Abstract

Progress in tissue engineering has led to the development of technologies allowing the reconstruction of autologous tissues from the patient's own cells. Thus, tissue-engineered epithelial substitutes produced from cultured skin epithelial cells undergo long-term regeneration after grafting, indicating that functional stem cells were preserved during culture and following grafting. However, these cultured epithelial sheets reconstruct only the upper layer of the skin and lack the mechanical properties associated to the connective tissue of the dermis. We have designed a reconstructed skin entirely made from human cutaneous cells comprising both the dermis and the epidermis, as well as a well-organized basement membrane by a method named the self-assembly approach. In this chapter, protocols to generate reconstructed skin and corneal epithelium suitable for grafting are described in details. The methods include extraction and culture of human skin keratinocytes, human skin fibroblasts as well as rabbit and human corneal epithelial cells, and a complete description of the skin reconstructed by the self-assembly approach and of corneal epithelium reconstructed over a fibrin gel.

Key words: Stem cells, keratin 19, epidermis, human, cornea, skin substitute, reconstructed skin, corneal epithelial cell, tissue engineering, fibrin gel.

Abbreviations alkDME-Ham: Air–liquid human keratinocyte culture medium, complete hcDME-Ham: Complete human corneal epithelial cell culture medium, complete hkDME-Ham: Complete human keratinocyte culture medium, complete rcDME-Ham: Complete rabbit corneal epithelial cell culture medium, DME: Dulbecco's modified Eagle's medium, DMSO: Dimethyl sulfoxide, EDTA: ethylenediaminetetraacetic acid, fDME: human fibroblast culture medium, Ham: Ham's F12 medium, iS3T3: Irradiated Swiss 3T3, K19: Keratin 19, PBS-P/G/F: Phosphate buffered saline-Penicillin G/Gentamicin/Fungizone, SHEM: Supplemented hormonal epithelial medium, T3: 3,3′,5′-triiodo-L-thyronine.

1. Introduction

The epidermis of the skin and the epithelium of the cornea are self-renewing tissues. Their homeostasis is ensured by the permanent pool of stem cells they house (*1–4*). Recent findings emphasize the importance of the specialized microenvironment of the stem cells, called the niche, for the establishment of their properties (*5–7*). For example, Keratin 19 (K19) - expressing cells of glabrous and hairy skin, which present stem cell properties, are in direct contact with the basement membrane (*8*). Thus, skin substitutes intended for long-term grafting or for basic research on stem cells were designed to mimic in vivo-like tissue architecture and phenotype.

The concept of the self-assembly approach is to reconstruct an organ in a fashion resembling its formation in vivo in which the use of appropriate culture and mechanical conditions induce cells to secrete a significant amount of extracellular matrix as during organogenesis (*9, 10*). The skin reconstructed by the self-assembly approach exhibits a well-developed epidermis (*see* **Fig. 15.1B**) that expresses differentiation markers and a well-organized basement membrane (*11, 12*). An adaptation of this method also enables the reconstruction of tissues with corneal epithelial cells (*13*). These models allowed us to study fundamental aspects of both the regeneration and differentiation of epithelia (*13–15*) and represent attractive candidates for the permanent coverage of wounds (*16, 17*).

This chapter first details the methods for extraction and culture of keratinocytes (*18*), fibroblasts (*11*), and corneal epithelial cells (*17, 19*), followed by a detailed description of the generation of reconstructed skin by the self-assembly approach (*11, 12*) and also of reconstructed corneal epithelium over a fibrin gel (*16, 17*).

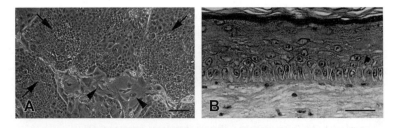

Fig. 15.1. (**A**) Phase contrast micrograph of newborn keratinocytes (arrows) grown to 80% confluency with an iS3T3 feeder layer (arrowheads). (**B**) Histological cross-section of human reconstructed skin after 21 days of culture at the air–liquid interface stained by hematoxylin and eosin. Scale bars: A: 250 μm; B: 36 μm.

2. Materials

2.1. Culture Media

2.1.1. Tissue Transport Medium

1. Preparation of tissue transport medium components (for final concentration *see* **Table 15.1**):

    a. DME-Ham. Dulbecco's modified Eagle's medium (DME) (cat. no. 12800, Invitrogen, Oakville, Ont., Canada): Ham's F12 medium (cat. no. 21700, Invitrogen), 3:1, 3.07 g/L NaHCO$_3$ (36.54 mM) (cat. no. S233, Fisher Scientific, Ottawa, Ont.), 24.3 mg/L adenine (0.18 mM) (cat. no. A2786, Sigma Chemicals, St Louis, MO), 312.5 μL/L 2 N HCl. Dissolve in apyrogenic ultrapure water. Adjust pH to 7.1. Sterilize by filtration through a 0.22 μm low-binding disposable filter. Aliquot and store at 4°C.

    b. Fetal calf serum (cat. no. SH30396, HyClone, Logan, UT). Thaw in cold water. Inactivate in hot water (56°C) for 30 min. Distribute in single use aliquots and store at −20°C.

    c. Penicillin G and gentamicin (cat. no. P3032 and G1264, respectively, Sigma Chemicals). Dissolve 50,000 IU/mL of penicillin G and 12.5 mg/mL of gentamicin sulfate in apyrogenic ultrapure water to make a 500× stock solution. Sterilize by filtration through a 0.22 μm low-binding disposable filter, distribute in single use aliquots, and store at −80°C.

    d. Fungizone (Amphotericin B, cat. no. A9528, Sigma Chemicals). Dissolve 0.25 mg/mL of amphotericin B (0.27 mM), in apyrogenic ultrapure water to make a 500× stock solution. Sterilize by filtration through a 0.22 μm low-binding disposable filter, distribute in single use aliquots, and store at −80°C.

Table 15.1
Tissue transport medium

Component	Quantity	Final concentration
DME-Ham	900 mL	90% (v/v)
Fetal calf serum	100 mL	10% (v/v)
Penicillin G-Gentamicin 500×	2 mL	Penicillin G 100 IU/mL Gentamicin 25 μg/Ml
Fungizone 500×	2 mL	0.5 μg/mL

2. Tissue transportation: Thaw all components at 4°C. To make 1 L, refer to **Table 15.1**. Store the transport medium at 4°C.

2.1.2. Complete Human Keratinocyte Culture Medium (Complete hkDME-Ham)

1. Preparation of human keratinocyte culture medium components (for final concentration *see* **Table 15.2**):
   a. DME-Ham (*see* **Section 2.1.1**; *1.a*).
   b. Fetal clone II serum (cat. no. SH30066, HyClone). Thaw in cold water. Inactivate in hot water (56°C) for 30 min. Distribute in single use aliquots and store at −20°C.
   c. Insulin (cat. no. I1882, Sigma Chemicals). Dissolve 250 mg in 50 mL 0.005 N HCl (125 μL 2 N HCl/50 mL apyrogenic ultrapure water) to make a 1,000× stock solution (0.87 mM). Sterilize by filtration through a 0.22 μm low-binding disposable filter, distribute in single use aliquots, and store at −80°C.
   d. Hydrocortisone (cat. no. 386698, Calbiochem, San Diego, CA). Dissolve 25 mg in 5 mL of 96% ethanol (4.8 mL 99% ethanol (Les Alcools de Commerce Inc., Brampton, Ont.)/0.2 mL apyrogenic ultrapure water). Complete to 125 mL with DME-Ham (*see* **Section 2.1.1**; *1.a*) to make a 500× stock solution (0.53 mM). Sterilize by filtration through a 0.22 μm low-binding disposable filter, distribute in single use aliquots, and store at −80°C.
   e. Cholera toxin (cat. no. C8052, Sigma Chemicals). Dissolve 1 mg in 1 mL of apyrogenic ultrapure water. Complete to 118.18 mL with DME-Ham (*see* **Section 2.1.1**; *1.a*) supplemented with 10% (v/v) fetal clone II (*see* **Section 2.1.2**; *1.b*) to make a 1,000× stock solution (10^{-7} M). Sterilize by filtration through a 0.22 μm low-binding disposable filter, distribute in single use aliquots, and store at −80°C.
   f. Epidermal growth factor (cat. no. GF-010-8, Austral Biologicals, San Ramon, CA). Dissolve 500 μg in 2.5 mL of 10 mM HCl. Complete to 50 mL with DME-Ham (*see* **Section 2.1.1**; *1.a*) supplemented with 10% (v/v) fetal clone II (*see* **Section 2.1.2**; *1.b*) to make a 1,000× stock solution. Sterilize by filtration through a 0.22 μm low-binding disposable filter, distribute in single use aliquots, and store at −80°C.
   g. Penicillin G-Gentamicin (*see* **Section 2.1.1**; *1.c*).
2. Thaw all components at 4°C. To make 1 L, refer to **Table 15.2** (*see* **Note 1**). Complete hkDME-Ham can be stored at 4°C for 10 days.

Table 15.2
Complete human keratinocyte culture medium (complete hkDME-Ham)

Component	Quantity	Final concentration
DME-Ham	950 mL	95% (v/v)
Fetal clone II	50 mL	5% (v/v)
Insulin 1,000×	1 mL	5 µg/mL
Hydrocortisone 500×	2 mL	0.4 µg/mL
Cholera toxin 1,000×	1 mL	10^{-10} M
Epidermal growth factor 1,000×	1 mL	10 ng/mL
Penicillin G-Gentamicin 500×	2 mL	Penicillin G 100 IU/mL Gentamicin 25 µg/mL

2.1.3. Air–Liquid Human Keratinocyte Culture Medium (alkDME-Ham)

1. Thaw all components described in **Section 2.1.2**; *1*, at 4°C, with the exception of epidermal growth factor. To make 1 L, refer to **Table 15.3** (*see* **Note 1**).
2. alkDME-Ham can be stored at 4°C for 10 days.

Table 15.3
Air–liquid human keratinocyte culture medium (alkDME-Ham)

Component	Quantity	Final concentration
DME-Ham	950 mL	95% (v/v)
Fetal clone II	50 mL	5% (v/v)
Insulin 1,000×	1 mL	5 µg/mL
Hydrocortisone 500×	2 mL	0.4 µg/mL
Cholera toxin 1,000×	1 mL	10^{-10} M
Penicillin G-Gentamicin 500×	2 mL	Penicillin G 100 IU/mL Gentamicin 25 µg/mL

Table 15.4
Human fibroblast culture medium (fDME)

Component	Quantity	Final concentration
DME	900 mL	90% (v/v)
Fetal calf serum	100 mL	10% (v/v)
Penicillin G-Gentamicin 500×	2 mL	Penicillin G 100 IU/mL Gentamicin 25 µg/mL

2.1.4. Human Fibroblast Culture Medium (fDME)

1. Preparation of fibroblast culture medium components (for final concentration see **Table 15.4**):
   a. DME. Dulbecco's modified Eagle's medium (DME) (cat. no. 12800, Invitrogen), 3.7 g/L (44 mM) NaHCO$_3$ (cat. no. S233, Fisher Scientific). Dissolve in apyrogenic ultrapure water. Adjust pH to 7.1. Sterilize by filtration through a 0.22 µm low-binding disposable filter, aliquot, and store at 4°C.
   b. Fetal calf serum (see **Section 2.1.1**; *1.b*).
   c. Penicillin G-Gentamicin (see **Section 2.1.1**; *1.c*).
2. Thaw all components at 4°C. To make 1 L, refer to **Table 15.4**. The fDME can be stored at 4°C for 10 days.

2.1.5. Complete Rabbit Corneal Epithelial Cell Culture Medium (Complete rcDME- Ham)

1. Preparation of rabbit corneal epithelial cell culture medium components (for final concentration, see **Table 15.5**):
   a. SHEM. Supplemented hormonal epithelial medium (SHEM). DME (see **Section 2.1.1**; *1.a*): Ham's F12 medium (see **Section 2.1.1**; *1.a*), 1:1, 3.07 g/L (36.54 mM) NaHCO$_3$ (cat. no. S233, Fisher Scientific), 24.3 mg/L (0.18 mM) adenine (cat. no. A2786, Sigma Chemicals), 1.25 mL/L 2 N HCL. Dissolve in apyrogenic ultrapure water. Ajust pH to 7.1. Sterilize by filtration through a 0.22 µm low-binding disposable filter, aliquot, and store at 4°C.
   b. Fetal clone II (see **Section 2.1.2**; *1.b*).
   c. Insulin (see **Section 2.1.2**; *1.c*).
   d. L-Glutamine (cat. no. 21051-024, Invitrogen). Dissolve 1.169 g in 40 mL of SHEM medium to make a 80× stock solution (200.17 mM). Agitation at 37°C can facilitate dissolution. Sterilize by filtration through a 0.22 µm low-binding disposable filter, distribute in single use aliquots and store at −80°C.

Table 15.5
Complete rabbit corneal epithelial cell culture medium (complete rcDME-Ham)

Component	Quantity	Final concentration
SHEM	900 mL	90% (v/v)
Fetal clone II	100 mL	10% (v/v)
Insulin 1,000×	1 mL	5 µg/mL
L-Glutamine 80×	12.5 mL	2.5 mM
Cholera toxin 1,000×X	5 mL	0.1 µg/mL
Epidermal growth factor 1,000×	1 mL	10 ng/mL
DMSO	5 mL	0.5% of 5 mg/mL
Penicillin G-Gentamicin 500×	2 mL	Penicillin G 100 IU/mL Gentamicin 25 µg/mL

  e. Cholera toxin (*see* **Section 2.1.2**; *1.e*).

  f. Epidermal growth factor (*see* **Section 2.1.2**; *1.f*).

  g. DMSO. Dimethyl sulfoxide (DMSO) (cat. no. D5879, Sigma Chemicals). Distribute the stock solution (99.7%) in single use aliquots and store at −20°C.

  h. Penicillin G-Gentamicin (*see* **Section 2.1.1**; *1.c*).

2. Thaw all components at 4°C. To make 1 L, refer to **Table 15.5**. *See* **Note 1**.

3. Complete rcDME-Ham can be stored at 4°C for 10 days.

2.1.6. Complete Human Corneal Epithelial Cell Culture Medium (Complete hcDME-Ham)

1. Preparation of human corneal epithelial cell culture medium components (for final concentration, *see* **Table 15.6**):

  a. Transferrin/T3. Transferrin (vial of 600 mg/20 mL, cat. no. 652202, Roche Diagnostics, Laval, Qc, Canada)/ 3,3′,5′-triiodo-L-thyronine (T3) (cat. no. T2752, Sigma Chemicals). Solution #1: Dissolve 6.8 mg of T3 in a small volume (about 3 mL) of 0.02 N NaOH (10 µL 10 N NaOH/4.99 mL apyrogenic ultrapure water). Complete to 50 mL with apyrogenic ultrapure water. Sterilize by filtration through a 0.22 µm low-binding disposable filter, distribute in single use aliquots, and store at 4°C. Mix the transferrin vial with 1.2 mL of solution #1 and complete to 120 mL with apyrogenic ultrapure water to make a 1,000× stock solution. Sterilize by filtration through a

Table 15.6
Complete human corneal epithelial cell culture medium (complete hcDME-Ham)

Component	Quantity	Final concentration
DME-Ham	900 mL	90% (v/v)
Fetal clone II	100 mL	10% (v/v)
Insulin 1,000×	1 mL	5 µg/mL
Hydrocortisone 500×	2 mL	0.4 µg/mL
Cholera toxin 1,000×	1 mL	10^{-10} M
Epidermal growth factor 1,000×	1 mL	10 ng/mL
Tranferrin/T3 1,000×	1 mL	Tranferrin: 5 µg/mL T3: 2×10^{-9} M
Penicillin G-Gentamicin 500×	2 mL	Penicillin G 100 IU/mL Gentamicin 25 µg/mL

0.22 µm low-binding disposable filter, distribute in single use aliquots, and store at −80°C.

2. Thaw all components described in **Section 2.1.2**; *1*, with the addition of transferrin/T3, at 4°C. To make 1 L, refer to **Table 15.6** (*see* **Note 1**). Complete hcDME-Ham can be stored at 4°C for 10 days.

2.1.7. Freezing Medium

1. Preparation of freezing medium components.
   a. Fetal calf serum (*see* **Section 2.1.1**; *1.b*).
   b. DMSO (*see* **Section 2.1.5**; *1.g*).
2. Thaw all components at 4°C. To make 10 mL, refer to **Table 15.7**. Keep on ice or at 4°C. *See* **Note 3**.

Table 15.7
Freezing medium

Component	Quantity	Final concentration
Fetal calf serum	9 mL	90% (v/v)
DMSO	1 mL	10% (v/v)

2.2. Monolayer Culture of Human Keratinocytes

2.2.1. Extraction and Culture of Human Keratinocytes

1. Source of human keratinocytes: normal adult skin or newborn foreskin removed by surgery.
2. Sterile container.
3. Transport medium (*see* **Section 2.1.1**).
4. 50 mL tubes (cat. no. 352070, BD Falcon, Franklin Lakes, NJ).
5. Phosphate buffered saline-Penicillin G/Gentamicin/Fungizone (PBS-P/G/F). 137 mM NaCl (cat. no. S271, Fisher Scientific), 2.7 mM KCl (cat. no. P285, Fisher Scientific), 6.5 mM Na_2HPO_4 (cat. no. S369, Fisher Scientific), 1.5 mM KH_2PO_4 (cat. no. P217, Fisher Scientific). Dissolve in apyrogenic ultrapure water to make a 10× stock solution. Store at room temperature. To dilute 10× PBS to 1×, add apyrogenic ultrapure water. Verify pH is 7.4. Store at room temperature. Before use, add Penicillin G-Gentamicin 500× stock solution (*see* **Section 2.1.1**; *1.c*) and Fungizone 500× stock solution (*see* **Section 2.1.1**; *1.d*) by diluting these additives to 1×.
6. Petri dish (size: 100 × 15 mm, cat. no. 08-757-12, Fisher Scientific).
7. HEPES (cat. no. 194549, MP Biomedicals Inc., Montreal, Qc, Canada). Make a 10× stock solution in apyrogenic ultrapure water: 0.1 M HEPES, 67 mM KCl (cat. no. P217, Fisher Scientific), 1.42 M NaCl (cat. no. S271, Sigma Chemicals). Adjust pH to 7.3. Protect from light and store at 4°C. HEPES 1×: dilute the 10× stock solution to 1× with apyrogenic ultrapure water and add 1 mM $CaCl_2$ (cat. no. C7902, Sigma Chemicals). Adjust pH to 7.45. Protect from light (*see* **Note 4**) and store at 4°C.
8. Thermolysin (cat. no. T7902, Sigma Chemicals). Dissolve 500 µg/mL in HEPES 1× (*see* **Section 2.2.1**; *7*). Sterilize by filtration through a 0.22 µm low-binding disposable filter and store at 4°C. See **Note 3**.
9. Trypsin/EDTA. 2.8 mM D-glucose (cat. no. DX0145, EMD Chemicals Inc., Gibbstown, NJ), 0.05% (w/v) trypsin 1–500 (cat. no. 7003, Intergen, Toronto, Ont., Canada), 0.00075% (v/v) phenol red (Phenol red solution 0.5%, sterile-filtered, cat. no. P0290, Sigma Chemicals), 100,000 IU/L penicillin G (cat. no. P3032, Sigma Chemicals), 25 mg/L gentamicin (cat. no. G1264, Sigma

Chemicals), 0.01% (w/v) ethylenediaminetetraacetic acid (EDTA, disodium salt, cat. no. 8993, J.T. Baker, Phillipsburg, NJ). Dissolve in 1× PBS. Adjust pH to 7.45. Sterilize by filtration through a 0.22 μm low-binding disposable filter, distribute in single use aliquots, and store at −20°C.

10. Dissecting curved forceps (cat. no. 08-953F, Fisher Scientific).
11. Scalpel (#4 cat. no. 08-917-5, Fisher Scientific) and blade (size: 22, cat. no. 73-0422, Personna Medical, Verona, VA).
12. Trypsinization unit (Celstir suspension culture flask, cat. no. 356875, Wheaton Sciences Products, Millville, NJ).
13. Parafilm (cat. no. 13-374, Fisher Scientific).
14. Tissue culture flask, 75 cm^2 (cat. no. 35310, BD Falcon).
15. Complete hkDME-Ham (*see* **Section 2.1.2**).
16. Irradiated Swiss 3T3 (iS3T3) (#CCL-92, ATCC, Manassas, VA). To obtain about 8–10 × 10^6 cells, seed 1 × 10^6 cells in a 75 cm^2 culture flask (Tissue culture flask, cat. no. 35310, BD Falcon) with 20 mL of DME. Incubate for 4 days in 8% CO$_2$, 100% humidity atmosphere at 37°C. Irradiate at 6,000 rads with a Gammacell irradiator (^{60}Co source). *See* **Note 5**.

2.2.2. Subculture of Human Keratinocytes (Passage)

1. Complete hkDME-Ham (*see* **Section 2.1.2**).
2. iS3T3 (*see* **Section 2.2.1**; *16*).
3. Trypsin/EDTA (*see* **Section 2.2.1**; *9*).
4. 50 mL tube (cat. no. 352070, BD Falcon).
5. Tissue culture flask, 75 cm^2 (cat. no. 35310, BD Falcon).

2.2.3. Cryopreservation of Human Keratinocytes

1. Freezing medium (*see* **Section 2.1.7**)
2. Sterile cryogenic vials (cat. no. 5000-020, Nalgene Labware, Rochester, NY).
3. Freezing container (cat. no. 5100-0001, Nalgene Labware), filled with 99% ethanol and pre-cooled at −20°C.

2.2.4. Thawing of Human Keratinocytes

1. Complete hkDME-Ham (*see* **Section 2.1.2**).
2. Tissue culture flask, 75 cm^2 (cat. no. 35310, BD Falcon).
3. iS3T3 (*see* **Section 2.2.1**; *16*).

2.3. Monolayer Culture of Human Fibroblasts

2.3.1. Extraction and Culture of Human Fibroblasts

1. Source of human fibroblasts: normal adult breast skin removed by surgery.
2. Sterile container.
3. Transport medium (*see* **Section 2.1.1**).
4. 50 mL tubes (cat. no. 352070, BD Falcon).
5. PBS-P/G/F (*see* **Section 2.2.1**; *5*).
6. Petri dish (size: 100 × 15 mm, cat. no. 08-757, Fisher Scientific).
7. Thermolysin (*see* **Section 2.2.1**; *8*).
8. Collagenase H (cat. no. 1074059, Roche Diagnostics, Laval, Qc, Canada). Dissolve 0.125 U/mL in fDME (*see* **Section 2.1.4**). Sterilize by filtration through a 0.22 μm low-binding disposable filter (*see* **Note 3**).
9. Dissecting curved forceps (cat. no. 08-953F, Fisher Scientific).
10. Scalpel (size: 4, cat. no. 08-917-5, Fisher Scientific) and blade (size: 22, cat. no. 73-0422, Personna Medical, Verona, VA).
11. Trypsinization unit (Celstir suspension culture flask, cat. no. 356875, Wheaton Sciences Products, Millville, NJ).
12. Parafilm (cat. no. 13-374, Fisher Scientific).
13. Tissue culture flask, 75 cm^2 (cat. no. 35310, BD Falcon).
14. fDME (*see* **Section 2.1.4**).

2.4. Monolayer Culture of Rabbit Corneal Epithelial Cells

2.4.1. Extraction and Culture of Rabbit Corneal Epithelial Cells

1. Source of rabbit corneal epithelial cells: rabbit post-mortem eyes from a local slaughterhouse.
2. 35 and 100 mm Petri dishes (size: 35 × 10 mm, cat. no. 351008, BD Falcon, and size: 100 × 15 mm, cat. no. 08-757-12, Fisher Scientific, respectively).
3. Sterile 3 in. × 3 in. gauzes (cat. no. A3120, AMD-Ritmed, Lachine, Qc, Canada).
4. Scalpel (#4 cat. no. 08-917-5, Fisher Scientific) and blades (size: 11, cat. no. 73-0411 and size: 22, cat. no. 73-0422, Personna Medical, Verona, VA).

5. Dissecting curved forceps (cat. no. 11272-30, Fine Science Tools, North Vancouver, B.C., Canada).
6. Dissecting curved scissors (cat. no. E3220, Storz, St Louis, MO).
7. 15 and 50 mL tubes (cat. no. 352096 and cat. no. 352070, respectively, BD Falcon).
8. PBS-P/G/F (*see* **Section 2.2.1**; *5*).
9. Dispase II (cat. no. 165-859, Roche Diagnostics). Dissolve 2 mg/mL dispase in HEPES 1× (*see* **Section 2.2.1**; *7*). Sterilize by filtration through a 0.22 μm low-binding disposable filter. Store at 4°C. See **Note 3**.
10. Complete rcDME-Ham (*see* **Section 2.1.5**).
11. Tissue culture flask, 75 cm^2 (cat. no. 35310, BD Falcon).
12. iS3T3 (*see* **Section 2.2.1**; *16*).
13. Dissecting microscope (SMZ-800, Nikon Canada Inc., Mississauga, Ont.).

2.4.2. Subculture of Rabbit Corneal Epithelial Cells (Passage)

1. Complete rcDME-Ham (*see* **Section 2.1.5**).
2. iS3T3 (*see* **Section 2.2.1**; *16*).
3. Trypsin/EDTA (*see* **Section 2.2.1**; *9*).
4. 50 mL tubes (cat. no. 352070, BD Falcon).
5. Tissue culture flask, 75 cm^2 (cat. no. 35310, BD Falcon).

2.4.3. Cryopreservation of Rabbit Corneal Epithelial Cells

1. Freezing medium (*see* **Section 2.1.7**).
2. Sterile cryogenic vials (cat. no. 5000-020, Nalgene Labware).
3. Freezing container (cat. no. 5100-0001, Nalgene Labware), filled with 99% ethanol and pre-cooled at −20°C.

2.4.4. Thawing of Rabbit Corneal Epithelial Cells

1. Complete rcDME-Ham (*see* **Section 2.1.5**).
2. Tissue culture flask, 75 cm^2 (cat. no. 35310, BD Falcon).
3. iS3T3 (*see* **Section 2.2.1**; *16*).

2.5. Monolayer Culture of Human Corneal Epithelial Cells

2.5.1. Extraction and Culture of Human Corneal Epithelial Cells

1. Source of human corneal epithelial cells: human post-mortem donor corneas unsuitable for transplantation (Banque Nationale d'yeux du CHUQ, Quebec, Canada).
2. 35 and 100 mm Petri dish (size: 35 × 10 mm, cat. no. 351008, BD Falcon and size: 100 × 15 mm, cat. no. 08-757-12, Fisher Scientific, respectively).

3. Sterile 3 in. × 3 in. gauzes (cat. no. A3120, AMD-Ritmed, Lachine, Qc, Canada).
4. Scalpel (#4 cat. no. 08-917-5, Fisher Scientific) and blades (size: 11, cat. no. 73-0411, and size: 22, cat. no. 73-0422, Personna Medical, Verona, VA).
5. Dissecting curved forceps (cat. no. 11272-30, Fine Science Tools, North Vancouver, B.C., Canada).
6. Dissecting curved scissors (cat. no. E3220, Storz, St Louis, MO).
7. 15 and 50 mL tubes (cat. no. 352096 and cat. no. 352070, respectively, BD Falcon).
8. PBS-P/G/F (*see* **Section 2.2.1**; *5*).
9. 7.5 mm diameter trephine (cat. no. 9711, Medtronic of Canada, Mississauga, Ont., Canada).
10. Dispase II (*see* **Section 2.4.1**; *9*).
11. Complete hcDME-Ham (*see* **Section 2.1.6**).
12. Tissue culture flask, 75 cm^2 (cat. no. 35310, BD Falcon).
13. iS3T3 (*see* **Section 2.2.1**; *16*).
14. Dissecting microscope (Nikon, SMZ-800).

2.5.2. Subculture of Human Corneal Epithelial Cells (Passage)

1. Complete hcDME-Ham (*see* **Section 2.1.6**).
2. iS3T3 (*see* **Section 2.2.1**; *16*).
3. Trypsin/EDTA (*see* **Section 2.2.1**; *9*).
4. 50 mL tubes (cat. no. 352070, BD Falcon).
5. Tissue culture flask, 75 cm^2 (cat. no. 35310, BD Falcon).

2.5.3. Cryopreservation of Human Corneal Epithelial Cells

1. Freezing medium (*see* **Section 2.1.7**).
2. Sterile cryogenic vials (cat. no. 5000-020, Nalgene Labware).
3. Freezing container (cat. no. 5100-0001, Nalgene Labware), filled with 99% ethanol and pre-cooled at −20°C.

2.5.4. Thawing of Human Corneal Epithelial Cells

1. Complete hcDME-Ham (*see* **Section 2.1.6**).
2. Tissue culture flask, 75 cm^2 (cat. no. 35310, BD Falcon).
3. iS3T3 (*see* **Section 2.2.1**; *16*).

2.6. Human Skin Reconstruction by the Self-Assembly Approach

1. Confluent fibroblasts between their second and eighth passages.
2. Keratinocytes at 80% confluency (*see* **Fig. 15.1A**) between their third and fifth passages (co-cultured with iS3T3, *see* **Section 2.2.1**; *16*).

3. Ascorbic acid (cat. no. A7631, Sigma Chemicals). Dissolve 10 mg/mL of ascorbic acid in DME (*see* **Section 2.1.1**; *1.a*) to make a 200× stock solution. Sterilize by filtration through a 0.22 µm low-binding disposable filter (*see* **Note 2**).

4. Culture media:

   a. fDME (*see* **Section 2.1.4**) containing 50 µg/mL ascorbic acid.

   b. Complete hkDME-Ham (*see* **Section 2.1.2**) containing 50 µg/mL ascorbic acid.

   c. alkDME-Ham (*see* **Section 2.1.3**) containing 50 µg/mL ascorbic acid.

5. Ingots (stainless steel grade # 316, Denmar, Quebec, Qc, Canada).

6. Merocel®, Medtronic (cat. no. 22-3620, Instruments Ophtalmiques INNOVA, Laval, Qc, Canada).

7. Anchoring ring (stainless steel grade # 316, Denmar, Quebec, Qc, Canada). Dimension: 3 cm diameter, 7/8 inch wide, 1/8 inch height.

8. Dissecting curved forceps (cat. no. 08-953F, Fisher Scientific).

9. Tissue culture flask, 25 cm^2 (cat. no. 35309, BD Falcon).

10. Cell culture dish (cat. no. 353003, BD Falcon).

11. Petri dish (size: 100 × 15 mm, cat. no. 08-757, Fisher Scientific).

12. Soldering iron (Weller Universal Dual Heat Soldering Gun, cat. no. 8200, CooperTools, Apex, NC; available in local hardwares).

13. Anchoring paper. Cut a circle with a 60 mm diameter in a Wathman sheet (cat. no. 09-825F, Fisher Scientific). Remove the concentric inside disk of 25 mm diameter after cutting.

14. Air–liquid stand (homemade acrylic stand).

2.7. Corneal Epithelium Reconstruction

1. Rabbit or human corneal epithelial cells at 80% confluency between their first and third passages (co-cultured with iS3T3, *see* **Section 2.2.1**; *16*).

2. iS3T3 (*see* **Section 2.2.1**; *16*).

3. Culture media:

   a. Complete rcDME-Ham (*see* **Section 2.1.5**)

   b. Complete hcDME-Ham (*see* **Section 2.1.6**)

4. Aprotinin (cat. no. A3428, Sigma Chemicals). Dissolve 5 mg/mL of aprotinin in DME (*see* **Section 2.1.1**; *1.a*) to make a 200× stock solution. Sterilize by filtration through a 0.22 µm low-binding disposable filter, distribute in single use aliquots, and store at −20°C.

5. Tisseel® kit VH (cat. no. B19312-40, Baxter Hyland Immuno, Mississauga, Ont., Canada). Solution A is obtained by diluting the original thrombin stock solution included in the kit to a final concentration of 3 IU/mL in saline solution 1 (1.1% (w/v) NaCl and 1 mM CaCl$_2$ in apyrogenic ultrapure water). Solution B is prepared by mixing 1 mL of the original fibrinogen stock solution included in the kit with 1.16 mL of saline solution 2 (2% (w/v) NaCl and 1.86 mM CaCl$_2$ in apyrogenic ultrapure water).

6. 30 mm diameter plastic ring (homemade HDPE (high-density polyethylene)) previously glued with silicone (Silicone Rubber Adhesive Sealant, 1200 Series Sealant, GE Silicones, Waterford, NY; available in local hardwares) in a 60 mm Petri dish (size: 60 mm × 15 mm, cat. no. 08-757-13A, Fisher Scientific). Sterilize with ethylene oxide.

3. Methods

3.1. Monolayer Culture of Human Keratinocytes

3.1.1. Extraction and Culture of Human Keratinocytes

1. Human keratinocytes are isolated from normal adult or newborn skin (*see* **Note 7**).

2. Transport and conservation: In the surgery room, put the skin specimen into a sterile container filled with cold (4°C) transport medium.
    All further manipulations are performed under a sterile laminar flow hood cabinet.

3. Wash the skin specimen in a 50 mL tube containing 30 mL PBS-P/G/F. Agitate vigorously. With sterile forceps, transfer the skin specimen in another PBS-P/G/F tube. Repeat this step three times.

4. Spread out the skin specimen, epidermis on the top, into a 100 mm Petri dish.

5. Cut the skin in 3 mm × 100 mm pieces with scalpel (blade 22).

6. Add 10 mL of cold (4°C) thermolysin. Seal the Petri dish with parafilm.

7. Incubate overnight at 4°C.

8. With two curved forceps, separate the epidermis from the dermis. Put epidermal pieces within a trypsinization unit

containing 20 mL of warm (37°C) trypsin/EDTA. Incubate under agitation during 15–30 min at 37°C.

9. Collect the cell suspension. Put into a 50 mL tube and add 20 mL of warm (37°C) DME-Ham supplemented with 5% (v/v) fetal clone II. Wash the trypsinization unit with 10 mL of DME-Ham 5% (v/v) fetal clone II and add it to the tube (complete the volume to a total of 40 mL).

10. Count the cells and measure the viability by trypan blue staining. The cell viability should be superior to 80%.

11. Centrifuge cell suspension at 300 g for 10 min at room temperature.

12. Seed 1×10^6 keratinocytes and 1.5×10^6 iS3T3 by 75 cm^2 culture flask with 20 mL of complete hkDME-Ham. Incubate in 8% CO_2, 100% humidity atmosphere at 37°C. Change culture medium three times a week.

13. When the keratinocytes reach 80% confluency (see **Fig. 15.1A**), subculture (see **Section 3.1.2**) or freeze (see **Section 3.1.3**) cells.

3.1.2. Subculture of Human Keratinocytes (Passage)

For a 75 cm^2 culture flask of keratinocytes:

All further manipulations are performed under a sterile laminar flow hood cabinet.

1. Remove medium.
2. Wash the culture flask with 2 mL of warm (37°C) trypsin/EDTA and remove it.
3. Add 8 mL of trypsin/EDTA. Incubate at 37°C until the cells are detached from the flask.
4. Add 8 mL of complete hkDME-Ham (37°C). Collect the cell suspension. Put into a 50 mL tube. Wash the flask with 2 mL of complete hkDME-Ham, collect the cell suspension, and add it to the 50 mL tube (total 18 mL).
5. Count the cells and measure the viability by trypan blue staining. The cell viability should be superior to 80%.
6. Centrifuge cell suspension at 300 g for 10 min at room temperature.
7. Resuspend cell pellet in a given volume of complete hkDME-Ham.
8. Seed 2×10^5 to 7×10^5 keratinocytes and 1.5×10^6 iS3T3 by 75 cm^2 culture flask with 20 mL of complete hkDME-Ham. Incubate in 8% CO_2, 100% humidity atmosphere at 37°C. Change culture medium three times a week.

3.1.3. Cryopreservation of Human Keratinocytes

All further manipulations are performed under a sterile laminar flow hood cabinet.

1. Resuspend cells in a given volume of cold (4°C) freezing medium in order to obtain 2×10^6 cells/mL (*see* **Note 6**).
2. On ice, aliquot in cryogenic vials. Put them in a freezing container filled with 99% ethanol that has previously been cooled at −20°C.
3. Freeze overnight at −80°C, in the freezing container.
4. Store in liquid nitrogen.

3.1.4. Thawing of Human Keratinocytes

1. Put the cryogenic vial in 37°C water until only a small cluster of ice remains.
   All further manipulations are performed under a sterile laminar flow hood cabinet.
2. Using a pipette, put 1 mL of cold (4°C) complete hkDME-Ham in the cryogenic vial. When the ice is melted, put the cryogenic vial content in a tube containing 9 mL of cold complete hkDME-Ham.
3. Count the cells and measure the viability by trypan blue staining.
4. Centrifuge cell suspension at 300 g for 10 min at room temperature.
5. Resuspend cell pellet in a given volume of warm (37°C) complete hkDME-Ham.
6. Seed 2×10^5 to 7×10^5 keratinocytes and 1.5×10^6 iS3T3 by 75 cm² culture flask with 20 mL of complete hkDME-Ham. Incubate in 8% CO_2, 100% humidity atmosphere at 37°C. Change culture medium three times a week.

3.2. Monolayer Culture of Human Fibroblasts

3.2.1. Extraction and Culture of Human Fibroblasts

1. Human fibroblasts are obtained from normal adult breast skin (*see* **Note 7**).
2. Transport and conservation: In the surgery room, put the skin specimen into a sterile container filled with cold (4°C) transport medium.
   All further manipulations are performed under a sterile laminar flow hood cabinet.
3. Follow steps 3–5 of **Section 3.1.1**.
4. Add 10 mL of warm (37°C) thermolysin. Seal the Petri dish with parafilm.
5. Incubate 2–3 hr at 37°C.
6. With two curved forceps, separate the epidermis from the dermis. Put dermal pieces within a trypsinization unit containing 20 mL of warm (37°C) collagenase H. Incubate under agitation, overnight at 37°C.

7. Collect the cellular suspension. Put into a 50 mL tube and add 20 mL of fDME. Wash the trypsinization unit with 10 mL of fDME and add it to the tube (total 40 mL).
8. Count the cells and measure the viability by trypan blue staining.
9. Centrifuge cell suspension at 300 g for 10 min at room temperature.
10. Seed 5×10^5 fibroblasts in a 75 cm^2 culture flask with 15 mL of fDME. Incubate in 8% CO_2, 100% humidity atmosphere at 37°C. Change culture medium three times a week.
11. When the fibroblasts reach confluency, subculture (*see* **Section 3.2.2**) or freeze (*see* **Section 3.2.3**) cells.

3.2.2. Subculture of Human Fibroblasts (Passage)

For a 75 cm^2 culture flask of fibroblasts:

All further manipulations are performed under a sterile laminar flow hood cabinet.

1. Remove medium.
2. Wash the culture flask with 2 mL of trypsin/EDTA and remove it.
3. Add 3 mL of trypsin/EDTA. Incubate at 37°C until the cells are detached from the flask.
4. Add 3 mL of fDME. Collect the cellular suspension. Put into a 15 mL tube. Wash the flask with 2 mL of fDME, collect the cellular suspension and add it to the 50 mL tube (total 8 mL).
5. Count the cells and measure the viability by trypan blue staining. The cell viability should be higher than 95%. Usually, a confluent culture flask contains 4×10^6 cells.
6. Centrifuge cell suspension at 300 g for 10 min at room temperature.
7. Resuspend cell pellet in a given volume of fDME.
8. Seed 5×10^5 fibroblasts in a 75 cm^2 culture flask with 15 mL of fDME. Incubate in 8% CO_2, 100% humidity atmosphere at 37°C. Change culture medium three times a week.

3.2.3. Cryopreservation of Human Fibroblasts

All further manipulations are performed under a sterile laminar flow hood cabinet.

1. Resuspend fibroblasts in a given volume of cold (4°C) freezing medium in order to obtain 2×10^6 fibroblasts/mL (*see* **Note 6**).
2. On ice, aliquot in cryogenic vials. Put them in a Nalgene freezing container filled with 99% ethanol that has previously been cooled at −20°C.
3. Freeze overnight at −80°C, in the freezing container.

4. Store in liquid nitrogen.

3.2.4. Thawing of Human Fibroblasts

1. Put the cryogenic vial in 37°C water until a small cluster of ice remains.
   All further manipulations are performed under a sterile laminar flow hood cabinet.
2. Using a pipette, put 1 mL of cold (4°C) fDME in the cryogenic vial. When the ice is thawed, put cryogenic vial content in a tube containing 9 mL of cold fDME.
3. Count the cells and measure the viability by trypan blue staining.
4. Centrifuge cell suspension at 300 g for 10 min at room temperature.
5. Resuspend cell pellet in a given volume of warm (37°C) fDME.
6. Seed 5×10^5 fibroblasts in a 75 cm^2 culture flask with 15 mL of fDME. Incubate in 8% CO_2, 100% humidity atmosphere at 37°C. Change culture medium three times a week.

3.3. Monolayer Culture of Rabbit Corneal Epithelial Cells

3.3.1. Extraction and Culture of Rabbit Corneal Epithelial Cells

1. Rabbit corneal epithelial cells are obtained from rabbit postmortem eyes. *See* **Note 8**.
   All further manipulations are performed under a sterile laminar flow hood cabinet.
2. Wash the eye in a 50 mL tube containing 30 mL PBS-P/G/F. Agitate gently 1–2 min. With sterile forceps, transfer the eye in another tube filled with PBS-P/G/F. Repeat this step three times.
3. Place the eye into a 100 mm Petri dish.
4. Surround the eye with a folded sterile gauze. This helps in holding the eye without having to touch it.
5. With the #11 scalpel blade, make a small opening of 2–3 mm.
6. With curved scissors, cut-out the cornea in order to obtain only the limbus and central cornea. Avoid leaving sclera to eliminate conjunctival epithelial cell contamination.
7. With two curved forceps, peel-off the iris. Do this step while holding the cornea in the air to avoid any damage to the epithelium during the procedure.
8. Place the Cornea into a 35 mm Petri dish.

9. Add 5 mL of cold (4°C) dispase II. Seal the Petri dish with parafilm.
10. Incubate overnight at 4°C.
11. With two curved forceps, mechanically separate the epithelium from the stroma under a dissecting microscope.
12. Put the epithelium in a 35 mm Petri dish and cut into small pieces with a #22 scalpel blade (*see* **Note 9**).
13. With a plastic pipette, collect the pieces with 2×5 mL of warm (37°C) complete rcDME-Ham. Put into a 15 mL tube.
14. Centrifuge at 300 g for 10 min at room temperature.
15. Seed 1×10^6 epithelial cells and 1.5×10^6 iS3T3 by 75 cm^2 culture flask with 20 mL of complete rcDME-Ham. Incubate in 8% CO_2, 100% humidity atmosphere at 37°C. Change culture medium three times a week.
16. When the rabbit corneal epithelial cells reach 80% confluency, subculture (*see* **Section 3.3.2**) or freeze (*see* **Section 3.3.3**) cells.

3.3.2. Subculture of Rabbit Corneal Epithelial Cells (Passage)

For a 75 cm^2 culture flask of corneal epithelial cells: Follow steps 1–8 of **Section 3.1.2** but use complete rcDME-Ham instead of complete hkDME-Ham.

3.3.3. Cryopreservation of rabbit corneal epithelial cells

Follow steps 1–4 of **Section 3.1.3**.

3.3.4. Thawing of Rabbit Corneal Epithelial Cells

Follow steps 1–6 of **Section 3.1.4** but use complete rcDME-Ham instead of complete hkDME-Ham.

3.4. Monolayer Culture of Human Corneal Epithelial Cells

3.4.1. Extraction and Culture of Human Corneal Epithelial Cells

1. Human corneal epithelial cells are obtained from human postmortem donor corneas unsuitable for transplantation (*see* **Note 7**).
2. Follow steps 1–16 of **Section 3.3.1** except for step 8. In fact, at step 8, place the limbal ring into a 35 mm Petri dish. The limbal ring is obtained by separating the limbus from the central cornea with a 7.5 mm diameter trephine.

3.4.2. Subculture of Human Corneal Epithelial Cells (Passage)

For a 75 cm^2 culture flask of corneal epithelial cells: Follow steps 1–8 of **Section 3.1.2** but use complete hcDME-Ham instead of complete hkDME-Ham.

3.4.3. Cryopreservation of Human Corneal Epithelial Cells	Follow steps 1–4 of **Section 3.1.3**.
3.4.4. Thawing of Human Corneal Epithelial Cells	Follow steps 1–6 of **Section 3.1.4** but use complete hcDME-Ham instead of complete hkDME-Ham.

3.5. Human Skin Reconstruction by the Self-Assembly Approach

3.5.1. Assembly of Fibroblast Sheets for Dermal Reconstruction	All further manipulations are performed under a sterile laminar flow hood cabinet. 1. Seed 2×10^5 fibroblasts by 25 cm² culture flask in 5 mL of fDME containing 50 µg/mL ascorbic acid. Incubate in 8% CO_2, 100% humidity atmosphere at 37°C. Change culture medium three times a week. 2. After 28–35 days, open the top of the flask with a soldering iron. 3. Using curved forceps, detach carefully one fibroblast sheet. Transfer it into a 100 mm cell culture dish. 4. Anchor peripherally the fibroblast sheet with ingots. Move one-by-one ingots towards the tissue periphery to flatten the fibroblast sheet. 5. Detach a second fibroblast sheet and transfer it on the first fibroblast sheet. Repeat step 4 of this section. 6. Place a Merocel® sponge (that has been cut to fit within the ingots) on top of the superimposed fibroblast sheets. Keep Merocel® sponge in place with ingots. 7. Add 25 mL of fDME containing 50 µg/mL ascorbic acid. Incubate in 8% CO_2, 100% humidity atmosphere at 37°C. 8. After 2 days, remove Merocel® sponge and sponge ingots, and put an anchoring ring on the fibroblast sheets to allow epithelial cell seeding. Incubate in 8% CO_2, 100% humidity atmosphere at 37°C. Change culture medium three times a week.
3.5.2. Assembly of Reconstructed Skin	All further manipulations are performed under a sterile laminar flow hood cabinet. 1. One week after the stacking of fibroblast sheets, remove culture medium and seed 8×10^5 keratinocytes within the anchoring ring (from a suspension of 1.6×10^6 cells/mL of complete hkDME-Ham). Incubate in 8% CO_2, 100% humidity atmosphere at 37°C.

2. Two hours later, add 25 mL of complete hkDME-Ham containing 50 μg/mL ascorbic acid. Incubate in 8% CO_2, 100% humidity atmosphere at 37°C. Change culture medium three times a week.

3. 24 hours later, anchoring ring can be removed.

3.5.3. Maturation of the Reconstructed Skin: Air–Liquid Interface Culture

All further manipulations are performed under a sterile laminar flow hood cabinet.

1. One week after the seeding of keratinocytes, remove culture medium and ingots.
2. Using curved forceps, detach carefully the reconstructed skin from the bottom of the cell culture dish.
3. Place the reconstructed skin on the anchoring paper; centering it in order that the area with keratinocytes is on top of the aperture.
4. Put an air–liquid stand in a Petri dish.
5. Lift reconstructed skin together with the anchoring paper and transfer it on the air–liquid stand.
6. Add 25 mL of alkDME-Ham containing 50 μg/mL ascorbic acid (*see* **Note 10**). Incubate in 8% CO_2, 100% humidity atmosphere at 37°C. Change culture medium three times a week.

   Note: The reconstructed skin can be cultivated at the air–liquid interface more than 28 days. A well-organized basement membrane is obtained after 21 days of culture at the air–liquid interface (11, 12).

3.6. Corneal Epithelium Reconstruction on Fibrin Gel

3.6.1. Fibrin Gel Production

All further manipulations are performed under a sterile laminar flow hood cabinet.

1. The fibrin gel is produced by adding 330 μL of each solution (solution A: thrombin and solution B: fibrinogen, *see* **Section 2.7; 5**) using the Duploject-System, included in the Tisseel® kit VH, in a 30 mm diameter plastic ring laid in a 60 mm Petri dish.
2. The Petri dish is left at room temperature for 10–15 min (until a complete substrate polymerization is obtained) and stored overnight at 37°C.

3.6.2. Corneal Epithelial Cells Seeding over the Fibrin Gel

All further manipulations are performed under a sterile laminar flow hood cabinet.

1. Seed 8.4×10^4 corneal epithelial cells (1.68 mL from a suspension of 5.0×10^4 cells/mL of complete rcDME-

Ham or hcDME-Ham) and 1.4×10^5 iS3T3 (1.4 mL from a suspension of 1.0×10^5 cells/mL of complete rcDME-Ham or hcDME-Ham) within the plastic ring. Incubate in 8% CO_2, 100 % humidity atmosphere at 37°C.

2. Change culture medium three times a week (3 mL of complete rcDME-Ham or hcDME-Ham containing 25 µg/mL aprotinin). *See* **Note 11**.

4. Notes

1. Serum must be added first followed by insulin. Insulin must be added with a sterile plastic pipette.
2. Ascorbic acid must be prepared immediately before use and protected from light.
3. This solution must be prepared the day of its use.
4. When exposed to light, HEPES buffer may undergo degradation and become toxic.
5. Irradiated S3T3 could be kept 1 week in 8% CO_2, 100% humidity atmosphere at 37°C. However, their number may fall by approximately 15% per day.
6. DMSO is an oxidative agent toxic for the cells, especially at temperature above 10°C. Thus, it must be used at 4°C.
7. Protocols were approved by the institution's committee for the protection of human subjects.
8. Protocols were approved by the institution's animal care and use committee (Comité de Protection des Animaux de Laboratoire de l'Université Laval).
9. Change scalpel blade and clean other instruments (forceps) between eye specimens.
10. The lower surface of the reconstructed skin must be in direct contact with culture medium.
11. A solution of aprotinin is added to the culture medium to prevent fibrinolysis. Aprotinin must be added to the culture medium just before use.

Acknowledgments

The authors would like to thank current and former members of the LOEX laboratory who have contributed to develop the foregoing protocols.

References

1. Benitah, S. A., Frye, M., Glogauer, M., and Watt, F. M. (2005) Stem cell depletion through epidermal deletion of Rac1 *Science* **309,** 933–935.
2. Bickenbach, J. R., McCutecheon, J., and Mackenzie, I. C. (1986) Rate of loss of tritiated thymidine label in basal cells in mouse epithelial tissues *Cell Tissue Kinet* **19,** 325–333.
3. Cotsarelis, G., Sun, T. T., and Lavker, R. M. (1990) Label-retaining cells reside in the bulge area of pilosebaceous unit: implications for follicular stem cells, hair cycle, and skin carcinogenesis *Cell* **61,** 1329–1337.
4. Kruse, F. E. (1994) Stem cells and corneal epithelial regeneration *Eye* **8** (Pt 2), 170–183.
5. Blanpain, C., Lowry, W. E., Geoghegan, A., Polak, L., and Fuchs, E. (2004) Self-renewal, multipotency, and the existence of two cell populations within an epithelial stem cell niche *Cell* **118,** 635–648.
6. Morris, R. J., Liu, Y., Marles, L., Yang, Z., Trempus, C., Li, S., Lin, J. S., Sawicki, J. A., and Cotsarelis, G. (2004) Capturing and profiling adult hair follicle stem cells *Nat Biotechnol* **22,** 411–417.
7. Tumbar, T., Guasch, G., Greco, V., Blanpain, C., Lowry, W. E., Rendl, M., and Fuchs, E. (2004) Defining the epithelial stem cell niche in skin *Science* **303,** 359–363.
8. Michel, M., Torok, N., Godbout, M. J., Lussier, M., Gaudreau, P., Royal, A., and Germain, L. (1996) Keratin 19 as a biochemical marker of skin stem cells in vivo and in vitro: keratin 19 expressing cells are differentially localized in function of anatomic sites, and their number varies with donor age and culture stage *J Cell Sci* **109** (Pt 5), 1017–1028.
9. Auger, F. A., Rémy-Zolghadri, M., Grenier, G., and Germain, L. (2000) The self-assembly approach for organ reconstruction by tissue engineering *e-biomed* **1,** 75–86.
10. Germain, L., Berthod, F., Moulin, V., Goulet, F., and Auger, F. A. (2004) Principles of living organ reconstruction by tissue engineering *in* "Tissue Engineering and Novel Delivery Systems" (Yaszemski, M. J., Trantolo, D. J., Lewandrowski, K.-W., Hasirci, V., Altobelli, D. E., and Wise, D. L., Eds.), pp. 197–228, Marcel Dekker, New-York.
11. Michel, M., L'Heureux, N., Pouliot, R., Xu, W., Auger, F. A., and Germain, L. (1999) Characterization of a new tissue-engineered human skin equivalent with hair *In Vitro Cell Dev Biol Anim* **35,** 318–326.
12. Pouliot, R., Larouche, D., Auger, F. A., Juhasz, J., Xu, W., Li, H., and Germain, L. (2002) Reconstructed human skin produced in vitro and grafted on athymic mice *Transplantation* **73,** 1751–1757.
13. Gaudreault, M., Carrier, P., Larouche, K., Leclerc, S., Giasson, M., Germain, L., and Guerin, S. L. (2003) Influence of sp1/sp3 expression on corneal epithelial cells proliferation and differentiation properties in reconstructed tissues *Invest Ophthalmol Vis Sci* **44,** 1447–1457.
14. Masson-Gadais, B., Fugere, C., Paquet, C., Leclerc, S., Lefort, N. R., Germain, L., and Guerin, S. L. (2006) The feeder layer-mediated extended lifetime of cultured human skin keratinocytes is associated with altered levels of the transcription factors Sp1 and Sp3 *J Cell Physiol* **206,** 831–842.
15. Robitaille, H., Proulx, R., Robitaille, K., Blouin, R., and Germain, L. (2005) The mitogen-activated protein kinase kinase kinase dual leucine zipper-bearing kinase (DLK) acts as a key regulator of keratinocyte terminal differentiation *J Biol Chem* **280,** 12732–12741.
16. Rama, P., Bonini, S., Lambiase, A., Golisano, O., Paterna, P., De Luca, M., and Pellegrini, G. (2001) Autologous fibrin-cultured limbal stem cells permanently restore the corneal surface of patients with total limbal stem cell deficiency *Transplantation* **72,** 1478–1485.
17. Talbot, M., Carrier, P., Giasson, C. J., Deschambeault, A., Guerin, S. L., Auger, F. A., Bazin, R., and Germain, L. (2006) Autologous transplantation of rabbit limbal epithelia cultured on fibrin gels for ocular surface reconstruction *Mol Vis* **12,** 65–75.
18. Germain, L., Rouabhia, M., Guignard, R., Carrier, L., Bouvard, V., and Auger, F. A. (1993) Improvement of human keratinocyte isolation and culture using thermolysin *Burns* **19,** 99–104.
19. Germain, L., Auger, F. A., Grandbois, E., Guignard, R., Giasson, M., Boisjoly, H., and Guerin, S. L. (1999) Reconstructed human cornea produced in vitro by tissue engineering *Pathobiology* **67,** 140–147.

Part V

Regeneration of the Musculoskeletal System

Chapter 16

Prospective Isolation of Mesenchymal Stem Cells from Mouse Compact Bone

Brenton J. Short, Nathalie Brouard, and Paul J. Simmons

Abstract

Bone marrow from numerous species, including rodents and man, has been shown to contain a rare population of cells known as marrow stromal cells or mesenchymal stem cells (MSC). Given the innate ability of these cells to give rise to multiple tissue types including bone, fat and cartilage, there is considerable interest in utilizing MSC in a broad repertoire of cell-based therapies for the treatment of human disease. In order for such therapies to be realized, a preclinical animal model in which to refine strategies utilizing MSC is required.

We have described methodology allowing for the prospective isolation by fluorescence activated cell sorting (FACS) of a highly purified population of MSC from murine compact bone (CB). These cells are multipotent and capable of extensive proliferation in vitro and thus represent an ideal source of cells with which to explore both the fundamental biology of MSC and their efficacy in a variety of cellular therapies.

Key words: Mesenchymal Stem Cell, MSC, CFU-F, multipotent, FACS, compact bone.

1. Introduction

In contrast to the defined phenotype of human MSC *(1, 2)*, neither the stem nor early progenitor cell populations of the murine stromal system have been well characterized either morphologically or at the molecular level. This, coupled with their extremely low incidence *(3)* and a lack of knowledge of their precise location within the marrow, has meant that much of our current understanding of stromal precursor cells has arisen from in vitro assays and culture manipulations. Pioneering studies conducted by Friedenstein and colleagues demonstrated that the explantation of single cell suspensions of bone marrow (BM) at appropriate densities (10^4–10^5 cells/cm^2) resulted in the outgrowth of adherent colonies of cells morphologically resembling fibroblasts (4). Originally termed fibroblast colony-forming cells, F-CFC, the clonogenic stromal precursor cells initiating these colonies have more recently been termed colony-forming unit-fibroblasts (CFU-F).

We have shown that the CB rather than the BM is the major source of MSC in the adult mouse. Enzymatic digestion of bone fragments followed by depletion of mature hematopoietic cells (Lin-) and subsequent FACS allows the resolution of a population of cells with the composite phenotype $Lin^-CD45^-CD31^-Sca-1^+$.

2. Materials

2.1. Mice

These protocols were developed from experiments performed using cells derived from specific pathogen-free (SPF) C57BL/6 J (Ly 5.2) mice. All animals were maintained in a clean, conventional animal facility and were fed sterile mouse pellets and acidified water *ad libitum*.

2.2. Isolation of CB-Derived Cells

1. Phosphate-buffered saline (PBS) supplemented with 2% (v/v) fetal bovine serum (FBS, JRH Biosciences, Lenexa, KS), sterile filtered, and stored at 4°C.
2. 70% (v/v) Ethanol solution (BDH Chemicals Ltd, Poole, England).
3. A 3 mg/ml solution of Type 1 collagenase (Worthington Biochemical Corporation, Lakewood, NJ). The collagenase solution should be prepared in PBS and sterile filtered prior to use and can be either prepared fresh or stored frozen. Collagenase should be warmed to 37°C prior to use.

2.3. Lineage Depletion of CB-Derived Cells

1. Phosphate buffered saline (PBS) supplemented with 2% fetal bovine serum (PBS2%).
2. Purified rat antibodies to murine CD3, CD4, CD5, CD8, CD11b (Mac-1), Gr-1, B220, and Ter-119 (BD Pharmingen, Franklin Lakes, NJ) diluted accordingly (*see* **Note 1**) into a single "lineage cocktail." Lineage antibody cocktail should be freshly prepared for each experiment.
3. Sheep-anti-rat IgG Dynabeads (Dynal Biotech ASA, Oslo, Norway)

2.4. Purification of CB-Derived MSC by FACS

1. Phosphate buffered saline (PBS) supplemented with 2% (v/v) fetal bovine serum.
2. FITC-conjugated rat anti-mouse stem cell antigen-1 (Sca-1), PE-conjugated rat antibodies to mouse-platelet endothelial

cell molecule-1 (PECAM-1/CD31), and CD45 (BD Pharmingen, Franklin Lakes, NJ) diluted accordingly (*see* **Note 2**).

3. FACS analysis buffer consisting of PBS2% containing a 1/100 (v/v) dilution of 7-amino actinomycin D (7AAD, Sigma, St Louis, MO).

2.5. Cell Culture

1. Alpha modification of Eagle's medium (α-MEM) (ICN Biomedicals, Inc, Aurora, OH) supplemented with 20% v/v FBS (Hyclone, South Logan, UT).

2. GLUTAMAX-1 (Invitrogen Corporation, Carlsbad, CA), purchased as a 100× stock solution and stored in 5 ml aliquots at −20°C.

3. Penicillin–streptomycin (Invitrogen Corporation, Carlsbad, CA), purchased as a 100× stock solution comprising 10,000 U/ml penicillin G sodium and 10,000 μg/ml streptomycin sulfate in 0.85% saline and stored at −20°C. Primary cultures are initiated and maintained in antibiotic-containing media for approximately 2 weeks.

4. Sodium pyruvate (Invitrogen Corporation, Carlsbad, CA), purchased as a 100×, 100 mM stock solution and stored in 5 ml aliquots at −20°C.

5. Trysin–EDTA (Invitrogen Corporation, Carlsbad, CA) was purchased as a 10× stock comprising 0.5% trypsin, 5.3 mM EDTA–4Na and stored frozen in 1 ml aliquots. A 1× working solution made by the addition of 9 ml of PBS can be stored for up to 1 week at 4°C.

3. Methods

3.1. Sampling of Bones

1. Kill mice by cervical dislocation and apply a 70% v/v EtOH solution liberally to the lower half of the animal.

2. Excise the tibiae, femurs, and iliac crests (*see* **Note 3**) and clean thoroughly with a sterile #11 scalpel to remove excess muscle tissue and tendons. Remove epiphyses and keep bones in an ice-cold PBS containing PBS2% prior to further processing.

3.2. Isolation of Compact Bone-Derived Cells

1. Place bones in a mortar containing approximately 10 ml of ice-cold PBS2% and crush with a pestle to free the marrow (*see* **Note 4**). Wash bone fragments with several changes of PBS2% with gentle agitation.

2. Once the majority of the marrow has been removed (as assessed by the color of the bones) transfer the bone fragments

to 100 mm petri dishes and cover in a 3 mg/ml solution of Type 1 collagenase for 5 min.

3. Chop bones into small fragments using a sterile # 22 scalpel (*see* **Note 5**) and transfer to a 50 ml polypropylene tube. Add further collagenase such that the final volume is approximately 2 ml per each mouse used for the experiment.

4. Seal the lid of the tube with parafilm and place the tube on a 37°C shaking platform at 240 rpm for 45 min.

5. Following digestion, make the volume up to 50 ml with ice-cold PBS2% and allow bone fragments to settle for 5 min prior to collection of supernatant. Repeat with a further 50 ml PBS2% to ensure maximal cell yield (*see* **Note 6**).

6. Filter the resultant cell suspension through a 70 µm nylon cell strainer (BD Falcon, Bedford, MA) and centrifuge at 400 g (all subsequent centrifugation steps should be performed at this speed). Transfer the supernatant to fresh tubes and spin again to ensure maximal cell recovery.

7. Discard supernatant and combine cells in a single sterile 5 ml polypropylene tube. Store on ice for further processing.

3.3. Lineage Depletion of Compact Bone-Derived Cells

1. Count viable cells obtained in **Section 3.2** using a viability dye such as trypan blue (*see* **Note 7**). Typical yields are between 1.5 and 2×10^6 CB cells per mouse.

2. Prepare the lineage depletion antibody cocktail described in **Section 2.3**. Add 50 µl of the aforementioned cocktail per 5×10^6 viable CB cells and incubate on ice for 25 min.

3. Wash cells twice with ice-cold PBS2%, aspirate supernatant and store on ice in a minimal volume.

4. Aliquot Dynabeads into a 5 ml polypropylene tube. The total number of beads used should be twice the total number of cells being depleted (Dynabead solution comprises 4×10^8 cells/ml). As the beads are added in two stages, the initial depletion step thus has a 1:1 bead:cell ratio with the subsequent step having a great excess of beads to remove cells expressing lower levels of lineage antigens.

5. Wash beads twice on a Dynal Magnetic Particle Concentrator (Dynal MPC-L, Dynal Biotech) with 3 ml PBS2%. Resuspend beads in 400 µl PBS2%.

6. Add half the beads (200 µl) to the CB cell pellet and incubate on ice with gentle agitation for 5 min.

7. Place cells on the Dynal MPC for 1 min to facilitate clearance of bead-bound lineage positive cells. Transfer non-bound cells to the tube containing the remaining 200 µl of beads.

8. Gently re-suspend the bead bound lineage positive cells in 2 ml PBS2% and place the cell suspension back on the Dynal MPC. Collect the non-bound cells and add to the tube containing the beads and lineage negative cells previously collected.

9. Parafilm the lid of the tube containing the depleted cells and place on a rotator at 4°C for 25 min.

10. Place tube on MPC for 1 min to remove bead-bound cells. Transfer non-bound cells to a fresh 5 ml polypropylene tube.

11. Add 2 ml PBS2% to the bead-bound lineage positive cells, and re-suspend gently. Repeat clearance of lineage positive cells and combine supernatant with that obtained in the first clearance.

12. Count viable lineage cells using a viability dye such as trypan blue (*see* **Note 8**) and store on ice.

3.4. Purification of Compact Bone-derived MSC by FACS

1. The lineage depleted CB cells are now prepared for FACS analysis. Aliquot an appropriate number of cells (25,000–30,000) into sterile polystyrene tubes for use as isotype controls (*see* **Note 9**), retaining the rest of the cells (the "sort sample") in the 5 ml polypropylene tube.

2. Centrifuge all cells and aspirate supernatant leaving approximately 250 µl of PBS2% in the tubes.

3. Re-suspend cells and add antibodies to appropriate control tubes (*see* **Note 2**) and store on ice under foil or in the dark.

4. Add test antibodies (Sca-1-FITC, CD45-PE, and CD31-PE) to the sort sample (*see* **Notes 2** and **10**) and store on ice under foil or in the dark.

5. Following a 25 min incubation, wash cells twice with 4 ml ice-cold PBS2%.

6. Re-suspend cells in either PBS2% (unstained control cells and all fluorochrome-conjugated controls) or FACS analysis buffer (7AAD control and sort sample), adding 500 µl to control tubes and 1 ml per 10^6 cells in the sort sample, which may need to be divided between multiple tubes.

7. Immediately prior to FACS analysis the sort sample should be passed through a 70 µm cell strainer to remove any clumps that may have formed and to prevent blockages.

8. Samples are now analyzed by FACS using a FACSDiva (Becton Dickinson, San Jose, CA). Unstained cells are used to set forward and side-scatter (FSC and SSC respectively), whilst the isotype negative controls are used to determine background fluorescence. Single color positive controls (CD45-PE/FITC and Sca-1-FITC) are used in setting the compensation.

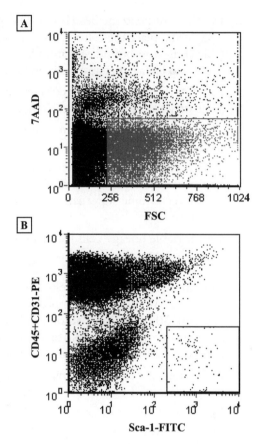

Fig. 16.1. Isolation of Lin⁻CD45⁻CD31⁻Sca-1⁺ MSC by FACS. Compact bone-derived cells were lineage depleted as described and stained with PE-conjugated anti-CD45 and CD31, FITC-conjugated anti-Sca-1, and 7AAD to exclude non-viable cells. (**A**) A dot-plot displaying forward scatter versus 7AAD fluorescence is used to select viable cells for analysis (*boxed region*). (**B**) Analysis of PE versus FITC fluorescence resolves a population of MSC lacking expression of both CD45 and CD31 and expressing high levels of Sca-1 (*boxed region*).

9. To set gates for sorting, a dot-plot of FSC versus the viability dye 7AAD is employed as shown in **Fig. 16.1A**. The gate should exclude both dead cells which have incorporated 7AAD and non-cellular debris (*see* **Note 11**).

10. Viable cells gated as described are subsequently displayed on a dot-plot showing PE versus FITC fluorescence. A gate selecting CD45/CD31⁻ Sca-1⁺ MSC is set (**Fig. 16.1B**) and these cells are collected into a tube containing α-MEM supplemented with 20% FBS.

3.5. Culture of Murine CB-Derived MSC

1. Following isolation of MSC by FACS as described, cells are counted prior to plating at appropriate densities (*see* **Note 12**) in tissue culture plates or flasks (Cellstar, Greiner Bio-One, Kremsmuenster, Austria) in α-MEM containing 20% v/v Hyclone FBS and supplemented with 1× sodium pyruvate, glutamax, and antibiotics.

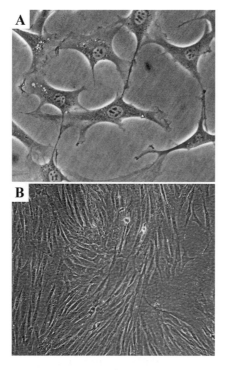

Fig. 16.2. Morphology of cultured Lin⁻CD45⁻CD31⁻Sca-1⁺ MSC. (**A**) Phase contrast analysis of clonal population of primary MSC after 7 days culture in α-MEM containing 20% v/v hyclone FBS in "triple mix" gas containing 5% O_2 (40× magnification). (**B**) Passage 1 MSC at 90% confluence (10× magnification).

2. Cells should be cultured in a 37°C incubator gassed with "Triple Mix" gas mixture comprising 5% O_2, 10% CO_2, and 85% N_2 (*see* **Note 13**).

3. For primary cultures, cells are cultured as colonies for 10–12 days prior to passaging by detachment with a 0.05% trypsin–EDTA solution. Subsequently, cultures should be split 1:4 once they reach 80–90% confluence (**Fig. 16.2B**). Antibiotic free media should be used if cells are to be passaged extensively.

4. Notes

1. We routinely use Pharmingen antibodies at a final dilution of 1/250, however you may wish to titrate each antibody individually on BM to ensure optimum efficiency.

2. Different antibody dilutions are needed following lineage depletion as there are far fewer cells to be stained. We routinely use 0.5 µl of fluorochrome-conjugated antibody per 10^6 lineage negative cells or a minimum of 1 µl if fewer than 10^6 cells are obtained. For isotype controls where few cells are

labeled, a 1/500 dilution (0.5 µl of antibody in 250 µl) is sufficient.

3. When harvesting bones, make an incision in the skin over the gut and carefully peel the skin from the lower half of the mouse. This removes the fur and helps prevent sample contamination. Collect all three bones together by cutting (with scissors) along the spine up from the tail then separate and clean individually. The iliac crest is difficult to clean; cut in half above the socket for the head of the femur and discard the lower portion. Tibias and femurs can be cleaned efficiently by holding one end of the bone with forceps and scraping along the bone with a scalpel blade.

4. When crushing bones, it is important to use the least amount of force possible, the objective being to crack the bones and release the marrow rather than to pulverize the bones. Do not use a circular "grinding" motion, rather apply downwards pressure to the pestle to fragment the bones.

5. When cutting bones, rather than using direct downwards force use the curved blade of the scalpel to rock back and forth across the bones. Continue cutting until the bones are in fragments 1–2 mm in size.

6. When collecting supernatant from digested bones, do not use vigorous agitation to break up any clumps that have formed during digestion. Digested matrix proteins and dead cells may be released that will contaminate the cell fraction.

7. Cell counts should be performed manually with a hemocytometer as bone particles may clog automatic cell counting machines as well as lead to inaccurate counts. Care should be taken to discriminate between cells and debris as bone fragments can look deceptively like cells.

8. Typically, lineage depletion will remove between 95 and 98% of the initial CB fraction so at this stage a 1:1 dilution of the cell solution in trypan blue may be sufficient to obtain an accurate count.

9. You will require several different control tubes which to ensure correct setup for FACS. You should have an unstained sample, one containing cells with the viability dye (7AAD) only, isotype matched control antibodies for Sca-1, CD45, and CD31 as well as positive control antibodies. For positive controls useful in compensation settings, you may wish to use CD45-PE and CD45-FITC, as the majority of lineage negative cells are primitive hematopoietic cells expressing CD45 and thus provide a strong signal. A tube containing cells labeled with Sca-1-FITC should be used to fine tune the compensation settings as the level of expression of this antigen is typically a log brighter than that of CD45.

10. If you wish to sort other CB-derived populations in addition to MSC, you may wish to use separate fluorochromes for CD31 and CD45. In this protocol, the two antigens are labeled with the same fluorochrome as an easy means of excluding cells expressing these markers. The Lin$^-$CD31$^+$ fraction represents vascular endothelial cells whilst the Lin$^-$CD45$^+$ cells are primitive hematopoietic cells.

11. When setting forward scatter, it is important to note that this protocol generates samples contaminated with bone fragments. FSC settings should be such that this debris is excluded as in general the bone debris is significantly smaller than the cellular fraction (see **Fig. 16.1A**).

12. Sorted Lin-CD45/CD31$^-$Sca-1$^+$ MSC exhibit a cloning frequency of approximately one in two cells plated and are clonogenic in the absence of feeder cells or exogenous mitogens. MSC can be cloned effectively using the automatic cell deposition function (ACDU) of the FACS or can be manually plated at clonal density and individual CFU-F subsequently collected.

13. Culture of murine MSC in low oxygen tension is a critical factor in the successful culture of these cells. Comparison of CFU-F formation between CB-derived cells cultured in a standard gas mixture containing 10% O_2 versus 5% O_2 reveals a seven-fold increase in the number of colonies when grown in low oxygen conditions. If you cannot dedicate an entire incubator to these conditions, an alternate method is to culture the cells in humidified, air-tight plastic containers gassed from a cylinder containing the "triple mix" gas. To do this, place some distilled water in a plastic container with a tight fitting lid and sit a wire rack in the bottom upon which your cell culture vessels can rest. Make two small holes in opposite corners of the lid into which a 21-gauge needle can fit. Connect the cylinder to a length of flexible tubing and secure a 0.22 µm syringe filter into the end of the tubing. Attach a 21-gauge needle to the filter and ensure that the gas is flowing at a low rate. Insert needle into one of the holes in the lid and gas for approximately 10 min. Seal the holes with tape (you may also need to tape around the lid to ensure the seal is air-tight) and place the sealed container in a 37°C incubator or oven.

Acknowledgements

The authors would like to thank Mr Ralph Rossi for his assistance with the fluorescent activated cell sorting (FACS).

References

1. Simmons, P.J. and B. Torok-Storb (1991). *Identification of stromal cell precursors in human bone marrow by a novel monoclonal antibody, STRO-1*. Blood, 78(1): 55–62.
2. Gronthos, S., et al., (2003). *Molecular and cellular characterisation of highly purified stromal stem cells derived from human bone marrow*. J Cell Sci, 116(Pt 9): 1827–35.
3. Phinney, D.G., et al., (1999). *Plastic adherent stromal cells from the bone marrow of commonly used strains of inbred mice: variations in yield, growth, and differentiation*. J Cell Biochem, 72(4): 570–85.
4. Friedenstein, A.J., J.F. Gorskaja, and N.N. Kulagina (1976). *Fibroblast precursors in normal and irradiated mouse hematopoietic organs*. Exp Hematol, 4(5): 267–74.

Chapter 17

Isolation, Propagation, and Characterization of Human Umbilical Cord Perivascular Cells (HUCPVCs)

Rahul Sarugaser, Jane Ennis, William L. Stanford, and John E. Davies

Abstract

Current sources of mesenchymal cells, including bone marrow, fat and muscle, all require invasive procurement procedures, and provide relatively low frequencies of progenitors. Here, we describe the non-invasive isolation, and characterization, of a rich source of mesenchymal progenitor cells, which we call human umbilical cord perivascular cells (HUCPVCs). HUCPVCs show a similar immunological phenotype to bone marrow-derived mesenchymal stromal cells (BM-MSCs), since they are non-alloreactive, exhibit immuno-suppression, and significantly reduce lymphocyte activation, in vitro. They present a non-hematopoietic myofibroblastic mesenchymal phenotype (CD45-, CD34-, CD105+, CD73+, CD90+, CD44+, CD106+, 3G5+, CD146+); with a 1:300 frequency at harvest, a short-doubling time, and a clonogenic frequency of >1:3 in culture. Furthermore, in addition to robust quinti-potential differentiation capacity in vitro, HUCPVCs have been shown to contribute to both musculo-skeletal and dermal wound healing in vivo.

Key words: Mesenchymal progenitor cell, mesenchymal stem cell, umbilical cord.

1. Introduction

Mesenchymal progenitor cells (MPCs), or mesenchymal stem/stromal cells (MSCs) have classically been obtained from bone marrow (BM), and more recently from adipose and muscle tissue. Although these cells have been shown to differentiate in vitro into various mesenchymal lineages, a major limitation is availability. The harvest of BM is a highly invasive procedure, and the number and maximal life span of BM-MSCs declines with the increasing age of the donor (1). The use of alternative MSC sources is thus required for the large-scale expansion necessary to produce the number of cells needed for clinical applications. Here we describe a unique, easily harvested, rapidly proliferating, **h**uman **u**mbilical cord perivascular cell (HUCPVC) population that provides high yields of MSCs (2). HUCPVCs are acquired through digestion of Wharton's jelly of the human umbilical cord. The unique dissection procedure specifically targets the perivascular cells, yielding a high-colony forming unit-fibroblast (CFU-F) frequency of 1:300, significantly higher than that observed in either neonatal bone marrow ($1:10^4$) (1) or umbilical cord blood ($1:5 \times 10^8$) (3). These highly proliferative cells exhibit a doubling time of 20 hr (serum-dependent), and a >1:3 clonogenic frequency in culture.

HUCPVCs present a myofibroblastic phenotype in culture, expressing high levels of α-smooth muscle-actin, desmin, vimentin, and the pericyte marker 3G5 (4). HUCPVCs have been characterized as CD45-, CD34-, CD105+, CD73+, CD90+, CD44+, CD106+, CD49e+, CD73+, CD146+, Oct4-, and telomerase negative, with expression levels similar to those observed in BM-MSCs (5). Also similar to BM-MSCs (6), HUCPVCs have an immunoprivileged phenotype in vitro, in that they do not cause lymphocyte proliferation in a lymphocyte co-culture. In addition, they are also immunomodulatory, reducing lymphocyte proliferation in one- and two-way mixed lymphocyte cultures (7). Finally, HUCPVCs maintain the potential to differentiate into all five mesenchymal lineages in culture: fibroblast, bone, cartilage, fat and muscle; and, due to their >1:3 clonogenic capacity, can be induced to do so at the clonal level.

2. Materials

2.1. Umbilical Cord Dissection

1. Umbilical cord (20–40 cm) from full term birth
2. α-MEM (VWR)
3. 4 oz sample container (VWR)
4. Antibiotics: Penicillin G (167 U/ml) (Sigma), Gentamicin solution (50 µg/ml) (Sigma), and Amphotericin B (0.3 µg/ml) (Sigma)
5. HUCPVC dissecting tray with sterile surgical instruments (scissors and forceps)
6. Sterile silk sutures (VWR)
7. 50 ml polypropylene tubes (VWR)
8. Collagenase Type I (Sigma)
9. Hyaluronidase Type II (Sigma)
10. D-PBS (VWR)
11. Rotator (VWR)

12. Ammonium chloride solution (StemCell Technologies)
13. DMSO (Sigma)
14. Fetal bovine serum (FBS)

2.2. HUCPVC Culture

1. Supplemented medium (SM): 95% α-MEM, 5% FBS, and antibiotics
2. 75 cm² tissue culture-treated flask (Falcon)
3. 0.25% trypsin solution (Gibco)

2.3. Characterization

1. D-PBS
2. FBS
3. Goat serum (Gibco)
4. Mouse–anti-human monoclonal antibodies: CD146, CD105 (SH2), CD73(SH3), CD90 (Thy1), CD44, CD117 (c-kit), CD34, CD45, STRO-1, 3G5, actin, desmin, vimentin, MHC I, MHC II, and Alexafluor 488 goat–anti-mouse secondary antibody (Molecular Probes)

2.4. Immuno-characterization

1. Supplemented medium (SM) (*see* **Section 2.2.1**)
2. 75 cm² tissue culture-treated flask (Falcon)
3. 0.25% trypsin solution (Gibco)
4. 20 mL syringe (BD)
5. 40 mL heparinized whole blood (venipuncture)
6. Heparin
7. Ficoll-Paque™
8. D-PBS
9. WBC media (WM): 90% RMPI-1640, 10% FBS, and antibiotics (*see* **Section 2.1.4**)
10. 96-well tissue culture-treated plates (Falcon)
11. 24-well tissue culture-treated plates (Falcon)
12. Transwell™ insert for 24-well plate
13. 5-Bromo-2-deoxyuridine, and antibody
14. CD25 (interleukin-2 receptor) antibody

2.5. Mesenchymal Differentiation and Characterization

1. Supplemented medium (SM) (*see* **Section 2.2.1**).
2. 75 cm² tissue culture-treated flask (Falcon).
3. 0.25% trypsin solution (Gibco).
4. Osteogenic induction medium: SM, 10 nM dexamethasone (Sigma-Aldrich), 5 mM β-glycerophosphate (Sigma-Aldrich), and 50 μg/ml l-ascorbic acid (Sigma-Aldrich).

5. Tetracycline stain: 9 µg/ml tetracycline (Sigma-Aldrich).
6. Alkaline phosphatase & Von Kossa staining: 10% NFB: 100 ml formalin/formaldehyde, 16 g Na$_2$HPO$_4$, and 4 g NaH$_2$PO$_4$.H$_2$O in 1 L distilled water.
7. 2.5% silver nitrate solution: 2.5 g AgNO$_3$ in 100 ml distilled water.
8. Naphthol AS MX-PO$_4$ (Sigma).
9. N,N-dimethylformamide (DMF) (Fischer Scientific, D1191).
10. 0.2 M Tris–HCl (pH 8.3).
11. Red violet LB salt (Sigma F1625).
12. Sodium carbonate formaldehyde: 25 ml formalin/formaldehyde, 5 g Na$_2$CO$_3$ in 100 ml distilled water.
13. Chondrogenic induction medium: SM, 10 ng/ml transforming growth factor-β 1 (TGF-β1) (Chemicon).
14. Collagen II staining: Mouse–anti-human collagen II antibody (Chemicon) and goat–anti-rabbit Alexa Fluor 488 secondary antibody (Molecular Probes).
15. Alcian blue solution (pH 2.5): 1 g 8GX Alcian blue, 3% acetic acid solution, and 0.1% nuclear fast red solution: (0.1 g nuclear fast red, 5 g ammonium sulfate, and 100 ml distilled water).
16. Adipogenic induction medium: 87% α-MEM, 10% antibiotics, 3% FBS, 33 µM biotin (Sigma-Aldrich), 17 µM pantothenate (Sigma-Aldrich), 5 µM Rosiglitazone (Cayman Chemical), 100 nM bovine insulin (Sigma-Aldrich), 1 µM dexamethasone (Sigma-Aldrich), and 200 µM isobutyl methylxanthine (Sigma-Aldrich).
17. Oil Red O (Sigma-Aldrich).
18. Myogenic induction medium: 88% α-MEM, 10% antibiotics, 2% horse serum (Gibco, Lot 480116), 1 nM dexamethasone (Sigma-Aldrich), and 2 µM hydrocortisone (Sigma-Aldrich).
19. Myogenic staining: rabbit–anti-human MyoD primary antibody (Santa Cruz biotech) and mouse–anti-human fast-skeletal-MLC primary antibody (Sigma-Aldrich), goat–anti-rabbit Alexa Fluor 488, and goat–anti-mouse Alexa Fluor 555 secondary antibodies (Molecular Probes).
20. 3.7% formalin.
21. RNA isolation and reverse transcription (RT): 0.5 ml Trizol reagent (Invitrogen); Primers: Runx2, Collagen IA1, Osteocalcin, Osteopontin, Sox9, Collagen II, Aggrecan, LPL, MyoD, Myf5, Desmin, Fast skeletal myosin light chain, and myosin heavy chain.

2.6. Clonal isolation of HUCPVCs	1. Supplemented medium (SM) (*see* **Section 2.2.1**). 2. 75 cm² tissue culture-treated flask (Falcon). 3. 96-well tissue culture-treated plates (Falcon). 4. 0.25% trypsin solution (Gibco).

3. Methods

3.1. Umbilical Cord Dissection for HUCPVC Extraction	1. A healthy umbilical cord is obtained from a full term delivery and placed in a 4 oz container containing α-MEM and antibiotics. The cord is then transported and stored at room temperature until dissected. The remainder of the protocol is carried out aseptically (*see* **Notes 1 and 2**). 2. The cord is placed on the HUCPVC dissecting tray (*see* **Fig. 17.1**), and clamped down at one end. 3. A scalpel is used to make a circumferential incision in the amniotic epithelium close to the clamped end of the cord. The epithelium is then separated from the bulk of the cord using blunt dissection, and removed with forceps. The epithelium is usually removed in one tubular piece, although if breakage does occur, the process can be repeated to ensure all epithelium is removed. 4. The cord is then severed from the clamped end, and the three vessels longitudinally separated using forceps. These individual

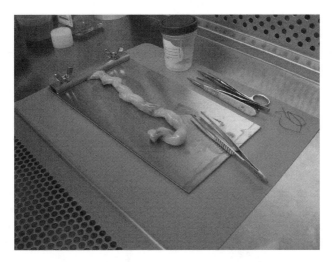

Fig. 17.1. Human umbilical cord clamped onto custom-designed dissecting table which allows easy removal of the amniotic epithelium and access to the umbilical vessels. The latter are surrounded by the perivascular tissue from which HUCPVCs are harvested.

vessels are then tied in a loop using silk sutures, and placed in 40 ml of 100 U/mL Type I collagenase and 0.01 U/mL hyaluronidase in a 50 mL Falcon tube and left to digest in a rotator for 3–5 hr (37°C oven), depending on the amount of matrix on the vessels.

5. After the digestion is completed, the looped vessels are removed from the suspension and the tubes are then centrifuged at 285 g for 10 min.

6. The supernatant is then aspirated and the cell pellets pooled. The cells are treated with 50 ml ammonium chloride (0.8%) and allowed to incubate at room temperature for 5 min to lyse the erythrocytes. The tube is then centrifuged for 10 min at 285 g. The supernatant is discarded, the cells are washed and counted, and plated in the T-75 tissue culture flasks in SM.

3.2. HUCPVC Culture

1. The cells are cultured on tissue-culture treated plastic, usually T-75 flasks. Once the cells reach 80–90% confluence, they are ready to be sub-cultured.

2. For sub-culture, the cells are washed twice with PBS, 5 ml of 0.25% trypsin is added, and the cells are incubated for 2–5 min. Once the cells are in suspension, they can be centrifuged for 5 min at 285 g before removing the supernatant. The cells are then resuspended in fresh SM and counted.

3. For optimal proliferation, the cells are then seeded at a density of 4×10^3 cells/cm^2 in SM, with the SM being replaced every 2–3 days until the cells are ready for sub-culture and seeding as required.

3.3. Characterization of HUCPVCs

1. The cells are washed, resuspended as described above, and placed on ice. The cells are then permeabilized with 1 ml of methanol (only for analysis of intracellular markers). The cells are then blocked on ice in 10% FBS in PBS for 1 hr.

2. After blocking, 1–2 μl (or respective working dilution) of the primary mouse–anti-human antibody is added to the cells in 0.5% goat serum in PBS, and incubated on ice for 20 min to overnight (for low-expressing markers). The cells are then washed twice in 2% goat serum in PBS, and then incubated with the Alexafluor 488 goat–anti-mouse secondary antibody for 20 min, and washed again in 2% goat serum in PBS.

3. The cells are analyzed by flow cytometry or immunohistochemistry as is shown in **Fig. 17.2** and Color Plate 13, for the α-actin, desmin, and vimentin.

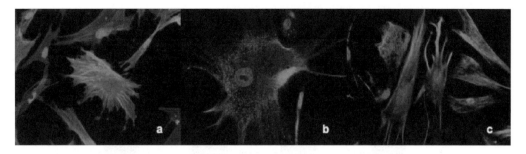

Fig. 17.2. HUCPVCs expressing α-actin (**a**), vimentin (**b**), and desmin (**c**). (*See* Color Plate 13)

3.4. Immuno characterization of HUCPVCs

3.4.1. Immunoprivilege

1. HUCPVCs are plated in 96-well plates at 10^4 cells/well.
2. Blood is obtained using heparinized syringes, and processed in a sterile biosafety cabinet.
3. Cells are separated using a Ficoll-Paque™ density centrifugation (20 mL blood layered slowly over 20 mL Ficoll), and the leukocytes are removed from the buffy coat and counted.
4. The cells are plated in the 96-well plate at 10^5 cells/well.
5. The two cell populations are co-cultured for 6 days before the leukocytes are assessed for BrdU expression using flow cytometry.

3.4.2. Immunomodulation

1. HUCPVCs are plated in 96-well plates at 10^4 cells/well.
2. Blood is obtained from two mismatched donors using heparinized syringes, and processed in a sterile biosafety cabinet.
3. Cells are separated using a Ficoll-Paque™ density centrifugation, and the leukocytes are removed from the buffy coat and counted.
4. Both populations of leukocytes are plated in the 96-well plate at 10^5 cells/well.
5. The three cell populations are co-cultured for 6 days before the leukocytes are assessed for proliferation or co-staining with BrdU and CD25 (*see* **Fig. 17.3**).
6. This assay can also be performed using 24-well plates, seeding the HUCPVCs on a Transwell™ insert, and placing the insert in a well with the leukocytes. The same endpoints are assessed.

3.5. Mesenchymal Differentiation and Characterization

1. For CFU-F frequency, cells should be seeded into six-well plates at low dilutions ranging from less than 1 cell/well to 12 cells/well. CFU-F frequency of HUCPVCs in culture should be observed as approximately 1:3.

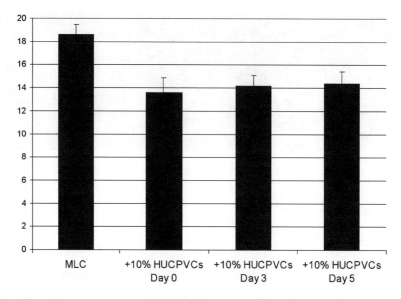

Fig. 17.3. HUCPVCs can differentiate into bone, cartilage, adipose, and muscle in vitro. Under induction, bone nodules are observed in culture (**a**). Cartilage pellet cultures of HUCPVCs express collagen II (**b**) and glycosaminoglycans that stained with Alcian blue (**c**). HUCPVC-derived adipocytes stain with Oil Red O (**d**), while myogenically-induced HUCPVCs expressed high levels of MyoD (**e**), and fast skeletal myosin light chain (FSMLC) (**f**) in multinucleated HUCPVC myotubes.

2. For mesenchymal differentiation analysis, the cells are subcultured and seeded into six-well plates as previously described (*see* **Fig. 17.4** and Color Plate 14).

3. For osteogenic induction, the cells are cultured in SM until they reach 60% confluence. The medium is then replaced with

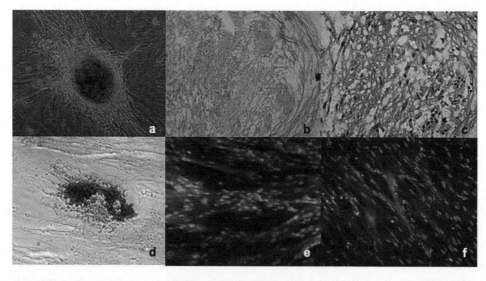

Fig. 17.4. HUCPVCs reduce lymphocyte proliferation, even if added 3 and 5 days into a 6 day culture. Addition of HUCPVCs showed a significant decrease in lymphocyte cell number compared to control (no HUCPVCs) over 6 days in a two-way MLC. There is no significant difference among HUCPVCs added on day 0, 3, or 5 ($n = 6$). This figure shows the average cell numbers, + standard deviations. (*See* Color Plate 14)

osteogenic induction medium, which is replaced every 2–3 days for 14 days or until mineralized bone nodules can be observed. Cells are analyzed for mineralization and alkaline phosphatase expression, while RNA can be extracted for RT-PCR analysis of osteogenic genes, Runx2, osteopontin, osteocalcin, and collagen I.

4. Nodules are analyzed by labeling with 9 µg/ml tetracycline overnight and examining the nodules for fluorescence. The nodules are then stained for alkaline phosphatase expression and mineralization by Von Kossa. First, the nodules are fixed in 10% cold NFB for 15 min. Fresh substrate is prepared: 0.005 g Naphthol AS MX-PO$_4$, 200 µl DMF, 25 ml 0.2 M Tris–HCl, 25 ml distilled water, and 0.03 g red violet LB salt, and filtered with Whatman's No. 1 filter paper immediately prior staining. Incubate for 45 min at room temperature and rinse in distilled water —three to four times. Stain with 2.5% silver nitrate for 30 min and rinse with distilled water three times. Alkaline phosphatase will appear as a red stain, while mineralized areas will appear black/grey.

5. For chondrogenic induction, the cells are resuspended as previously described, and centrifuged at 285 g for 5 min to obtain a cell pellet. The supernatant is removed and replaced with chondrogenic induction medium, which is replaced every 2–3 days for 21 days. The chondrogenic pellets are analyzed by histology for expression of collagen II and glycosaminoglycans, or by RT-PCR for expression of Sox 9, collagen II, and aggrecan mRNA.

6. Immunohistochemical analysis of collagen II expression is done as previously described. For analysis of glycosaminoglycans, alcian blue staining is used. Stain in alcian blue solution (1 g 8GX Alcian blue, 3% acetic acid adjusted to pH 2.5) for 30 min, and wash in running tap water for 2 min. Rinse in distilled water and counterstain in nuclear fast red solution for 5 min. Wash in running tap water for 1 min. Glycosaminoglycans will stain blue.

7. For adipogenic induction, the cells are cultured in SM until they reach 60% confluence. The SM is then replaced with adipogenic induction medium, which is replaced every 2–3 days for 21 days or until cells appear with large lipid droplets. Cells can be stained with 0.5% oil red O solution or analyzed by RT-PCR for expression of lipoprotein lipase (LPL).

8. For myogenic induction, the cells are cultured in SM until they reach 95% confluence. The SM is then replaced with myogenic induction medium, which is replaced every 2–3 days for 28 days or until multinucleated myotubes can be observed. Analysis of myotubes can be performed by immunohistochemical

staining for MyoD and fast-skeletal-myosin light chain, or by RT-PCR analysis of MyoD, Myf5, desmin, myosin heavy chain, and fast-skeletal-myosin light chain.

3.6. Clonal isolation of HUCPVCs

1. HUCPVCs are obtained in a single cell suspension as previously described and passed through a 70 μm filter to ensure single cell separation.

2. The cell suspension is then diluted to 1 cell/250 μl in SM. Using a multichannel micropipette, 50 μl of the suspension is deposited into individual wells of 96-well tissue culture plates, which are then incubated overnight. The following day, a further 50 μl of SM is added to the wells. The SM is then replaced every 5 days for 15 days.

3. The plates are then observed by light microscopy for the presence of cells in individual wells. Only wells with cells are continued in culture with SM replacement every 2–3 days until they reach 80–90% confluence. The cloned cells are then sub-cultured and transferred into individual wells of six-well plates and then into T75 flasks for further expansion. The cloned cells can then be plated as required for analysis of mesenchymal differentiation potential as previously described.

4. Notes

1. All umbilical cord dissections are done aseptically in a biological safety cabinet.

2. If the umbilical cord obtained is by natural birth, it is recommended to wipe the cord with 70% alcohol three times and place in fresh αMEM/antibiotic solution.

Acknowledgments

The authors would like to thank the staff of the neonatal and birthing suite at Sunnybrook and Women's College Hospital, and Mrs. Elaine Cheng for her expertise at dissecting umbilical cords.

References

1. Caplan, A.I., *Mesenchymal stem cells.* J. Orthop. Res, 1991. 9(5): 641–650.
2. Sarugaser, R., et al., *Human umbilical cord perivascular (HUCPV) cells: a source of mesenchymal progenitors.* Stem Cells, 2005. 23(2): 220–229.
3. Kogler, G., et al., *A new human somatic stem cell from placental cord blood with intrinsic*

pluripotent differentiation potential. J. Exp. Med., 2004. 200(2): 123–135.
4. Nayak, R.C., et al., *A monoclonal antibody (3G5)-defined ganglioside antigen is expressed on the cell surface of microvascular pericytes.* J. Exp. Med., 1988. 167(3): 1003–1015.
5. Baksh, D., et al., *Comparison of proliferative and multilineage differentiation potential of human mesenchymal stem cells derived from umbilical cord and bone marrow.* Stem Cells, 2007. 45(67): 99–106.
6. Le Blanc, K., *Immunomodulatory effects of fetal and adult mesenchymal stem cells.* Cytotherapy, 2003. 5(6): 485–489.
7. Ennis, J, et al., *In vitro immunologic properties of human umbilical cord perivascular cells.* Cytotherapy, 2008. 10(2): 174–181.

Chapter 18

Bone Marrow-Derived Mesenchymal Stem Cells: Isolation, Expansion, Characterization, Viral Transduction, and Production of Conditioned Medium

Massimiliano Gnecchi and Luis G. Melo[†]

Abstract

Mesenchymal stem cells (MSCs) are defined as self-renewing and multipotent cells capable of differentiating into multiple cell types, including osteocytes, chondrocytes, adipocytes, hepatocytes, myocytes, neurons, and cardiomyocytes. MSCs were originally isolated from the bone marrow stroma but they have recently been identified also in other tissues, such as fat, epidermis, and cord blood. Several methods have been used for MSC isolation. The most common method is based on the ability of the MSCs to selectively adhere to plastic surfaces. Phenotypic characterization of MSCs is usually carried out using immunocytochemical detection or fluorescence-activated cell sorting (FACS) analysis of cell surface molecule expression. However, the lack of specific markers renders the characterization of MSCs difficult and sometimes ambiguous. MSCs posses remarkable expansion potential in culture and are highly amenable to genetic modification with various viral vectors rendering them optimal vehicles for cell-based gene therapy. Most importantly, MSC plasticity and the possibility to use them as autologous cells render MSCs suitable for cell therapy and tissue engineering. Furthermore, it is known that MSCs produce and secrete a great variety of cytokines and chemokines that play beneficial paracrine actions when MSCs are used for tissue repair. In this chapter, we describe methods for isolation, ex vivo expansion, phenotypic characterization, and viral infection of MSCs from mouse bone marrow. We also describe a method for preparation of conditioned and concentrated conditioned medium from MSCs. The conditioned medium can be easily tested both in vitro and in vivo when a particular paracrine effect (i.e., cytoprotection) is hypothesized to be an important mechanism of action of the MSCs and/or screened to identify a target paracrine/autocrine mediator.

Key words: Mesenchymal stem cells, cell surface markers, flow cytometry, gene transfer, conditioned medium, soluble factors, paracrine effect, drug discovery.

1. Introduction

The bone marrow stroma was originally thought to function mainly as a structural support for the hematopoietic stem and

---

[†]Luis G. Melo passed away suddenly on September 26, 2007. This work is dedicated to his courage.

progenitor cells in the bone marrow (*1*). It was in the 1960s that Ernest A. McCulloch and James E. Till first revealed the clonal nature of marrow stromal cells (*2*, *3*). In the 1970s, Friedenstein and colleagues reported an ex vivo assay for examining the clonogenic potential of stromal marrow cells (*4*, *5*). In this assay, stromal cells were referred to as colony-forming unit fibroblasts (CFU-F). Subsequent experiments revealed that marrow stromal cells possess self-renewal and multilineage differentiation capacity (*6–8*), features typical of stem cells. Consequently, many investigators started to refer to cultured stromal cells as mesenchymal stem cells (MSCs) (*9*).

MSCs are a rare population of fibroblast-like cells resident in the stroma of the bone marrow. In addition, MSCs have been isolated from adipose tissue, epidermis, vessel wall, and cord blood (*10*). Furthermore, some recent evidence suggests that MSCs can also be mobilized into the circulation (*11*). In the bone marrow, MSCs represent approximately 0.001–0.01% of the nucleated cells. Even though MSCs are rare, they are readily separated from the hematopoietic stem cells (HSCs) in culture by their preferential attachment to plastic surfaces (*12*) and can be easily expanded ex vivo. The multipotency of the MSCs has now been convincingly confirmed and a variety of terminally differentiated cell lines, such as osteocytes, chondrocytes, adipocytes, hepatocytes, myocytes, neurons, and cardiomyocytes have been derived from the MSCs (*12–17*). Furthermore, several studies have suggested that this cell population may play a significant role in endogenous tissue repair (*18–20*) and recently MSCs have been evaluated as a potential cellular substrate for tissue engineering and cell therapy for tissue repair and regeneration (*21–23*). Indeed, the ease of handling and enormous expansion potential of MSCs, together with their low immunogenicity (*24*) provides a nearly unlimited number of cells suitable for a wide range of syngeneic and allogeneic therapeutic applications.

The phenotypic characterization of MSCs relies on cell surface molecule marker detection (*10*, *17*). However, many of these epitopes are shared by hematopoietic stem cells and by adult mesenchyme-derived cells, rendering accurate characterization of MSCs quite challenging. Furthermore, it has been shown that MSCs from different species do not necessarily express the same markers. Nevertheless, general agreement exists regarding a limited number of cell surface markers that are accepted when defining a MSC population. In particular, undifferentiated cultured MSCs typically express CD29, SH2 (CD105), SH3 and SH4 (CD73), CD44, CD90, and CD166 while fail to express common hematopoietic and endothelial markers, such as CD11b, CD14, CD31, CD34, and CD45.

Cultured MSCs are amenable to genetic manipulation using replication-deficient viral vectors. The overexpression of certain genes may then be used to enhance the cell function and survival or to deliver proteins of interest (*25*). Another important property of the MSCs is represented by their capacity to produce and secrete a great variety of cytokines and chemokines (*26*) that play important beneficial actions in tissue repair. For example, it has been shown that intramyocardial injection of MSCs overexpressing the gene Akt limits infarct size in mice and rats mainly by releasing cytoprotective factors (*27–29*). The same cardioprotective effect has been recently confirmed with human MSCs (*30*). Furthermore, a cytoprotective effect exerted by the MSCs has been demonstrated also in kidney and brain injury models (*31, 32*). The identification of specific protective factors secreted by MSCs using genomic and/or proteomic approaches may provide the opportunity for the discovery of therapeutic targets for the development of novel drug and molecular therapies.

2. Materials

2.1. Marrow Collection

1. Wild-type male C57 BL/6 mice, 8–10 weeks old (The Jackson Laboratory, Bar Harbor, ME, USA).
2. Appropriate surgical and cell culture facilities.
3. A sterilized instrument pack consisting of the following items: two small hemostats, a pair of surgical scissors, two scalpel blades, a bone cutter, two forceps, one 10 ml syringe, one hypodermic 27G needle, drapes, and sterile gloves.
4. Tissue culture supplies including a motorized pipettor, 10 ml pipette, 100 × 20 mm dishes, 50 ml centrifuge tubes.
5. Povidone–iodine U.S.P. Prep Pad (#B40600, PDI, Orangeburg, NY).
6. 70% Alcohol Prep Pads (#B15901, PDI).
7. Hanks' Balanced Salt Solution (HBBS) (#14025-092 Invitrogen, Carlsbad, CA, USA).
8. Ketamine hydrochloride injection U.S.P. 100 mg/ml store at 4–8°C protected from light (Pfizer, Kirkland, Qc, Canada).
9. Xylazine hydrochloride injection U.S.P. 20 mg/ml store at 4–8°C protected from light (Bayer, Toronto, Ont., Canada).

2.2. Isolation, Expansion, and Viral Transduction

1. Tissue culture supplies including a motorized pipettor, 10 and 25 ml pipettes, 12 and 50 ml centrifuge tubes, and 100 × 20 mm dishes.
2. Ficoll-Paque™ Plus store at 4–8 °C protected from light (#17-1440-02 Amersham Biosciences, Uppsala, Sweden).

3. Hanks' Balanced Salt Solution (HBBS) (#14025-092 Invitrogen).
4. Phosphate-buffered saline (PBS) (#70011-044 Invitrogen).
5. Fetal bovine serum (FBS) heat inactivated (#SH30071.03HI, Hyclone, Logan, UT, USA).
6. Penicillin–streptomycin (P/S) (#P4333, Sigma-Aldrich, St Louis, MO, USA).
7. Alpha minimal essential medium (α-MEM) (#32571-036, Invitrogen).
8. Trypsin 0.25% in EDTA (#SV30031.01, Hyclone).
9. Hexadimethrine bromide (Polybrene) (#H9268, Sigma-Aldrich).
10. Recombinant retrovirus encoding IRES-GFP (Harvard Viral Core Facility, Boston, MA, USA).

2.3. Characterization — Immunophenotyping

1. CD29-PE antibody (#102207, Biolegend, San Diego, CA, USA).
2. CD44-APC Antibody (#BD-559250, BD Biosciences, San Jose, CA, USA).
3. CD105 Alexa Fluor® 488 Antibody (#120405, Biolegend).
4. CD11b-APC Antibody (#101211, Biolegend).
5. CD14-FITC Antibody (#BD-553739, BD Biosciences).
6. CD45-FITC Antibody (#BD-553079, BD Biosciences).
7. CD73-PE Antibody (#BD-550741, BD Biosciences).
8. CD166-PE Antibody (#BD-559263, BD Biosciences).
9. CD31-APC Antibody (#BD-551262, BD Biosciences).
10. Sca-1-FITC Antibody (#BD-557405, BD Biosciences).
11. IgG1-FITC Isotype Control (#BD-552916, BD Biosciences).
12. IgG2b-FITC Isotype Control (#BD-556923, BD Biosciences).
13. IgG2b-APC Isotype Control (#BD-556924, BD Biosciences).
14. IgG2a-APC Isotype Control (#BD-554690, BD Biosciences).
15. IgG2a-Alexa Fluor® 488 Isotype Control (#400525, Biolegend).
16. IgG2a-PE Isotype Control (#400507, Biolegend).
17. IgG1-PE Isotype Control (#BD-349043, BD Biosciences).
18. IgG-PE Isotype Control (#400907, Biolegend).
19. IgG2a-FITC Isotype Control (#BD-400505, BD Biosciences).

2.4. Production of Conditioned Medium

1. Tissue culture supplies including a motorized pipettor, 10 ml pipettes, 12 ml centrifuge tubes, 100×20 mm dishes, manual pipettor and 200 µL tip with filter (for retentate recovery), 1.5 ml microcentrifuge sterile tubes, and 2 ml cryovials.
2. PBS (#14190144, Invitrogen).
3. α-MEM (#32571-036, Invitrogen).
4. Amicon Ultra-15 centrifugal filter device (#UFC9 005 24, BD Biosciences).

3. Methods

3.1. Bone Marrow Collection

1. Place 10 ml of pre-warmed (at 37°C) α-MEM in a 50 ml tube and place a 100 mm dish with sterile HBSS on ice for bone dissection.
2. Anesthetize the animal with ketamine/xylazine (*see* **Notes 1** and **2**).
3. Shave fur from the hind limbs, back, and belly with electric clippers.
4. Wear Sterile Goves and, using the alcohol pads, clean the skin with unidirectional movements from the abdomen to the foot (*see* **Note 3**).
5. After the skin dries, scrub it using the povidone–iodine pads.
6. Euthanize the animal by cervical dislocation.
7. Wear new sterile gloves and make an incision with a scalpel blade to the midline of the leg. Firmly grip skin and pull rostrally and caudally to expose the muscles.
8. Using a pair of surgical scissors dissect muscles, tendons, and connective tissue to expose the bones. Keep cuts as close to the bones as possible to allow a clean dissection.
9. With the scalpel blade cut through the hip joint, separating the head of the femur from the acetabulum.
10. Using a forcep elevate the tibial–femoral complex and cut with the scissors the distal part of the tibia and fibula.
11. The femur and tibia are now separated from the animal; place the bones into the HBSS filled dish on ice.
12. Continue the dissection process for all the animals (*see* **Note 4**).
13. Continue the marrow collection under a laminar flow hood.
14. Wear a new pair of sterile gloves. Using the bone cutter carefully cut each bone at the epiphysis level paying attention not to break the bones (*see* **Note 5**).

15. Attach a 27G needle to a 10 ml syringe and draw 10 ml of α-MEM into the syringe.
16. Hold the femur (or the tibia) with a forcep and use the needle to drill a hole in the distal end of the bone. Slowly inject the medium into the bone flushing the marrow into a 50 ml centrifuge tube. Repeat this step twice.
17. When the marrow from all the bones has been ejected, add additional 7 ml of 37°C α-MEM and 17 ml of HBSS into the 50 ml tube and proceed to the next step (**Section 3.2.1**).

3.2. MSC Isolation, Culture, and Viral Transduction

3.2.1. Isolation

1. With a 25 ml serological pipette aspirate 15 ml of room-temperature Ficoll-Paque™. Place the pipette tip at the bottom of the 50 ml centrifuge tube, below the marrow-medium-HBBS mixture, and underlay the Ficoll while slowly raising the pipette.
2. Centrifuge at 400 rcf (with brake off) for 30 min at 20°C (*see* **Note 6**).
3. During centrifugation, reconstitute the growth medium supplementing α-MEM with 10% FBS and 1% P/S (*see* **Note 7**).
4. Remove the 50 ml tube from the centrifuge. Four layers are clearly visible after centrifugation: (1) the bottom layer is predominately formed by red blood cells, (2) the second layer is Ficoll and appears transparent, (3) the third is a slightly pink hazy layer containing the majority of the mononuclear cells, and (4) the uppermost layer is predominately HBSS/α-MEM.
5. With a 10 ml pipette aspirate and discard the uppermost layer.
6. Then aspirate the mononuclear layer and transfer it into a 10 ml tube.
7. Centrifuge at 1,000 rcf for 10 min at room-temperature.
8. Aspirate supernatant and resuspend pellet with HBBS. Centrifuge at 500 rcf for 5 min. Repeat this washing step twice.
9. Resuspend the cells in 10 ml of growth medium passing them several times with a pipette to ensure adequate separation.
10. Plate the cells in a 100 mm cell culture dish and place in a CO_2 incubator (*see* **Note 8**).
11. Change the medium for the first time 48 h after plating and then every 72 h.
12. Colonies of MSCs should appear 6–8 days after initial plating (*see* **Note 9**).

3.2.2. Expansion

1. Check the cells daily using an inverted phase contrast microscope.
2. When colonies are ~80% confluent, split the cells.
3. Aspirate the growth medium and wash the cells twice using PBS (*see* **Note 10**).
4. Aspirate all the PBS and add 2 ml of 37°C pre-warmed trypsin.
5. Place the dish in the CO_2 incubator for 3 min.
6. Add 8 ml of growth medium and collect all the cells in a 12 ml tube (*see* **Note 11**).
7. Centrifuge at 500 rcf for 5 min at room temperature.
8. Aspirate the supernatant and resuspend the pellet with PBS. Centrifuge at 500 rcf for 5 min. Repeat this washing step twice and count the cells using a hemocytometer.
9. Resuspend the pellet in growth medium passing the cells several times with the pipette to ensure adequate separation.
10. Plate at 2×10^3 cells/cm^2 in 10 ml of growth medium.
11. These are passage one (P1) cells.
12. Check the cells daily and change the medium every 72 h. Split the cells when 90% confluent (*see* **Note 12**).

3.2.3. Viral Transduction

1. Use P1 or P2 cells when they are 30–40% confluent.
2. Aspirate the growth medium and gently wash three times with PBS (*see* **Note 10**).
3. Incubate the cells in 7 ml of growth medium containing 6 µg/ml of polybrene and 100 multiplicities of infection (MOI) of virus (*see* **Note 13**).
4. Incubate the cells for 6 h in the CO_2 incubator at 37°C.
5. Aspirate the medium containing the virus and gently wash thrice with PBS (*see* **Note 10**).
6. Add 10 ml of growth medium and place the dish in the CO_2 incubator.
7. After 3 h, check the viability of the cells which should be greater than 60–70%.
8. 24 h after the first transduction cycle, repeat steps 2–7.
9. 96 h after the second transduction cycle, by fluorescent microscopy and flow cytometry determine transduction efficiency by fluorescent microscopy and flow cytimetry (*see* **Fig. 18.1** and **Color Plate 15**).

3.3. Characterizatio—Immuno phenotyping

1. Detach P5 cells using trypsin (see **Section 3.2.2**, steps 3–7).
2. Aspirate supernatant and resuspend pellet with 5% fetal bovine serum (serum-PBS). Centrifuge at 500 rcf for 5 min.

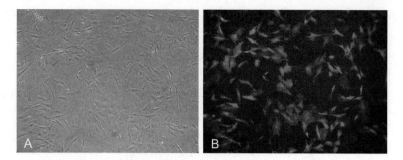

Fig. 18.1. Transduction efficiency by fluorescent microscopy. (**A**) Phase contrast microphotograph of P5 bone marrow-derived MSCs. (**B**) The same microscopic field observed under fluorescent light shows that the majority of the cells express GFP, meaning that they were successfully transduced. In our hands, this protocol allows a transduction efficiency higher than 80%, as confirmed also by FACS analysis. (*See* Color Plate 15)

Repeat this washing step one more time and resuspend in serum-PBS.

3. Count the cells and aliquot 5×10^5 cells for each 12 ml tube.
4. Pellet cells, gently resuspend in 90 μl serum-PBS the 5×10^5 cells and transfer them in a 1.5 ml microcentrifuge tube (*see* **Note 14**).
5. Add 10 μl of antibody/isotype control to 5×10^5 cells.
6. Shake on ice protected from light for 1 h.
7. Centrifuge at 800 rcf for 5 min and resuspend in serum-PBS. Repeat this washing step two to three times.
8. Resuspend each sample in 500 μl serum-PBS.
9. Perform flow cytometry using forward scatter and side scatter analysis to gate for live cells.
10. Use isotype controls to set photo-multiplier tube (PMT), compensation, and analysis gates (*see* **Note 14**).
11. MSC population will be positive for CD29, CD44, CD105, CD73, CD166, and Sca-1, and negative for CD11b, CD14, CD45, and CD31.

3.4. Production of Conditioned Medium

3.4.1. Conditioned Medium

1. Use P5, 90% confluent cells.
2. Aspirate the growth medium and wash three times with PBS.
3. Add 8 ml of α-MEM containing neither FBS nor P/S.
4. Incubate the cells in the CO_2 incubator.
5. Incubate at the same time 8 ml of α-MEM containing neither FBS nor P/S in a dish where no cells are seeded. This will serve as control medium.

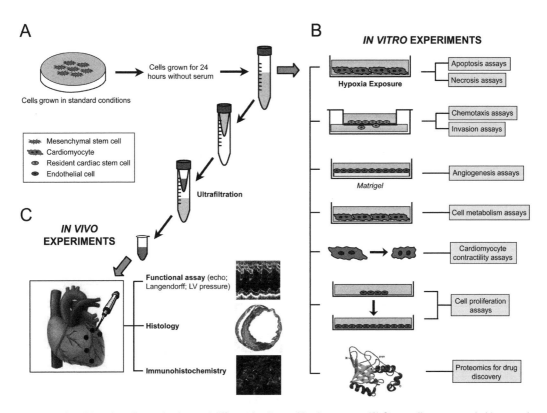

Fig. 18.2. Conditioned media production and different in vitro and in vivo assays. (**A**) Stem cells are expanded in normal conditions until they are 90% confluent. The growth medium is then exchanged with medium not containing serum and the cells are left for 24 h in a CO_2 incubator. The medium is then collected and tested either in vitro or in vivo. (**B**) For the in vitro experiments, the conditioned medium is transferred into culture dishes containing different kind of cells according to the goal of the specific experiment. Several different properties of the medium can be tested in vitro. For example, the cytoprotective effects exerted by the conditioned medium can be tested on murine cardiomyocytes. After exposing the cardiomyocytes to hypoxia in the presence of control medium or conditioned medium, apoptosis and necrosis assays are performed and the results compared. To verify if the conditioned medium contains chemotactic factors, a specific cell type (i.e., endothelial cell or cardiac stem cell) is seeded on the membrane of the upper chamber of a dual chamber dish. The number of cells migrating into the lower chamber, containing either conditioned medium or control medium, is then counted.

To test the pro-angiogenetic properties of conditioned medium, endothelial cells are seeded on a matrigel and the number of capillaries is quantified after exposure to conditioned medium or control medium. To verify if cell metabolism is influenced by factors present in the medium, the cell type of interest (i.e., murine cardiomyocytes) is exposed to the conditioned and to the control medium and then collected to perform metabolic assays. Another example: cardiomyocyte contractility may be assessed in the presence of control or conditioned medium; if inotropic factors are present, then cell contractility will be increased in the presence of conditioned medium. Proliferation assay may also be performed. Finally, proteomic analysis of conditioned medium may allow the discovery of new therapeutic molecules and targets. (**C**) Conditioned medium may be tested also in vivo using different experimental disease models. For example, the effects of conditioned medium on ischemic myocardium may be assessed in a murine model of myocardial infarction. Small volumes of concentrated conditioned medium obtained by ultrafiltration are injected at the infarct border zone after left coronary ligation. At established time points, heart function is analyzed by echo (an M-mode image of the left ventricle is depicted in the figure) or other methods. The heart may also be collected for histology to determine, for example, the infarct size (a cross-section of mouse heart stained with Masson Trichrome is depicted in the figure). Furthermore, immunohistochemistry staining may performed to determine different parameters, such as cardiac regeneration, neoangiogenesis, apoptosis. (*See* Color Plate 16)

6. After 24 h of incubation, aspirate the medium and place into a 12 ml tube.
7. Centrifuge at 2,500 rcf for 5 min at 20°C in case cell debris are present in the medium.
8. Transfer the conditioned medium into a new 12 ml centrifuge tube, paying attention not to aspirate the cell debris at the bottom of the tube.
9. The conditioned medium can be tested immediately in different assays (*see* **Fig. 18.2** and Color Plate 16) or stored at −80°C until use (*see* **Note 15**).

3.4.2. Concentrated Conditioned Medium

1. Repeat steps 1–7 reported in **Section 3.4.1**.
2. Transfer the conditioned medium and the control medium in two centrifugal filter devices.
3. Centrifuge at 4,000 rcf for 30 min at 4°C (*see* **Note 16**).
4. Immediately recover the concentrate using a 200 µl pipette (*see* **Note 17**).
5. Proceed by desalting the concentrate (*see* **Note 18**).
6. The concentrated conditioned medium (ultrafiltrate) can be placed in 1.5 ml microcentrifuge tubes on ice and used immediately (*see* **Fig. 18.2**) or stored in 2 ml cryovials at −80 °C until use (*see* **Notes 15 and 19**).

4. Notes

1. All animal procedures should be conducted in accordance with animal care policies and must be approved by the institutional animal care committee.
2. Use the following dose: ketamine 100 mg/kg and xylazine 10 mg/kg by intraperitoneal injection (IP). Mix 0.5 ml (50 mg) ketamine + 0.25 ml (5 mg) xylazine + 4.25 ml sterile water or saline. Mouse dose is 0.1 ml/10 g body weight.
3. Unless stated otherwise, all procedures should be conducted with the utmost adherence to aseptic technique.
4. Eight to ten mice should provide adequate number of marrow cells for isolation.
5. Before cutting the epiphysis, use a scalpel blade to scrape off any remaining soft tissue from the bones.
6. It is very important to allow the centrifuge to accelerate/decelerate slowly to avoid disrupting proper separation.
7. All compounds must be pre-warmed at 37°C.

8. Set the incubator temperature at 37°C and create a humidified atmosphere containing 95% air and 5% CO_2.

9. Colonies appear as a central core of round cells surrounded by more elongated cells at the periphery. The number of colonies is an index of MSC functional capacity.

10. Be very gentle when adding the 10 ml of PBS needed to wash the cells. Shake orbitally the dish two to three times to ensure a more thorough wash.

11. Before adding the medium, rock the plate gently three to four times on the side of the bench to help dislodge the cells. The cells that remain attached to the dish are discarded.

12. Under the conditions described above, cultures usually remain morphologically heterogeneous until P3 or P4, presenting a mix of round- and spindle-shaped cells. The proportion of spindle-shaped cells increases gradually with time and starting from P5, the cultures become quite homogeneous. When passaging the cells, from P1 to P4 plate at 2×10^3 cells/cm^2, from P5 decrease plating density to 10^3 cells/cm^2.

13. The protocol described here has been developed for recombinant retrovirus. We use retrovirus produced by the Harvard Viral Core Facility. Briefly, for the generation of this specific retrovirus, the expression vector pMSCV encoding IRES-GFP, and the retroviral vector pVSV-G (both from Clontech, Mountain View, CA, USA) were co-transfected into a packaging cell line (GP2-293).

14. An extensive description of flow cytometry immunophenotyping cannot be reported here due to space limitation. However, we would like to provide some suggestions. We recommend a minimum of 5×10^5 cells per antibody or isotype control and a minimum of 50,000 events for characterization. It is best to first run cells with isotype control; then a second set of cells with only one antigen. When performing multiple antigen-labeling, these controls can be used to verify if there is spectral overlap in multiparameter measurements.

15. To compare media conditioned in the presence of two different cell types (e.g., MSCs and fibroblasts), we suggest normalizing the volumes by the number of cells. To do so, count the cells after collecting the conditioned medium and adjust the volume of the conditioned medium according to the cell number.

16. This centrifugation will collect a 200 μL retentate volume starting from a 15 mL sample; adjust the centrifugation time accordingly.

17. To recover the concentrated solute, insert a manual pipettor with a 200 µL pipette tip into the bottom of the filter unit and withdraw the sample using a side-to-side sweeping motion to ensure total recovery. For optimal recovery, remove concentrated sample immediately after centrifugation.

18. Desalting is accomplished by reconstituting the concentrate to the original sample volume with any desired solvent (we suggest PBS) and concentrate again. Repeat this process twice.

19. The concentrated conditioned medium can be used for animal experiments since the volumes are suitable for vein or intraparenchymal injection. In mouse myocardial infarction models, we suggest using 20 µl of medium normalized for a number of cells ranging from 1×10^5 and 1×10^6 for each heart. For dilutions we suggest using PBS.

Acknowledgments

Massimiliano Gnecchi is founded by the Fondazione IRCCS Policlinico San Matteo Pavia, Italy, the Ministero Italiano dell'Università e della Ricerca (MIUR), the Ministero Italiano della Sanità, and the Fondazione Cariplo. Luis G. Melo is founded by the Canadian Institutes of Health Research (CIHR) and the Heart and Stroke Foundation of Ontario.

References

1. Dexter, T.M., Allen, T.D., Lajtha, L.G. (1977) Conditions controlling the proliferation of haemopoietic stem cells in vitro. *J Cell Physiol* **91**, 335–344.
2. Becker, A.J., McCulloch, E.A., Till, J.E. (1963) Cytological demonstration of the clonal nature of spleen colonies derived from transplanted mouse marrow cells. *Nature* **197**, 452–454.
3. Siminovitch, L., McCulloch, E.A., Till, J.E. (1963) The distribution of colony-forming cells among spleen colonies. *J Cell Physiol* **62**, 327–336.
4. Friedenstein, A.J., Chailakhjan, R.K., Lalykina, K.S. (1970) The development of fibroblast colonies in monolayer cultures of guinea pig bone marrow and spleen cells. *Cell Tissue Kinet* **3**, 393–403.
5. Friedenstein, A.J., Deriglasova, U.F., Kulagina, N.N., Panasuk, A.F., Rudakowa, S.F., Luria, E.A., Ruadkow, I.A. (1974) Precursors for fibroblasts in different populations of hematopoietic cells as detected by the in vitro colony assay method. *Exp Hematol* **2**, 83–92.
6. Friedenstein, A.J., Chailakhyan, R.K., Gerasimov, U.V. (1987) Bone marrow osteogenic stem cells: in vitro cultivation and transplantation in diffusion chambers. *Cell Tissue Kinet* **20**, 263–272.
7. Aston, B.A., Allen, T.D., Howlett, C.R., Eaglesom, C.C., Hattori, A., Owen, M. (1980) Formation of bone and cartilage by marrow stromal cells in diffusion chambers in vivo. *Clin Orthop Relat Res* **151**, 294–307.
8. Owen, M. (1988) Marrow stromal stem cells. *J Cell Sci Suppl* **10**, 63–76.
9. Caplan, A.I. (1991) Mesenchymal stem cells. *J Orthop Res* **9**, 641–650.

10. Minguelll, J.J., Erices, A., Conget, P. (2001) Mesenchymal stem cells. *Exp Biol Med* **226**, 507–520.
11. He, Q., Wan, C., Li, G. (2007) Concise review: multipotent mesenchymal stromal cells in blood. *Stem cells* **25**, 69–77.
12. Prockop, D.J. (1997) Marrow stromal cells as stem cells for nonhematopoietic tissues. *Science* **276**, 71–74.
13. Nuttall, M.E., Patton, A.J., Olivera, D.L., Nadeau, D.P., Gowen, M. (1998) Human trabecular bone cells are able to express both osteoblastic and adipocytic phenotype: implications for osteopenic disorders. *J Bone Miner Res* **13**, 371–382.
14. Wakitani, S., Saito, T., Caplan, A.I. (1995) Myogenic cells derived from rat bone marrow mesenchymal stem cells exposed to 5-azacytidine. *Muscle Nerve* **18**, 1417–1426.
15. Kopen, G.C., Prockop, D.J., Phinney, D.G. (1999) Marrow stromal cells migrate throughout forebrain and cerebellum, and they differentiate into astrocytes after injection into neonatal mouse brains. *Proc Natl Acad Sci U S A* **96**, 10711–10716.
16. Makino, S., Fukuda, K., Miyoshi, S., Kodama, H., Pan, J., Sano, M., Takahashi, T., Hori, S., Abe, H., Hata, J., Umezawa, A., Ogawa, S. (1999) Cardiomyocytes can be generated from marrow stromal cells in vitro. *J Clin Invest* **103**, 697–705.
17. Pittenger, M.F., Mackay, A.M., Beck, S.C., Jaiswal, R.K., Douglas, R., Mosca, J.D., Moorman, M.A., Limoneti, D.W., Carig, S., Marshak, D.R. (1999) Multilineage potential of adult human mesenchymal stem cells. *Science* **284**, 143–147.
18. Bianco, P., Riminucci, M., Kuznetsov, S., and Robey, P.G. (1999) Multipotential cells in the bone marrow stroma: regulation in the context of organ physiology. *Crit Rev Eukaryot Gene Exp* **9**, 159–173.
19. Liechty, K.W., MacKenzie, T.C., Shaaban, A.F., Radu, A., Moseley, A. M., Deans, R., Marshak, D. R., and Flake, A. W. (2000) Human mesenchymal stem cells engraft and demonstrate site-specific differentiation after in utero transplantation in sheep. *Nat Med* **6**, 1282–1286.
20. Pochampally, R.R., Neville, B.T., Schwarz, E.J., Li, M. M., and Prockop, D. J. (2004) Rat adult stem cells (marrow stromal cells) engraft and differentiate in chick embryos without evidence of cell fusion. *Proc Natl Acad Sci USA* **101**, 9282–9285.
21. Horwitz, E.M., Prockop, D.J., Fitzpatrick, L.A., Koo, W.W., Gordon, P.L., Neel, M., (1999) Transplantability and therapeutic effects of bone marrow-derived mesenchymal cells in children with osteogenesis imperfecta, *Nat Med* **5**, 262–264.
22. Caplan A.I., Bruden, S.P. (2001) Mesenchymal stem cells: building blocks for molecular medicine in the 21th century. *Trends Mol Med* **7**, 259–264.
23. Melo, L.G., Pachori, A.S., Kong, D., Gnecchi, M., Wang, K., Pratt, R. E., and Dzau, V. J. (2004) Molecular and cell-based therapies for protection, rescue, and repair of ischemic myocardium: reasons for cautious optimism. *Circulation* **109**, 2386–2393.
24. Ryan, J.M., Barry, F.P., Murphy, J.M., Mahorn, B.P. (2005) Mesenchymal stem cells avoid allogeneic rejection. *J Inflamm* **26**, 2–8.
25. Dzau, V.J., Gnecchi, M., Pachori, A.S. (2005) Enhancing stem cell therapy through genetic modification. *J Am Coll Cardiol* **46**, 1351–1353.
26. Caplan, A.I., Dennis, J.E. (2006) Mesenchymal Stem Cells as Trophic Mediators. *J Cell Biochem* **98**, 1076–1084.
27. Gnecchi, M., He, H., Liang, O.D., Melo, L. G., Morello, F., Mu, H., Noiseux, N., Zhang, L., Pratt, R. E., Ingwall, J. S., et al. (2005) Paracrine action accounts for marked protection of ischemic heart by Akt-modified mesenchymal stem cells. *Nat Med* **11**, 367–368.
28. Gnecchi, M., He, H., Noiseux, N., Liang, O. D., Zhang, L., Morello, F., Mu, H., Melo, L. G., Pratt, R. E., Ingwall, J. S., Dzau, V. J. (2006) Evidence supporting paracrine hypothesis for Akt-modified mesenchymal stem cell-mediated cardiac protection and functional improvement. *FASEB J* **20**, 661–669.
29. Noiseux, N., Gnecchi, M., Lopez-Ilasca, M., Zhang, L., Solomon, S.D., Deb, A., Dzau, V.J., Pratt, R.E. (2006) Mesenchymal stem cells overexpressing Akt dramatically repair infarcted myocardium and improve cardiac function despite infrequent cellular fusion or differentiation. *Mol Ther* **14**, 840–850.
30. Iso, Y., Spees, J.L., Serrano, C., Bakondi, B., Pochampally, R., Song, Y.H., Sobel, B.E., Delafontaine, P., Prockop, D.J. (2007)

Multipotent human stromal cells improve cardiac function after myocardial infarction in mice without long-term engraftment. *Biochem Biophys Res Commun* **354**, 700–706.

31. Honma, T., Honmou, O., Iihoshi, S., Harada, K., Houkin, K., Hamada, H., Kocsis, J. D. (2005) Intravenous infusion of immortalized human mesenchymal stem cells protects against injury in a cerebral ischemia model in adult rat. *Exp Neurol* **199**, 56–66.

32. Togel, F., Weiss, K., Yang, Y., Hu, Z., Zhang, P., Westenfelder, C. (2007) Vasculotropic, paracrine actions of infused mesenchymal stem cells are important to the recovery from acute kidney injury. *Am J Physiol Renal Physiol* **292**, F1626–F1635.

Chapter 19

Template DNA-Strand Co-Segregation and Asymmetric Cell Division in Skeletal Muscle Stem Cells

Vasily Shinin, Barbara Gayraud-Morel, and Shahragim Tajbakhsh

Abstract

Stem cells are present in all tissues and organs, and are crucial for normal regulated growth. How the pool size of stem cells and their progeny is regulated to establish the tissue prenatally, then maintain it throughout life, is a key question in biology and medicine. The ability to precisely locate stem and progenitors requires defining lineage progression from stem to differentiated cells, assessing the mode of cell expansion and self-renewal and identifying markers to assess the different cell states within the lineage. We have shown that during lineage progression from a quiescent adult muscle satellite cell to a differentiated myofibre, both symmetric and asymmetric divisions take place. Furthermore, we provide evidence that a sub-population of label retaining satellite cells co-segregate template DNA strands to one daughter cell. These findings provide a means of identifying presumed stem and progenitor cells within the lineage. In addition, asymmetric segregation of template DNA and the cytoplasmic protein Numb provides a landmark to define cell behaviour as self-renewal and differentiation decisions are being executed.

Key words: 'Immortal' template DNA, asymmetric cell division, Numb, videomicroscopy, label retaining cell.

1. Introduction

Skeletal muscle is a terminal tissue, where the functional entity that generates force and permits movement, is post-mitotic. Throughout life, therefore, a reservoir of cells, called satellite cells in the adult, assures continued growth and repair (1, 2). Like other tissues and organs, this dependence on stem cells is crucial for survival. However, information is lacking on the true nature of the stem cell entity and how these cells are distinguished from their daughters, which often are morphologically similar. Thus, it has become critical to identify markers and properties

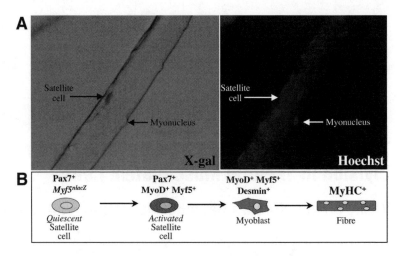

Fig. 19.1. Lineage progression of an adult skeletal muscle satellite cell to a differentiated cell. (**A**) Isolated fibre from the Tibialis Anterior muscle of a *Myf5$^{nlacZ/+}$* mouse showing nuclear β-galactosidase activity by X-gal staining. Hoechst staining reveals myonuclei inside the fibre as well as the nucleus of the satellite cell on the fibre. (**B**) Lineage progression of a quiescent to an activated satellite cell, hallmarked by MyoD expression, to a myoblast. Myoblasts fuse homotypically or to pre-existing differentiated fibres or myotubes after leaving the cell cycle. Differentiated fibres are characterised by the expression of myosin heavy chain (MyHC). The expression of different commonly used markers is indicated. (*See* Color Plate 17)

that can distinguish different cell states within a lineage progression of a stem to a differentiated cell (**Fig. 19.1** and Color Plate 17). It is likely that lineage specific and universal properties that can distinguish stem from progenitor cells will be identified. Stem cells are thought to undergo symmetric and asymmetric divisions to generate specialised daughters and to allow for self-renewal of the stem cell pool. How this is done, particularly in vertebrates is largely unknown *(3)*.

We have investigated asymmetric cell divisions by monitoring the selective segregation of the homologue of a Drosophila cell fate-determinant Numb in adult mouse satellite cells *(4)*. In addition, using a pulse-chase approach with the nucleotide analogue BrdU to identify stem cells and their niche, we recently identified a sub-population of cells which are label-retaining cells (LRCs) in skeletal muscles. Intriguingly, some of these LRCs co-segregate template DNA strands to only one daughter cell after cell division *(4)*. This finding suggests that the retention of 'immortal' DNA strands in stem cells may be one strategy that stem cells have adopted to escape the accumulation of errors arising during successive rounds of DNA replication *(5, 6)*. This hypothesis, which has not been universally accepted, was postulated for adult stem cells *(5)*. It is important test this model in other tissues, and establish a link with asymmetric cell divisions using high resolution techniques and appropriate controls.

This chapter provides detailed protocols for the detection of template DNA strands in adult muscle satellite cells. In addition, we describe the approaches which we used to observe asymmetric cells divisions, using principally the *Drosophila* cell fate determinant Numb as a readout.

2. Materials

2.1. Cell Culture Medium Components

1. Dulbecco's modified Eagle's medium (DMEM) (41966-029, Invitrogen, Bethesda, MD).
2. MCDB-201 medium (Sigma).
3. Fetal calf serum (Invitrogen).
4. Ultroser SF (steroid-free) (BioSepra, France), final concentration in the medium is 1–2%.
5. Insulin–transferrin–selenium (100× ITS-G: Insulin, 1 g/L; sodium selenite, 0.67 mg/L; transferrin, 0.55 g/L; Invitrogen).
6. Glutamine (Invitrogen). 100× concentrate.
7. Sodium pyruvate (Invitrogen) 100× concentrate.
8. Penicillin–streptomycin (Invitrogen) 100× concentrate.
9. Matrigel (BD Biosciences) 1 mg/ml solution in DMEM for coating of culture dishes.
10. Poly-D-lysine (Sigma) diluted in distilled water at 0.1 mg/ml and stored at −20°C. After thawing filter through 0.22 μm Millipore.
11. Tissue culture plastic dishes.

2.2. Reagents for Skeletal Muscle Fibre Dissociation and Satellite Cell Isolation

1. Freshly prepared solution of 0.1% collagenase type I (C0130, Sigma) in DMEM filtered through 0.22 μm Millipore filter, then pre-warmed to 37°C (*see* **Note 1**).
2. Freshly prepared solution of 0.5% dispase I (Sigma) in DMEM filtered through 0.22 μm Millipore filter pre-warmed to 37°C.
3. Solution of trypsin (0.25%) and ethylenediamine tetraacetic acid (EDTA) (1 mM) pre-warmed to 37°C before use from Invitrogen.
4. Horse serum (HS) (30G1348D, Invitrogen).
5. Matrigel (354234, Becton Dickinson). Stored at −20°C (*see* **Note 2**).

6. DMEM with 2% L-glutamine (Invitrogen) and 1% pen-strep (Invitrogen).
7. MCDB-201 (Sigma): DMEM (Invitrogen) 1:1 used in some experiments.
8. Coating medium: 10–20% horse serum/DMEM filtered with a 0.22 µm filter.
9. Chick embryo extract (CEE) (6-650-05, Imperial).
10. 50×18 mm non-tissue culture petri dishes (Sterilin, UK).
11. 24-well plates (Falcon).
12. 0.22 µm filters for sterilisation.
13. 100 µm (or 40, 70 µm) nylon mesh (Becton Dickinson) for filtering away myofibres from cells.
14. Pasteur pipettes: these are cut with a diamond knife and adapted to different bore sizes for the particular muscle. Smooth the cut edge by polishing using a flame to prevent damaging the fibres.
15. Tissue culture incubator: 37°C and 5% CO_2.
16. Carrageenan (C3889, Sigma).

2.3. BrdU Solutions

1. For labelling cells in vitro, we often used a final concentration of about 10^{-7} M of BrdU (Sigma Ultra) diluted in PBS from a 10^{-3} M stock solution, stored in aliquots at −20°C.
2. For in vivo injections, the BrdU (Sigma Ultra) is dissolved in PBS and stored as aliquots of 30 mg/ml at −20°C. The stock solution is diluted 10× in PBS before use (3 mg/ml final) and pre-warmed to room temperature or 37°C before injections are done. Intraperitoneal injections are done using a syringe equipped with a 27–30 gauge needle.

2.4. Immuno-fluorescence

1. Phosphate-buffered saline (PBS, without bivalent cations Ca^{2+}, Mg^{2+}): Prepared from 10× stock solution (Invitrogen) at room temperature.
2. Methanol (−20°C) (Merck).
3. Permeabilisation buffer solution: 0.2% (v/v) Triton X-100, 0.5% bovine serum albumin fraction V (BSA, Sigma), 1 mM sodium azide in PBS.
4. Washing buffer solution: 0.001% (v/v) Tween-20, 0.1% BSA, and 1 mM sodium azide in PBS (*see* **Note 3**).
5. Blocking solution for isolated muscle fibres: 10–20% filtered heat inactivated goat serum in PBS with 1 mM sodium azide in PBS.
6. 4% solution of paraformaldehyde (PFA) in PBS prepared freshly for each experiment, or thawed and used once from a −20°C stock.

7. 70% acid ethanol (Merck) solution in water, prepared by dissolving of 96% ethanol with distilled water and addition of two drops of 12 M (37%) hydrochloric acid (HCl) and cooled to 4°C before use.

8. DNA denaturation buffer: 2 N HCl solution in distilled water freshly prepared before use. Do not store more than one week.

9. Post-HCl neutralisation buffer: 0.1 M sodium borate solution in distilled water.

10. DNAse buffer: crude DNAse I (Roche) is dissolved in ice-cold DMEM on ice, and used immediately at 2,000U/2–5 × 10^5 cells in DMEM (150 µl total) for 30 min in a water bath at 37°C with agitation.

2.5. Constructs and Plasmids

Mouse p66 Numb cDNA was engineered and linked with EGFP as a fusion protein (fused portion shown below) separated by a flexible GlySer linker *(7)*. The EGFP contains a Val164Ala mutation in pEGFP-C1 (Clontech; a gift from H. LeMouellic). The EGFP sequence terminates with a 'strong' translation stop sequence containing opal (TGA), ochre (TAA) and amber (TAG) stop codons that are randomised, in triplets, and in three reading frames (TGA TAA TAG G TGA TAG TAA G TAA TAG TGA). We found that Numb overexpression may have deleterious effects on cell behaviour during mitosis. An H2B-mRFP fusion was made by PCR amplification of mouse H2B (H2B-F1: 5′-CGCG AGA TCT GAT ATC ACC ATG CCA GAG CCA GCG AAG TCT GCT-3′; H2B-R1: 5′-TCA CCA AGT ACA CCA GCG CTA AG-3′). This PCR fragment was fused in frame at the BglII site to linker-EGFP (see below) and linker-mRFP (PCR amplified from mRFP, which was kindly provided by R. Tsien *(8)*.

Numb-EGFP fusion junction:

Glu Ile Glu Leu Arg Ser Gly Gly Gly Gly Ser Gly Gly Gly Gly Ser Gly Gly
GAA ATA GAA CTT <u>AGA TCT</u> GGG GGT GGA GGC TCC GGC GGG GGT GGA TCT GGA GGT
　　　　　　　　　BglII

Gly Gly Ser Gly Gly Gly Gly Leu Ala Leu Pro Arg Ala Thr Met Ser Lys
GGG GGC AGC GGC GGA GGT GG<u>G CTA GC</u>G CTA CCG CGG GCC ACC ATG AGC AAG
　　　　　　　　　　　　　　　NheI

Numb; Linker ([Gly₄Ser]X3, Gly₄); GFP

H2B-mRGFP fusion junction:

Thr Ser Ala Lys Gly Ser Gly Gly Gly Gly Ser Gly Gly Gly Gly Ser Gly Gly
ACC AGC GCT AAG <u>GGA TCT</u> GGG GGT GGA GGC TCC GGC GGG GGT GGA TCT GGA GGT
　　　　　　　　　BglII

```
Gly Gly Ser Gly Gly Gly Gly Leu Ala Cys Arg Asn Ser Ile Ser Ser Leu
GGG GGC AGC GGC GGA GGT GGG CTA GCC TGC AGG AAT TCG ATA TCA AGC TTA
                                NheI
Asp Pro Met Ala Ser
GAT CCG ATG GCC TCC
```

H2B; Linker ([Gly₄Ser]X3, Gly₄); mRFP

2.6. Videomicroscopy

1. Glassware: 35 mm diameter glass-bottomed dishes (MatTek), glass thickness no. 0 (0.085–0.013 mm), coated with Matrigel or poly-D-lysine. In experiments involving tracking of LRCs, glass-bottom dishes with grided glass coverslips (thickness no. 2) are used to simplify localisation of recently divided cells after staining with BrdU antibodies. The dish is filled with medium and covered with a large coverslip (50 mm; Menzel-Glaser, In vitro diagnostics, Menzel GmbH & Co.) and sealed with vacuum grease while minimising bubbles in the sealed dish.

2. Spinning disc confocal microscope (Perkin Elmer) controlled by PC computer running original software from (Perkin Elmer) was used for imaging satellite cell derived myoblasts expressing Numb-GFP fusion protein. 20×, 40× oil objectives.

3. An inverted microscope (Ziess Axiovert 200 M), equipped with a cooled CCD camera (Photometrics CoolSNAP monochrome), controlled by PC computer, running Axiovision software v.4.5 (Zeiss). 10× air Ph1 objective.

4. Heating stage (Zeiss).

5. Image and cell tracking analysis: Power Mac G5 (Apple) station, MacOS × 10.4, running Adobe Photoshop and ImageJ v1.37 (NIH) and Imaris (Bitplane) softwares and Improvision Volocity.

3. Methods

Skeletal muscle fibres are routinely obtained from the Extensor Digitorium Longus (EDL), Tibialis Anterior (TA), and Soleus muscles located below the knee. Generally, the cleanest preparations of satellite cells are obtained from the Extensor Digitorium Longus. The dissections are carried out using a dissecting microscope and care is taken to keep the preparations sterile particularly during the muscle fibre dissociation and washings. The single fibres are subsequently transferred to the tissue culture incubator.

For most of the experiments, the satellite cells are obtained from single fibre preparations.

3.1. Dissection of Muscles below the Knee

For all dissections, it is necessary to remove the hair of the hindlimb by shaving with a scalpel or directly removing the entire skin from the leg (like a glove). All of the manipulations should be performed without touching the muscle, to prevent any damage that would compromise myofibre integrity.

1. Dissection of the Extensor Digitorium Longus (EDL): The mouse is positioned on its side. The tail is pinned in a distal position, and the leg is pinned through the toes into cork surface (**Fig. 19.2** and Color Plate 18). Dissection is done under a dissecting microscope. The lower four tendons

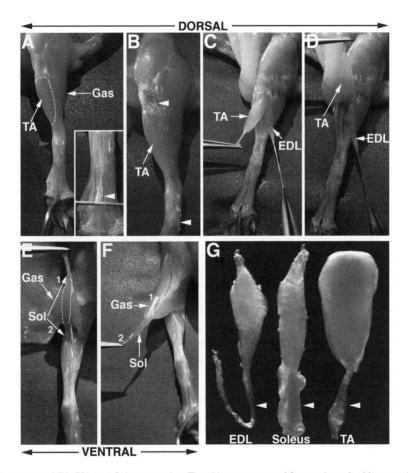

Fig. 19.2. Dissections of TA, EDL and Soleus muscles. The skin was removed from a 4-week-old mouse to reveal the underlying muscles. (**A–D**) Removal of TA and EDL muscles. (**E-F**) Removal of Soleus muscle. Lower tendon (1) is sectioned first, the muscles are lifted, then the upper Soleus tendon (2) is sectioned. Note in older mice, the soleus appears more red and distinguishable. G) Actual sizes of dissected muscles with associated tendons (white arrowheads). Muscle attachment proximal to the knee is at the top of the photo. See main text for details. Red arrowhead, cartilage bridge in foot under which tendons are transit. White arrowheads, tendons. (*See* Color Plate 18)

connecting the EDL to the four digits are cut with fine scissors (*see* **Fig. 19.2A**). Cut in the most distal part from the muscle to leave long tendons easy to manipulate with the forceps without touching the muscle. For the EDL and the TA, both of the lower tendons are released from a small cartilage bridge present close to the base of the muscles (**Fig. 19.2A**, red arrowhead). The connective tissue is cut at the base of the TA and EDL, to facilitate later extraction of the EDL. The upper tendon of the EDL, which is located close to the external face of the knee, is cut with scissors. Two close tendons are present in this area; section both of them and check that they are released from any attachment of connective tissue (**Fig. 19.2B**). Once both tendons are severed, pull gently the lower EDL tendon and the muscle should slide under the TA. If the entire bundle is not released, check that both of the tendons at the level of the knee are severed. The relative positions of these muscles appear in **Fig. 19.2C, D**.

2. Dissection of the Tibialis Anterior (TA): Follow instructions for the removal of the EDL. The lower tendon is cut at the most distal part from the muscle. The muscle is gently pulled upward away from the leg by releasing simultaneously the connective tissue on both sides of the TA. Since the upper tendon of the TA is not always visible and the TA is attached with a broad face at the knee, when arriving at the level of the knee, using fine tip scissors, cut as close as possible the base of the TA by applying the scissors against the knee bone (**Fig. 19.2D**).

3. Dissection of the Soleus: The mouse is positioned on its belly, with the back of the leg facing up, and the toes are pinned to the cork. Section the large white bundle of tendons in the lowest part of the Gastrocnemius. By holding this bundle of ligaments with forceps, the muscle mass is pulled backwards gently while sectioning the connective tissue on both sides with fine scissors. As the Gastrocnemius is lifted from its position, the Soleus becomes apparent on the underside of the Gastrocnemius, and it is distinguishable by its red colour (less so in younger mice). The upper tendon of the Soleus is localised and sectioned, while holding the Gastrocnemius and Soleus muscles firmly with forceps (**Fig. 19.2E**). While holding the lower tendon, pull the Soleus towards you and downwards very gently, releasing the muscle mass from its connective tissue attachments to the Gastrocnemius. Since the lower tendon of the Soleus is tightly linked to that of the Gastrocnemius, both are sectioned and these muscles are lifted to reveal the Soleus within the underside of the Gastrocnemius (**Fig. 19.2E**). The upper tendon of the Soleus is then sectioned and this muscle is removed (**Fig. 19.2F**).

3.2. Isolation of Live Skeletal Muscle Myofibres

For the isolation of highly pure satellite cells (*see* **Note 4**), it is important to obtain undamaged single myofibres free of cellular debris and interstitial contaminating cells. A number of precautions are therefore taken to ensure that myofibres remain intact and do not hypercontract (*see* **Note 5**). We usually isolate myofibres from the muscles below the knee, however, single fibres can be isolated from virtually any muscle mass if precautions are taken not to touch or damage the muscle itself. The protocol developed in the laboratory of T. Partridge (with help from L. Heslop and P. Zammit) was used *(9)*, with some modifications. Mice (often 4–18 weeks old) were killed by cervical dislocation.

1. Petri dishes 50 × 18 mm in size are coated by adding 15 ml of coating medium to dishes successively, then immediately removing it. Pasteur pipettes are coated with coating medium. This prevents isolated muscle fibres from attaching to the dish during the disaggregation and cleaning procedure.

2. Most frequently the EDL, Soleus or TA muscles are used (*see* **Note 6**). These are located below the knee. The distal tendons are cut without touching the muscle itself. This is critical as touching the muscle itself will cause the fibres to contract.

3. The dissected muscle is placed in 0.1% collagenase (5 ml) in sterilin 7 ml bijou container white cap plastic tubes and incubated in a shaking water bath at 37°C for 60–90 min with gentle agitation. Larger muscles require longer incubation times (about 20–30 min more). The mouse EDL requires about 60 min. Muscle from older mice requires longer to digest due to an increased amount of connective tissue.

4. When individual myofibres are observed to be coming away from the edge of the muscle bundle itself, remove the sample from 37°C.

5. Gently suck the muscle up and down repeatedly with the adapted, coated Pasteur pipette. This action gradually releases individual myofibres from the main muscle mass. Depending on the muscle type, considerable debris and interstitial cells can be released during the tituration of the muscle (*see* **Note 7**).

6. Wash single myofibres by serial passage into fresh Sterilin dishes containing pre-warmed medium. As myofibres are purified, they can be stored in the incubator until the isolation procedure is terminated.

7. At this stage, purified myofibres can be fixed for immunostaining (*see* **Note 8**) or cultured to release satellite cells.

8. For culturing of single myofibres, place individual fibres in Matrigel coated dishes (*see* **Note 9**).

9. Under a dissecting microscope, transfer single myofibres with a 1 ml pipettman carrying a coated tip, to a new Petri dish

leaving behind any debris to reduce contamination from other cell types (e.g., fibroblasts, adipocytes).

10. Remove the 24-well plate from the incubator and take off the lid to allow any moisture to evaporate.

11. Add one fibre to the center of each well. Place the fibre in the well, rather than squirting it in, to prevent fibre damage.

12. Add 0.5 ml of plating media per well, slowly, to prevent dislodging the fibre.

13. Place the plate in the incubator to allow satellite cells to leave the fibre. The first cell division takes place between 30 and 36 h, but satellite cells can leave the myofibres well before mitosis.

3.3. Satellite Cell Cultures and BrdU Labelling In Vitro

To grow satellite cells for analysis, and for pulse-chase experiments using the nucleotide analog BrdU, primary satellite cells are routinely derived from purified single myofibres.

1. The myofibres and associated satellite cells are cultured in medium consisting of a 1:1 ratio of MCDB-201:DMEM, with 20% fetal bovine serum (FBS; Invitrogen), 2% steroid-free Ultroser, ITS, and 1% penicillin/streptomycin at 37°C in 5% CO_2.

2. Freshly isolated myofibres (400–600) from EDL muscle are resuspended in 10 ml of freshly prepared growth medium and plated on Matrigel-coated T35–T75 flasks. In some experiments, on the following day, BrdU is added to a final concentration of 2×10^{-7} M for up to 96 h to label proliferating cells. The medium is changed after 48 h and fresh BrdU is added at the same concentration. To avoid dislodging the fibres, for the first addition of BrdU, a 2× stock in medium is prepared and one volume in added to the dish.

3. The BrdU is removed after 96 h and the cells are trypsinised and filtered through a 100 μm nylon mesh to remove the myofibres.

4. The cells are then centrifuged and resuspended in fresh growing medium without BrdU (start of chase period) at the desired plating densities (high) of 18,000–20,000 cells/cm². Mitotic cells are collected after 24–96 h of chase by mitotic shake-off (*see* **Section 3.4**).

3.4. Collection of Dividing Cells by Shake-Off

1. Satellite cells are grown on T75/T35 ventilated flasks.

2. Wash 3- to 7-day-old satellite cell cultures once with pre-warmed medium and leave in 5–6 ml of medium for 30 min.

3. The low adhering mitotic cells are released from the surface of the flask into the medium by gentle shake-off.

4. The medium with detached cells is removed and placed into a 50 ml collecting tube. The flasks with attached cells are rinsed

briefly with 1–2 ml of fresh medium to collect the rest of the dividing cells, and these cells are add to the 50 ml tube collection tube.

5. The flasks with attached cells are filled with 4–5 ml of fresh medium and placed in the incubator for 30–90 min for the next round of mitotic shake off.

6. The cells collected in the 50 ml tubes are fixed by adding pre-warmed 4% PFA in PBS (37°C) in the medium (2% PFA final), and stored at room temperature (30 min for staining with Numb antibody). *See* **Note 10**.

7. Mitotic cells are collected again and fix in the same manner as above, and this is repeated several times until the 50 ml tube is filled. After 30–60 min incubation at room temperature, the fixed cells are stored at 4°C.

8. The fixed cells in 50 ml tubes are centrifuged for 15 min, 1,600–1,800 rpm (about 400 g) at 4°C.

9. The supernatant is discarded and the fixed mitotic cells are resuspended very gently by dissociating in cold washing buffer (0.1% BSA, sodium azide, 0.001% Tween-20 in PBS) with 1 ml pipettman until all clumps dissociate.

10. The cell suspensions are transferred from all of the 50 ml collecting tubes into a 15 ml tube.

11. The cells are incubated on ice for at least 10 min and centrifuged for 15 min, 1,600–1,800 rpm (about 400 g) at 4°C. The supernatant is discarded by aspiration and the cells are resuspended carefully in fresh cold washing buffer. After the second round of centrifugation in the 15 ml tubes, the cells are resuspended in 200 µl of washing buffer. Centrifugation in the 15 ml tubes (second round) allows the concentration of the cell suspension. Although it is a washing step, and it is preferred to use large volumes of washing buffer and to reduce the PFA concentration to minimum.

3.5. Immunostaining with Anti-Numb Antibody

Numb protein has a cortical localisation with some protein located in the cytoplasm. Using standard immunostaining protocols, the asymmetric staining of Numb can be lost (*see* **Note 11**). Co-staining with other antigens, particularly those in the nucleus is therefore problematic. To obtain reliable results, the staining protocol for Numb is done first and subsequently other antigens are revealed (**Fig. 19.3** and Color Plate 19). Since staining for BrdU antigen required antigen exposure from the DNA often by HCl treatment, in some cases when double stainings are done, cells are treated with DNAseI rather than HCl to expose the BrdU antigen.

1. For mitotic shake-off, *see* **Section 3.4**; for antibody blocking solutions, *see* **Section 2.4**.

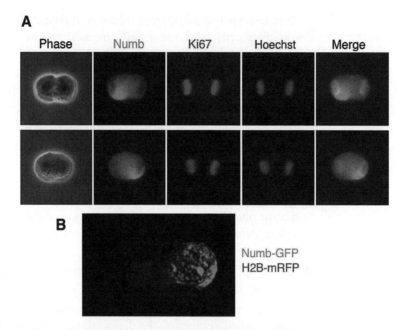

Fig. 19.3. Asymmetric segregation of Numb in non-differentiating satellite cell-derived myoblasts. (**A**) Two examples of asymmetric distribution of endogenous Numb protein in mitotic satellite cell-derived myoblasts after 4 days in culture; immunocytochemistry with anti-Numb and anti-Ki67 antibodies; (**B**) p66Numb-EGFP and H2B-mRFP fusion proteins were overexpressed in satellite cell-derived myoblasts using pCAG-p66Nb-EGFP and pCAG-H2B- mRFP plasmids (transfected at 96 h after plating of myofibres). At 4 days after plating, cells were harvested by mitotic shake-off, re-plated at low density and grown for several hours on poly-D-lysine-coated dishes. The cells were fixed and stained with anti-Numb antibody (green), and visualised using EGFP and mRFP epifluorescence. Phase contrast on left; reconstituted confocal stack on right was rendered with Imaris software. Note asymmetric distribution of the majority of the Numb-GFP protein to one pole of this cell. (*See* Color Plate 19)

2. The cells are incubated in primary anti-Numb rabbit polyclonal antibody overnight at 4°C.

3. The primary anti-Numb antibody is removed by centrifugation in 14 ml washing buffer for 15 min, 1,600–1,800 rpm (about 400 g) in a 15 ml tube. Repeat for a total of two times.

4. The cells are stained with secondary fluorescent antibodies (1/300 final) diluted in washing buffer for 3 hr with or without agitation, at room temperature. For certain fluorescent-coupled secondary antibodies, it is important to clarify them by brief centrifugation in a microcentrifuge at 10,000 rpm (about 10,000 g) at 4°C in the staining buffer. This reduces unspecific fluorescent clumps in the solution.

5. The secondary antibodies are washed away by centrifugation in the washing buffer at 4°C for 15 min at 1.600–1.800 rpm (about 400 g).

6. The secondary antibodies are fixed in 4% PFA in PBS (total volume 1 ml) for at least 30 min at room temperature with frequent but gentle and slow agitation every 10 min.

7. The post-fixation is stopped by resuspending the fixed cells in 1–3 ml permeabilisation buffer containing 0.5–0.2% Triton-X100, 0.1% BSA in PBS on ice for 10–15 min.

8. The cells are washed twice in the washing buffer by centrifugation for 15 min, 1,600 rpm at 4°C.

3.6. DNase Treatment and anti-BrdU Staining: Mitotic Shake-Off

1. The cells are collected by mitotic shake-off and fixed as described above (*see* **Section 3.4**). The cell pellet is resuspended in about 0.5 ml of permeabilisation buffer and incubated for 10–25 min on ice.

2. To wash the cells, 14 ml of 0.1% BSA in cold DMEM is added to the tube and the cells are centrifuged for 15 min, 1,600–1,800 rpm (about 400 g) at 4°C. Repeat for a total of two times.

3. The cell pellet is dissociated gently in freshly prepared solution of DNAse I (Roche) in cold (4°C) DMEM 2,000 U/2–5 × 10^5 cells (150 µl total) and the cells are incubated for 30 min at 37°C in a waterbath with gentle agitation.

4. The cells are resuspended in 14 ml of washing buffer and centrifuged for 15 min, 1,600–1,800 rpm (about 400 g) at 4°C.

5. After the wash, the cell pellet is resuspended in 0.1 ml of primary antibody and the cells are incubated at 4°C overnight. Several anti-BrdU antibodies were tested (sheep polyclonal, Abcam 1/500; or rabbit polyclonal, Megabase, 1/5,000–1/10,000; or mouse monoclonal antibody, Sigma, BU-33, 1/200–1/1,000).

6. The cells are washed by centrifugation for 15 min, 1,600–1,800 rpm (about 400 g) at 4°C and stained with appropriate secondary antibodies (AlexaFluor 350, 488, donkey anti-sheep biotinylated; Cy3-strepavidin) in washing buffer on ice for 1–2 h.

7. The cells are washed in staining buffer by centrifugation and the cell pellet is resuspended in the mounting medium (70% glycerol/PBS, in 50 mM Tris (pH 7.5)).

8. For microscopic observation, 50–100 µl of stained cell suspension is mounted on glass slides and covered with a glass coverslip. The preparation is examined with an upright fluorescent microscope or spinning disk confocal microscope.

3.7. BrdU Staining on Cells Attached to the Culture Dish

1. The dishes or culture flasks with attached satellite cells (poly-D-lysine) are washed very gently at room temperature in serum-free DMEM.

2. The medium is then aspirated and the cells are fixed with cold methanol (−20°C) for 5 min in the −20°C freezer.

3. After removal of the fixative, the cells are air-dried for several minutes at room temperature.

4. The cells are reconstituted in 2–4 ml of PBS for 10 min at room temperature.

5. The PBS was washed out two to three times with freshly prepared 2 M HCl to change the pH as quickly as possible, and the cells were incubated in an excess volume of 2 M HCl solution for 30 min at room temperature.

6. The HCl treatment was stopped by performing several washes with PBS or freshly prepared 0.1 M sodium borate (pH 8.5) and the cells are incubated in PBS for 10–20 min at room temperature.

7. The cells are washed two to three times in washing buffer (with BSA) for 10 min and stained with anti-BrdU antibodies for 2 h at 4°C or at room temperature.

8. The primary antibodies are washed out twice with washing buffer for at least 10 min and the cells are incubated with secondary antibodies for 1–2 h at room temperature.

9. The cells are then washed in the staining buffer three times for 5 min and covered with the mounting medium (70% glycerol/PBS, in 50 mM Tris (pH 7.5)).

3.8. Double Immunofluorescence with anti-Pax7 and anti-BrdU Antibodies on Cells Attached and Grown on poly-D-Lysine

1) The cells are washed with serum-free DMEM and fixed with 2% paraformaldehyde in PBS for 10 min.

2) The cells are washed with washing buffer twice for 5 min and permeabilised with permeabilisation buffer for 5–10 min on ice.

3) The cells are washed again several times with washing buffer for 15 min and incubated with anti-Pax7 mouse antibody (1/5 in washing buffer) overnight at 4°C.

4) The cells are washed twice (5 min each) with washing buffer and incubated with secondary antibodies for 2–3 h with AlexaFluor 594 anti–mouse secondary antibodies diluted in washing buffer at room temperature.

5) The cells are washed twice in PBS and post-fixed for 30 min in 4% PFA in PBS at room temperature.

6) The cells are washed twice in PBS and permeabilised in 0.1–0.2% Triton-X100 in PBS for 10 min.

7) The cells are washed in PBS three times for 5 min each.

8) To prepare for the anti-BrdU antibody staining, the cells are fixed with cold methanol (−20°C) for 5 min in the freezer.

9) The fixative is then removed and the cells are air-dried for several minutes at room temperature.

10) The cells are reconstituted in PBS for 10 min at room temperature.

11) The PBS is washed out two to three times with freshly prepared 2 M HCl and the cells are incubated in an excess volume of 2 M HCl solution for 30 min at room temperature.

12) The HCl treatment is stopped by performing several washes with PBS or freshly prepared 0.1 M sodium borate (pH 8.5) and the cells are incubated in PBS for 10–20 min at room temperature.

13) The cells are washed two to three times in washing buffer (with BSA) for 10 min and stained with polyclonal rabbit anti-BrdU antibody (1/2,500 dilution in washing buffer, Megabase, USA) for 2 hr at 4°C or at room temperature.

14) The primary antibody is washed out twice with washing buffer for at least 10 min and the cells are incubated with anti-rabbit secondary antibody diluted at 1/300 in washing buffer for 1–2 h at room temperature.

15) The cells are then washed in the staining buffer three times for 5 min and covered with the mounting medium (70% glycerol/PBS, in 50 mM Tris (pH 7.5)).

3.9. Pulse-Chase Labelling In Vivo with BrdU

A) **Pulse-chase during early postnatal growth**

1. Mouse pups at early postnatal stages (P3) are injected with single doses of BrdU at a dose of 25 mg/kg, intraperitoneally twice a day for six consecutive days. The period of chase starts at the end of the last injection. The volume of BrdU injected depends on the weight of the pups.

2. The mice are chased for 4–10 weeks. When genetically modified mice are to be used, such as Myf5$^{nlacZ/+}$ to mark satellite cells, they are phenotyped by X-gal staining or genotyped by PCR (9, 10) after or during the chase period. Some growth retardation and hair loss can occur if too many doses of BrdU are administered.

B) **Pulse-chase after cryogenic injury of muscle**

1. Mice that are 6–8 weeks old are anesthetised with 0.5% Imalgene/2% Rompun and left hindlimbs are immobilised with tape. The skin overlying the limb is shaved and opened with sterile scalpel under aseptic conditions.

2. The TA muscle is freeze-injured by application of liquid-nitrogen chilled copper twice for 14 s.

3. The wound is closed with a silk suture and the mouse is left overnight in a cage pre-warmed to 37°C cages for proper recovery.

4. Starting from 20 hr post-injury, the mouse receives intraperitoneal injections of BrdU at 30 mg/kg three times a day (9:00 am, 3:00 and 9:00 pm) for 3 days.

5. The mouse is killed at a desired time point. The chase period can last up to 2–3 months. Contralateral control TA muscles are also collected together with regenerating counterparts to assess if the injury provokes systemic effects which may influence muscle regeneration.

6. The muscles are digested with Collagenase I in DMEM to isolate muscle fibres (**Sections 3.1 and 3.2**). The resident satellite cells are stained with anti-BrdU antibody or cultured before analysis. In some experiments, isolated muscles are fixed, sectioned and reacted with the appropriate antibody.

3.10. BrdU Staining of Freshly-Isolated Myofibres

1. 20–100 undamaged myofibres are placed in 2 ml U-shaped Eppendorf tubes and the fibres are collected by gravity sedimentation (leave longer for fibres from very young mice). Sedimentation is monitored by eye, then the medium is removed, leaving 1 ml of the medium with sedimented fibres in the bottom of the tube.

2. The fibres are then fixed by adding pre-warmed 4% PFA in PBS (final 2% PFA) for 10 min at room temperature. The fibres are then washed three times with washing buffer and permeabilised with 0.2–0.5% Triton-X100 (final concentration 0.1–0.2%) for 5–10 min at room temperature.

3. The fibres are washed two times with washing buffer and equilibrated in serum-free DMEM on ice. The supernatant is removed and fibres are incubated in freshly prepared DNAseI (1,000 U/ml; Roche) solution in DMEM at 37°C for 30 min.

4. The fibres are then washed twice with cold washing buffer and blocked with 10–20% filtered goat serum in 1–2 ml of PBS for 30 min at room temperature.

5. The fibres are incubated with primary antibodies (1/500 monoclonal anti-BrdU (Sigma) and 1/1,000–1/2,000 rabbit anti-β-gal antibody to identify satellite cells from the Myf5$^{n\text{-}lacZ/+}$ mice) diluted in 0.35% lambda-Carrageennan in PBS overnight at 4°C.

6. After incubation with the primary antibody, the fibres are washed twice with PBS at room temperature for 20 min then incubated with secondary antibodies diluted in 0.35% lambda-Carrageennan for 2 h room temperature.

7. The fibres are washed in PBS three times, stained with Hoechst (1/500 of 1 mg/ml) for 20 min and mounted on a glass slide in mounting medium (*see* **Section 3.8.15**; **Fig. 19.4A** and Color Plate 20).

3.11. Cloning of Freshly Isolated Satellite Cells

1. Freshly isolated myofibres (200–1,000) are incubated in a freshly prepared and filtered solution of 0.2% Collagenase I/0.5% Dispase I in DMEM for 30–60 min at 37°C.

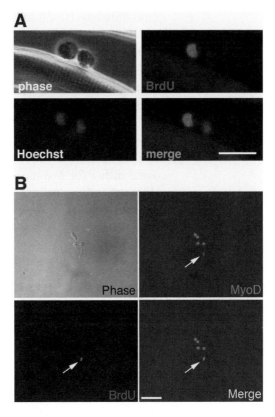

Fig. 19.4. Asymmetric segregation of template DNA strands in label retaining cells. (**A**) Asymmetric segregation of BrdU label to only one daughter satellite cell after mitosis, on a freshly isolated EDL fibre. $Myf5^{nlacZ/+}$ mice were pulsed with BrdU from P3-P7; chase for 7 days. (**B**) Clonal analysis of LRCs after 4 weeks chase in vivo, 72 h in culture. Antibody stainings for detecting MyoD protein and BrdU in the nuclei of satellite cell derived myoblasts. The LRC in this case has a lower expression of MyoD. Scale bars: A, 15 μm; B, 40 μm. (*See* Color Plate 20)

2. The fibres are dissociated by pipetting and the satellite cell suspension is filtered through 100 μm nylon mesh to remove fibre fragments.

3. The satellite cell suspension is centrifuged in the presence of 10% FCS in DMEM in 15–50 ml tubes for 20 min, 1,600–1,800 rpm (about 400 g) at room temperature.

4. The satellite cell pellet is diluted in growth medium and plated on poly-D-lysine coated 35 cm Petri dishes at a density of 300 cells per dish. At this plating density after 48–72 h, we observe 15–20 small, compact and well-spaced myoblast colonies. These are fixed and stained for the presence of myogenic markers and BrdU. If the concentration of satellite cells in the digestion buffer is high enough, they can be diluted in the growth medium and plated directly, thereby eliminating the centrifugation step. The medium in this case should be changed as soon as the cells attach to poly-D-lysine coated dish

(approximately 10–30 min). The assessment of clonogenicity should be done soon after plating by direct observation under a microscope.

5. Individual satellite cells can also be plated as single cells using a mouth pipette onto poly-D-lysine coated Terasaki or 96-well plates. Although the plating efficiency using this technique is low (approx. 25%; higher in 96-well) and requires practice as well as patience, the clonogenic argument is strengthened further.

6. Alternatively, single fibre can be plated and grown on Matrigel coated dishes for 3–6 days. They are then trypsinised, filtered (100 μm) and the cells are plated at clonal density on poly-D-lysine coated 3.5 cm tissue culture dishes or flasks at the density 10–100 cells/cm^2. The cells are fixed after 24–72 h with PFA or methanol for analysis (**Fig. 19.4B**).

3.12. Videomicroscopy of LRC Cell Divisions In Vitro

1. Several hundred freshly isolated myofibres and associated satellite cells are cultured for 3–4 days in Matrigel-coated flasks in the growth medium in the incubator.

2. The cultures are trypsinised and approximately 1,000 cells are plated on the Matrigel-coated MatTek grided glass-bottomed dishes (*see* **Note 12**).

3. The cells are seeded and allowed to attach on the surface of the coverslips (3–4 h). The medium is then changed and the cells are overlayed with sterile 30×30 coverslips, thereby generating a chamber for cells which is filled with fresh medium.

4. The dishes are then filled completely with fresh medium and sealed with another thick and large coverslip (50 mm) covered with vacuum silicone grease at the periphery. The coverslip is gently pressed from the top to remove the excess medium and air bubbles, thus creating a hermetic chamber completely filled with the medium, where gas composition does not change significantly for about 48 h.

5. The dishes are placed on a conventional heating stage used for time-lapse imaging on an inverted microscope (Zeiss Axiovert with a Zeiss heating chamber) and left for 1 h to equilibrate. The temperature of the stage is adjusted to 37.4–38.4°C.

6. The cells are filmed using a 10× phase contrast air objective and time-lapse images are routinely acquired every 4–8 min for 12–18 h or longer. To increase the chances of recording mitotic events, several fields are recorded using the MosaiX module of Axiovision software v.4.5 (Zeiss).

7. Immediately following the last frame, the chamber is opened, some medium is removed and 4% PFA pre-warmed to room temperature is added to the dish to fix the cells (2% PFA final) for 20 min at room temperature.

8. The cells on the grided coverslips are stained with anti-BrdU antibody and the positions of BrdU-positive cells are identified, imaged and recorded. Some cells are lost during the antibody staining.

9. Consecutive series of time lapse images are tilled and combined using Axiovision software and exported as a sequential series of tiff images to ImageJ software (NIH) for further analysis using the manual tracking module plug-in. The dividing cells are tracked and cell pairs are located, marked by colour and their positions are recorded.

10. Cell pairs with asymmetric BrdU staining are identified and the corresponding time-lapse sequences of cell divisions are exported to Avi and QuickTime movie formats using ImageJ software.

3.13. Live Videomicroscopy of Satellite Cell Divisions Overexpressing Numb-GFP

1. Single fibre derived satellite cells are plated on glass-bottomed dishes (MatTek) and transfected with Numb-GFP and H2B-mRFP expression vector using Lipofectamine 2000, or infected with lentivirus expressing Numb-GFP.

2. The dishes are sealed as mentioned above, mounted on the heating stage and filmed using a spinning-disk confocal microscope (Perkin Elmer) 48 hr post-transfection. Stacks of 20–40 optical sections are taken with 0.5–1 μm intervals, each 2–5 min for 12–18 h, using a 25× oil immersion objective. EGFP fluorescence is visualised using an FITC filter set (450–490 nm) and the images are acquired using a Hamamatsu digital CCD camera (binning mode 2×2).

3. The time-lapse stack series are exported and rendered using Velocity software (Improvision). Maximum intensity projection time-lapse tiff series are exported and combined as Avi and QuickTime time-lapse movies using ImageJ software.

4. Notes

1. Due to variability, collagenase must be batch tested; concentration ranges between 0.1 and 0.3% final.

2. Matrigel is handled as recommended by the manufacturer to avoid polymerisation and solidification. Briefly, the vial of approximately 10 mg/ml stock Matrigel is thawed on ice overnight in the cold room at 4–8°C. If the temperature increases during this step, the highly concentrated Matrigel will solidify and will be unusable. When thawed, dilute Matrigel to 1 mg/ml with ice-cold DMEM. At this point, diluted Matrigel can be filtered, aliquoted and stored at −20°C. Coat 24-well, Lab-tek (Nunc), or the culture dish of choice with

diluted Matrigel by adding and removing the solution with a 1 ml pipette. Place dishes in the 37°C incubator for at least 30 min to allow the Matrigel to gel. Do not leave longer than 60 min as the Matrigel dries and the fibres no longer adhere. The Matrigel aliquot is thawed on the bench just before use for coating dishes. The unused Matrigel is discarded.

3. In some experiments with single fibres or isolated cells, blocking reagents can be increased by using combinations of heat inactivated goat serum (2–20% final); BSA (up to 2% final) and/or Carrageenan (0.35% final; more often used for myofibres) in 1× PBS. Antibodies are diluted in washing buffer, centrifuged for up to 15 min at 2,000 rpm (about 450 g) and can be stored at 4°C. Sodium azide can be added to 1 mM in PBS.

4. This procedure can be performed in the laboratory under reasonably sterile conditions. The coated tips are changed often after transferring the fibres to the culture plate. Culture medium with high serum is added in the culture hood. The purified myofibres can be accumulated and stored in the incubator until all of the muscles are processed.

5. During release of the myofibres with the Pasteur pipette, care must be taken not to pinch the myofibres between the pipette and the dish. Any physical damage causes hypercontraction of the fibres.

6. Between the EDL, Soleus and TA, generally the cleanest myofibre preparations are obtained with the EDL muscle.

7. To obtain highly purified single fibre preparations, single myofibres are selected individually, transferred to a new dish containing fresh pre-warmed medium and serially cleaned by transferring into new dishes containing pre-warmed medium. Alternately, the transfer can be done once, and debris is then removed from the dish containing the fibres. Care is taken to remove as much debris as possible, including microvessels and connective tissue, and fresh medium is added at the same time. Selecting clean myofibres is a tedious enterprise, but an important step to ensure the purity of the satellite cell preparation. Fibre bundles, doublets and contracted fibres are not chosen. The fibres should be pipetted from the extremities to prevent damage and hypercontraction.

8. For immunostaining: Fibres are fixed for 5 min using warmed (room temperature) 4% paraformaldehyde (PFA). Add the PFA directly to the well. Do not remove the medium first as this will cause the fibres to contract. Rinse with PBS thoroughly. Generally, permeabilisation is done with 0.5% Triton X-100. To reduce background staining, add primary antibodies using 0.35% Carrageenan in PBS as a carrier. In some experiments involving filming of cells or taking pictures, it is

desirable to filter the Matrigel and the medium (0.45 μm) due to the amount of debris present in the preparation. When large numbers of satellite cells are required, a richer medium is used containing 20% fetal calf serum in 50% DMEM/50% MCDB-201, with 2% Ultroser. Under these conditions, differentiation is also delayed. Satellite cells can also be grown in 10% horse serum + 0.5% chick embryo extract in DMEM without Ultroser *(9)*. Under these conditions, satellite cells grow more slowly and differentiate sooner.

9. Matrigel should be handled with care to avoid precocious polymerisation and solidification. Do not leave longer than about 60 min as the Matrigel dries and the fibres no longer adhere.

10. Collected cells should be washed and stained with Numb antibody the same day of fixation. Usually five to seven mitotic shakes off are necessary to collect 200,000 or more cells, which can be aliquoted for separate stainings. For Numb staining, the cells should be fixed for at least 30 min in PFA at room temperature and then stored at 4°C in fixative for several hours before centrifugation.

11. While doing routine immunocytochemistry of fixed mitotic satellite cells in vivo or after culture with anti-Numb antibody, we noted that the staining pattern depends on: (A) fixative used; (B) time of fixation; (C) concentration of nonionic detergents (e.g., Triton-X100, Tween-20) used for permeabilisation. In actively proliferating cultures, when cells are fixed with PFA for 30 min and incubated with as low as 0.001% Tween-20 on ice for 20–30 min, many dividing cells exhibit asymmetric staining. This staining in characterised by a dotted, probably vesicular (100–200 nm) pattern at high (100×) magnification (**Fig. 19.3A**). This pattern was reproduced using Numb-GFP overexpression (**Fig. 19.3B**) and live videomicroscopy in interphase cells. As low as 5 min permeabilisation with 0.01% Tween-20 or Triton X-100 can dramatically reduce anti-Numb antibody staining. In addition, we often observe that after permeabilisation with high concentrations of Triton or Tween detergents, the punctate or vesicular staining pattern was partially lost. This observation suggests that permeabilisation with high concentrations of detergents may not be adequate for the assessment of asymmetric division events and may be artefactual for certain antibody stainings. However, the use of low permeabilisation conditions for antibody staining to reveal localisation of Numb protein, which is cortical and cytoplasmic, does not permit the simultaneous revelation of nuclear antigens routinely and with confidence. To reveal nuclear antigens, as well as BrdU labelled DNA, modifications were made to the protocol: cells stained with Numb

antibodies were post-fixed with PFA for at least 30–60 min. This crosslinks primary and secondary antibodies used to reveal Numb protein, and permits permeabilisation required for subsequent steps. Although after permeabilisation and subsequent washing steps, immunofluorescence of Numb is slightly reduced, this does not affect asymmetrical and the speckled-like pattern of immunostaining. Best results of such an approach are achieved when cells are collected by mitotic shake-off and each washing step is followed by centrifugation. Staining of cells attached to the culture dish can result in the loss of mitotic cells, which are loosely attached to the culture dish. We note that during each centrifugation step, there is a significant loss of cells from the preparation. This should not be compensated by increasing the g-force and time of centrifugation significantly since cells fixed with PFA are fragile.

12. Cell density should not be high to ensure proper tracking of individual cells (ca. 500–1,000 cells/cm^2).

Acknowledgements

We thank T. Partridge, P. Zammit and L. Heslop for kindly demonstrating how to isolate single myofibres, and members of the Tajbakhsh lab for helpful comments. We also thank S. Shorte, P. Roux and E. Perret from the Pasteur Imaging Center for their support. This work was funded by grants from the Pasteur Institute, AFM, ARC, Pasteur GPH 'Cellules Souches' programme, MyoRes (EU Framework 6 project LSHG-CT-2004-511978) and EuroStemCell (EU Framework 6 project LHSB-CT-2003-503005).

References

1. Zammit, P.S., Partridge, T.A., Yablonka-Reuveni, Z. (2006) The skeletal muscle satellite cell: the stem cell that came in from the cold. *J Histochem Cytochem* 54, 1177–1191.
2. Tajbakhsh, S. (2005) keletal muscle stem and progenitor cells: reconciling genetics and lineage. *Exp Cell Res* 306, 364–372.
3. Morrison, S.J., Kimble, J. (2006) Asymmetric and symmetric stem-cell divisions in development and cancer. *Nature* 441, 1068–1074.
4. Shinin, V., Gayraud-Morel, B., Gomes, D., and Tajbakhsh, S. (2006) Asymmetric division and cosegregation of template DNA strands in adult muscle satellite cells. *Nat Cell Biol* 8, 677–682.
5. Cairns, J. (1975) Mutation selection and the natural history of cancer. *Nature* 255, 197–200.
6. Cairns, J. (2002) Somatic stem cells and the kinetics of mutagenesis and carcinogenesis. *Proc Natl Acad Sci U S A* 99, 10567–10570.
7. Mack, M., Riethmuller, G., Kufer, P. (1995) A small bispecific antibody construct expressed as a functional single-chain molecule with high tumor cell cytotoxicity. *Proc Natl Acad Sci U S A* 92, 7021–7025.

8. Campbell, R.E., Tour, O., Palmer, A.E., Steinbach, P. A., Baird, G. S., Zacharias, D. A., and Tsien, R. Y. (2002) A monomeric red fluorescent protein. *Proc Natl Acad Sci U S A* 99, 7877–7882.

9. Beauchamp, J.R., Heslop, L., Yu, D. S., Tajbakhsh, S., Kelly, R. G., Wernig, A., Buckingham, M. E., Partridge, T. A., and Zammit, P. S. (2000) Expression of CD34 and Myf5 defines the majority of quiescent adult skeletal muscle satellite cells. *J Cell Biol* 151, 1221–1234.

10. Tajbakhsh, S., Rocancourt, D., Cossu, G., and Buckingham, M. (1997) Redefining the genetic hierarchies controlling skeletal myogenesis: Pax-3 and Myf-5 act upstream of MyoD. *Cell* 89, 127–138.

11. Kassar-Duchossoy, L., Gayraud-Morel, B., Gomès, D., Rocancourt, D., Buckingham, M., Shinin, V., and Tajbakhsh, S. (2004) Mrf4 determines skeletal muscle identity in Myf5:Myod double-mutant mice. *Nature* 431, 466–471.

Chapter 20

Isolation and Grafting of Single Muscle Fibres

Charlotte A. Collins and Peter S. Zammit

Abstract

Satellite cells are mononucleate muscle precursor cells resident beneath the basal lamina, which surrounds each skeletal muscle fibre. Normally quiescent in adult muscle, in response to muscle damage satellite cells are activated and proliferate to generate a pool of muscle precursor cells, which subsequently differentiate and fuse together to repair and replace terminally differentiated muscle fibre syncytia. Cells prepared by enzymatic digestion of whole muscle tissue are likely to contain myogenic cells derived both from the satellite cell niche and from other populations in the muscle interstitium and vasculature. Single muscle fibre preparations, in which satellite cells retain their normal anatomical position beneath the basal lamina, are free of interstitial and vascular tissue and can therefore be used to investigate satellite cell behaviour in the absence of other myogenic cell types. Here, we describe methods for the isolation of viable muscle fibres and for grafting of muscle fibres and their associated satellite cells into mouse muscles to assess the contribution of satellite cells to muscle regeneration.

Key words: Satellite cell, stem cell, skeletal muscle, muscle fibre, muscle regeneration, single fibre graft, self-renewal.

1. Introduction

Adult skeletal muscle is a normally stable, low-turnover tissue that nevertheless retains a remarkably good ability to regenerate following injury (*1–3*). The main cellular constituent of skeletal muscle is the muscle fibre, an elongate syncytial cell that contracts in response to nervous stimuli and is supported by several hundred postmitotic myonuclei within a continuous cytoplasm. During development, muscle fibre syncytia are formed by the terminal differentiation and fusion of mononucleate muscle precursor cells, and at least in the mammal, remain incapable of further mitosis. The ability of postnatal muscle to regenerate is mainly attributable to a population of undifferentiated precursor cells which are resident beneath the basal lamina that surrounds each fibre, and according to this anatomical criterion, termed satellite cells (*4–6*).

Satellite cells are predominantly quiescent in adult muscle and are formally defined by their anatomical location beneath the basal lamina. Unambiguous identification of satellite cells by anatomical criteria alone requires electron microscopy. However, quiescent satellite cells can also be readily identified by immunofluorescence for a range of molecular markers, including the paired-box transcription factor Pax7, the membrane glycoprotein CD34 (*7*) and the adhesion protein M-cadherin (**Fig. 20.1**). None of these markers is wholly specific to satellite cells, but Pax7 is arguably the most useful since it is expressed by >95% of the population and a good antibody is available (*8, 9*). Injury to muscle results in the rapid activation of satellite cells, which enter mitosis, migrate out of their anatomical niche and proliferate to generate a pool of muscle precursor cells that subsequently differentiate and fuse to repair or replace damaged fibres. Some satellite cell progeny opt out of immediate differentiation and instead become quiescent and reoccupy the sublaminal satellite cell niche, thus maintaining the population of satellite cells (*9–11*).

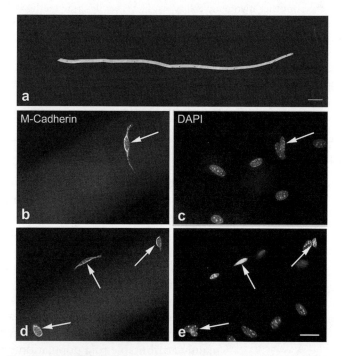

Fig. 20.1. Visualisation of satellite cells on isolated muscle fibres by M-cadherin immunolabelling. (**a**) Isolated mouse muscle fibre. (**b** and **d**) Muscle fibre-associated satellite cells labelled with an antibody recognising the surface antigen M-cadherin. (**c** and **e**) DAPI counterstain of satellite cell nuclei and of the myonuclei within the muscle fibre. Scale bars=200 μm (**a**) and 10 μm (**b–d**).

The ability of satellite cells to both self-renew and generate differentiated progeny satisfies the basic criteria for a tissue-specific stem cell.

In postnatal muscles, the satellite cell population accounts for just 1–5% of total muscle nuclei, with variability between muscles of different age and fibre type. This rarity, combined with the presence of other myogenic cell types within muscle tissue (*12–14*), and a paucity of specific cell surface markers, makes satellite cells a challenge to isolate as a pure population. Recently, it has been demonstrated that pure populations of myogenic cells can be isolated from whole muscle tissue on the basis of small size, non-granularity and CD34 expression (*15*), albeit the satellite cells present within the population have necessarily lost the anatomical criteria which define their compartment. Here, we describe a method for isolating satellite cells within their endogenous anatomical niche, such that the normal association of satellite cell and myofibre is preserved.

A method for isolating single muscle fibres, with their associated populations of sublaminal satellite cells, was first developed by Dr Richard Bischoff, and Dr. Irwin Konigsberg and colleagues. (*16, 17*). In the laboratory of Prof. Terence Partridge, the method was further developed, chiefly by Dr David Rosenblatt, to allow isolation and culture of single muscle fibres from the longer, more fragile *extensor digitorum longus* (EDL), *soleus* and *tibialis anterior* (TA) muscles (*18*). Indeed, this method can be effectively utilised for almost any given muscle. We and others have used the single fibre isolation technique to generate models of satellite cell behaviour in culture (*9, 11, 19, 20*), and in vivo after transplantation of viable muscle fibres into irradiated mouse muscles (*10, 21*).

2. Materials

2.1. Isolation of Viable Single Fibres from Mouse Muscles

1. sDMEM Medium: Dulbecco's modified Eagle's medium (DMEM) with sodium pyruvate and high glucose (GibcoBRL, 41966) (see **Note 1**) supplemented with l-glutamine (G7513, Sigma) to a final concentration of 2% v/v and penicillin and streptomycin solution (P0781, Sigma) to a final concentration of 1% v/v. To 10 mls of sDMEM, add filtered (0.22 μm) horse serum (GibcoBRL 16050-098) to a final concentration of 10% v/v (see **Note 2**).

2. Collagenase solution: Immediately before the dissection, weigh out collagenase type 1 (Sigma C-0130) (see **Note 3**) and prepare a 0.2% (2 mg/ml) solution in sDMEM. About 2 ml solution per muscle is sufficient. Filter-sterilise the

collagenase solution using a sterile syringe with a 0.22 μm filter. Aliquot 2 ml of collagenase solution into each of several 7 ml tubes (one per muscle) and label tubes with an alcohol-resistant marker. Collagenase should be prepared as soon as possible before use and then kept at room temperature until the addition of muscles to tubes.

3. Serum-coated Petri dishes: Take six 50 × 18 mm sterile Petri dishes per muscle and rinse with horse serum. Remove the excess and add 8 ml of sDMEM per dish. Place the dishes in the incubator for at least 30 min before use.

4. Heat-polished, serum-coated Pasteur pipettes: Score around a glass Pasteur pipette using a diamond pen and then snap away the end to create a range of cut pipettes with diameters at the mouth of between 1 and 6 mm. Heat polish the cut ends of the Pasteur pipettes in the hottest part of a Bunsen flame, to partially melt the glass at the mouth and so remove the sharp and jagged edges. Putting a slight bend in the centre of the finer mouthed pipettes using the flame makes manipulation of fibres easier. Coat the heat polished glass Pasteur pipettes with 10% horse serum/DMEM just before use to prevent the fibres from adhering.

2.2. Host and Donor Mouse Strains

1. Host mice: The C57Bl/10ScSn *mdx*-nu/nu (*mdx*-nude) mouse model can be used to generate an optimised environment for donor-derived muscle regeneration. The *mdx* mouse is a genetic and biochemical homologue of Duchenne's (dystrophin deficient) muscular dystrophy that spontaneously arose within a C57Bl/10ScSn mouse colony. The *mdx* mutation in the *dystrophin* gene results in the near-absence of dystrophin protein. *mdx* mice are bred onto a homozygous nude background. Nude mice fail to develop a mature immune system as a result of absent thymus glands and are therefore immunotolerant of grafted cells (22). To provide an optimal environment, host mice should be grafted at 24 days of age and their hindlimbs treated with γ-irradiation 3 days prior to this.

2. Donor mouse strains: The 3F-*nlacZ*-2E transgenic mouse contains seven copies of a construct consisting of 2 kb upstream of the myosin light chain (MLC)-3F transcriptional start site, *nlacZ*-SV40 poly(A) in frame in the second MLC3F-specific exon, 1 kb of MLC3F sequence 3' of *nlacZ*, and a 260 bp 3' MLC1F/3F enhancer (23). β-Gal expression faithfully recapitulates fast myosin expression in the nuclei of all fast muscle fibres, and is absent in the nuclei of satellite cells (23, 7). This mouse is therefore a useful donor strain for investigating the contribution of satellite cells to differentiated muscle. The $Myf5^{nlacZ/+}$ mouse has *nlacZ*-SV40poly(A)

RNA*polII*/*Neo* targeted to the first exon of the *Myf5* gene such that β-gal is produced as a fusion protein with the first 13 amino acids of Myf5 (*24*). Myf5-β-gal fusion protein is expressed by both myonuclei and satellite cell nuclei during the perinatal period. In adults, Myf5-β-gal protein is expressed in the majority of satellite cell nuclei and also in the myonuclei of recently regenerated muscle fibres (*24, 7*), making it a useful donor strain for investigating the contribution of grafted cells to the satellite cell compartment and to recently regenerated muscle.

2.3. Irradiation of Host Mouse Muscles and Grafting

1. Anaesthetic: Mixture of one part Hypnorm® (79 µg/ml fentanyl citrate, 2.5 mg/ml fluanisone, Janssen-Cilag Ltd), one part Midazolam (1.25 mg/ml midazolam, CP Pharmaceuticals Ltd) and two parts sterile water.
2. 4 cm-thick lead jigs designed to shield body of mouse but allow exposure of hindlimbs to irradiation.
3. Glass needles: prepare glass needles from PCR glass Micropipets (10 µl) with plungers (Drummond Scientific Company, 5-000-1001-X10) by drawing them out to a fine point in a hot bunsen flame, and then cutting with a glass knife to give a tip diameter of about 0.5 mm. Fit each needle with a wire plunger and sterilise before use.

2.4. Analysis of Donor Muscle Regeneration in the Graft Site

1. Detergent solution: Potassium-buffered saline (PBS) containing 2 mM $MgCl_2$, 0.02% (v/v) IGEPAL CA-630 (Sigma) and 0.01% (w/v) sodium deoxycholate.
2. X-gal stock: dimethyl sulphoxide containing 40 mg/ml of 5-bromo-4-chloro-3-indolyl-B-d-galactopyronoside (BDH).
3. X-gal diluant: PBS containing 2 mM $MgCl_2$, 0.02% (v/v) IGEPAL CA-630, 0.01% (w/v) sodium deoxycholate, 5 mM $K_3Fe(CN)_6$ and 5 mM $K_4Fe(CN)_6$.
4. Primary anti-dystrophin antibody and an appropriate secondary antibody.
5. Silane-coated glass slides.

3. Methods

3.1. Dissection of TA, EDL and Soleus Muscles from the Mouse Hindlimb

1. Kill mouse by cervical dislocation.
2. Saturate the hind limbs and lower half of body with 70% EtOH and closely shave the hindlimbs using a scalpel. Rinse with 70% EtOH to remove any loose hairs.

3. Pin the mouse onto a dissecting board, with one forelimb pinned through the palm. Pin the contralateral hind limb through the dorsal side of the paw. Place the tail to one side and the other hindlimb to the opposite.

4. Make an incision with the scalpel through the skin from above the knee to the paw.

5. Free the skin from the underlying muscle on both sides of the incision, by gripping the skin with forceps and using another pair of forceps to ease the skin away from the underlying musculature to expose the TA (**Fig. 20.2a**).

6. Locate the four tendons of the extensor digitorum longus on the dorsal side of the paw, and cut them proximal to their insertions on the base of the third phalanx of digits two to five.

7. Locate the EDL tendons (lateral to the tendon of the TA), promixal to the extensor retinaculum (annular ligament) and gently loop them out, thus pulling the cut ends of the tendons through the extensor retinaculum. Ensure that all four tendons are present.

8. Cut the TA tendon proximal to its insertion on the first cuneiform and proximal end of the first metatarsal.

9. Gently cut the epimysium overlying the TA muscle, avoiding damage to the underlying muscle.

10. Firmly grip the tibialis anterior tendon and gently ease it away from the underlying musculature and bone, using forceps to gently disrupt connections (**Fig. 20.2b**).

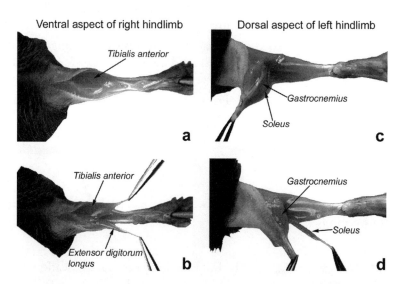

Fig. 20.2. Location of the TA, EDL and soleus muscles in the mouse hindlimb. (**a** and **b**) Partial dissection of the TA and EDL muscles. (**c** and **d**) Partial dissection of the gastrocnemius and soleus muscles.

11. Cut the proximal TA tendons and where unavoidable, the TA muscle itself, to free it from the margin of the lateral condyle, tuberosity and ventral crest of the tibia. Handle the muscle carefully, gripping it only by its distal tendon to avoid damage and place it into a tube of collagenase solution. Do not allow dissected muscles to get cold.

12. Firmly grip all four distal tendons of the EDL muscle and ease the muscle away from the underlying musculature and bone (**Fig. 20.2b**).

13. Expose and cut the proximal tendons where the muscle arises from the lateral epicondyle of the femur and carefully pull the muscle free. Handle only by the distal tendons to avoid damage and immediately transfer the muscle to a tube of collagenase solution. To rearrange the mouse for dissection of the soleus muscles, turn it over so that it is lying flat on its ventral surface and pin the forelimbs through the paws.

14. Cut the skin above the ankle and then pull it back over the leg to expose the gastrocnemius muscle (**Fig. 20.2c**).

15. Cut the Achilles tendon close to the ankle and gently raise the gastrocnemius free of the underlying musculature and bone (**Fig. 20.2c and d**). The soleus is a long thin muscle running along the centre of the anterior surface of the gastrocnemius. In adult animals it can be readily located by its dark red colour which contrasts with surrounding muscles.

16. Whilst holding the Achilles tendon, use a No. 11 scalpel to gently free the gastrocnemius from the underlying tissue until the thin tendon that connects the soleus muscle to the head of the fibula becomes apparent. Gently free the thin tendon with forceps and cut.

17. Carefully grip the proximal tendon of the soleus and gently draw it towards the Achilles tendon, whilst freeing it from the gastrocnemius using a No. 11 scalpel. Avoid any application of tension.

18. Cut the Achilles tendon, and gastrocnemius muscle where necessary, to free the entire soleus and place it into a tube of collagenase.

3.2. Isolation of Myofibres by Enzymatic Digestion and Trituration

1. Incubate the muscles in tubes of collagenase solution in a shaking water bath at 35°C for 60–90 min. As a general guide, an EDL or soleus muscle from a 6–8-week-old mouse requires approximately 60 min of digestion, whilst the larger TA muscle takes approximately 90 min. The precise time depends on both the age and size of the mouse and the activity of the batch of enzyme used, and should therefore be determined empirically. Digestion is complete when

the muscle looks less defined and slightly swollen, with hair-like single fibres seen coming away from the edge of the muscle.

2. When digestion is complete, remove tubes from water bath, wipe dry, and then wipe with 70% EtOH before placing in the culture hood.

3. Remove several Petri dishes of warmed sDMEM from the incubator and place in the culture hood.

4. Fit the largest diameter Pasteur pipette with a rubber pipette filler, rinse with DMEM/horse serum medium, and use to carefully remove the muscle into one of the Petri dishes.

5. Place the muscles in the Petri dishes back into the incubator and remove one dish at a time to isolate myofibres under the microscope.

6. Ideally, place the stereo dissecting microscope directly in the tissue culture hood or lamina flow cabinet; if this is not possible, place on a clean, draft free bench.

7. Place the Petri dish under the stereo dissecting microscope. The muscle should look swollen and much less defined with myofibres starting to emanate from the edges.

8. Using the largest-diameter heat-polished Pasteur pipette (pre-rinsed with DMEM/horse serum medium) take up each muscle and triturate repeatedly in the sDMEM. This procedure will result in highly refractive, hair-like myofibres being liberated from the muscle (**Fig. 20.1a**). In addition, hypercontracted dead fibres, together with fat droplets, tendon and other debris may be seen.

9. Using the largest diameter Pasteur pipette, carefully remove the remaining muscle to a new Petri dish with warmed sDMEM and label accordingly.

10. Using successively smaller diameter heat-polished Pasteur pipettes, continue to gently titurate the muscle to liberate myofibres. Remove any obvious debris.

11. Using a small diameter pipette, collect intact liberated myofibres and place them in a fresh Petri dish of sDMEM, being careful to avoid collecting hypercontracted myofibres and debris.

12. Check the myofibres are smooth and so free of associated endothelium. Serially transfer isolated myofibres through 2–3 further Petri dishes of sDMEM to ensure that any contaminating endothelium, cells and collagenase are removed.

13. Store dishes of prepared myofibres in the incubator at 37 °C with 5% CO_2.

3.3. Grafting of Single Myofibres into Mouse Muscles

3.3.1. Irradiation of Host Mouse Muscles

1. Anaesthetise mice by subcutaneous injection of 40 μl Hypnorm/midazolam preparation.
2. Place anaesthetised mice in lead jigs and arrange such that only the hindlegs are exposed to the radioactive source with the remainder of the body (including tail) shielded by the lead barrier. Tape legs in position. Wrap mice in cotton wool to prevent them from becoming chilled during the procedure.
3. Expose mice to 18 Gray of γ-irradiation delivered at 0.7 Gray/min (*25*) (see **Note 4**).
4. Remove mice from irradiator promptly and place back in cage on a heated pad until fully recovered from the effects of the anaesthetic.

3.3.2. Insertion of Grafts

1. Prepare viable muscle fibres on the day of grafting and maintain in the incubator at 37 °C/5% CO_2 until use.
2. Grafts should be inserted 3 days following irradiation. Mice are anaesthetised using either hypnorm/midazolam, or if preferred, an inhalable anaesthetic such as isoflurane.
3. Cover a heated pad with a sterile drape and place underneath a dissection microscope. Place each mouse under the microscope and tape one hindlimb in position such that the TA muscle is uppermost.
4. Swab each hindlimb with 70% EtOH and use a sterile No. 11 scalpel to make a 1 mm incision at the base of the TA muscle.
5. Place a dish of prepared muscle fibres under the stereo dissecting microscope. Select an intact muscle fibre that is free from damage or adherent debris and carefully draw into a PCR pipette needle in 3–4 μl medium.
6. Hold the needle up to a bright light to check that the muscle fibre is lying straight and undamaged in the barrel of the needle, and has not become folded over on itself.
7. Insert the needle into the incision at the base of the TA muscle, and slowly withdraw whilst gently expelling the myofibre under close observation.
8. Under the stereo dissection microscope, flush out the needle with medium to confirm that the myofibre has been expelled (see **Note 5**).
9. Repeat the procedure with the contralateral hindleg.
10. Keep mice warm on a heat pad until fully recovered from the effects of the anaesthetic.

3.3.3. Analysis of Donor Muscle Regeneration in the Graft Site

1. The graft site is typically analysed 3–5 weeks following insertion of the graft.
2. Kill mice by cervical dislocation and then pin out on a cork dissection board.
3. Dissect engrafted TA muscles and bisect transversely. Embed the pieces of muscle together in gum tragacanth on a cork disc, with the cut sides uppermost and level with each other.
4. Freeze muscles in isopentane cooled in liquid nitrogen. Once completely frozen, remove the muscles and store at −80 °C until analysis.
5. Using a cryostat, cut three or four 7 μm sections at 100 μm intervals throughout the entire length of each muscle. Collect serial sections onto silanised slides and air dry for 30 min before storing at −80 °C.
6. For X-gal staining, sections should be fixed with 0.5% gluteraldehyde, washed in 2 mM $MgCl_2$ and then incubated in detergent solution for 10 min on ice. A solution of the chromogenic β-gal substrate, X-gal, is prepared by diluting X-gal stock to 1 mg/ml in X-gal diluant. Slides should be incubated in X-gal solution overnight at 37 °C, rinsed three times with distilled water, briefly air-dried and then mounted with aqueous mounting medium and coverslips. The bright blue reaction product of X-gal localises β-gal activity to donor-derived nuclei (**Fig. 20.3a**).
7. For dystrophin immunolabelling, unfixed sections should be blocked with serum appropriate to the origin of the secondary antibody, and then incubated with a polyclonal antibody against dystrophin for 1 hr at room temperature, followed by washing and incubation with a fluorescent or peroxidase labelled secondary antibody. The details of the protocol will depend on which primary antibody is used (**Fig. 20.3b**).

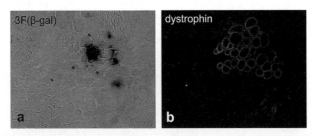

Fig. 20.3. Identification of donor-derived muscle regeneration in an engrafted host muscle. The images show serial sections of an irradiated *mdx*-nude muscle that had been engrafted with a single muscle fibre 3 weeks prior to killing. Satellite cells associated with the donor 3F-*nLacZ*-2E muscle fibre gave rise to a cluster of new muscle fibres identified by X-gal staining to reveal 3F(β-gal) activity (**a**) and in an adjacent section, immunolabelling for dystrophin (**b**).

4. Notes

1. Viability of muscle fibres may be reduced if media is not freshly prepared.
2. Batch testing is required to identify horse serum which is optimal for muscle fibre survival.
3. Collagenase should be batch tested for optimal yield and survival of isolated muscle fibres.
4. It has been demonstrated that the rate of radiation delivery is an important determinant of the subsequent efficiency of graft-derived muscle regeneration (*24*).
5. Muscle fibres that have hypercontracted (appearing short, dark and optically dense) can become trapped in the tip of the needle. In this case it is better to replace the needle.

Acknowledgements

The authors thank Dr David Rosenblatt for the images shown in **Fig. 20.1** and Dr Jennifer Morgan for useful discussions. The lab of PSZ is supported by the Medical Research Council, Muscular Dystrophy Campaign, Association of International Cancer Research, Association Française contre les Myopathies and the MYORES Network of Excellence contract 511978

References

1. Studitsky, A.N. (1964) Free auto- and homografts of muscle tissue in experiments on animals. *Ann N Y Acad Sci* 120, 789–801.
2. Chargé S.B., Rudnicki, M.A. (2004) Cellular and molecular regulation of muscle regeneration. *Physiol Rev* 84, 209–238.
3. Zammit, P.S., Partridge, T.A., Yablonka-Reuveni, Z. (2006) The skeletal muscle satellite cell: the stem cell that came in from the cold. *J Histochem Cytochem* 54, 1177–1191.
4. Mauro, A. (1961) Satellite cells of skeletal muscle fibres. *J Biophys Biochem Cytol* 9, 493–496.
5. Moss, F.P., Leblond, C.P. (1971) Satellite cells as a source of myonuclei in the muscles of growing rats. *Anat Rec* 170, 421–436.
6. Snow, M.H. (1978) An autoradiographic study of satellite cell differentiation into regenerating myotubes following transplantation of muscles in young rats. *Cell Tissue Res* 186, 535–540.
7. Beauchamp, J.R., Heslop, L., Yu, D.S., Tajbakhsh, S., Kelly, R.G., Wernig, A., Buckingham, M.E., Partridge, T.A. and Zammit, P.S. (2000) Expression of CD34 and Myf5 defines the majority of quiescent adult skeletal muscle satellite cells. *J Cell Biol* 151, 1221–1234.
8. Seale, P., Sabourin, L.A., Girgis-Gabardo, A., Mansouri, A., Gruss, P. and Rudnicki, M.A. (2000) Pax7 is required for the specification of myogenic satellite cells. *Cell* 102, 777–786.
9. Zammit, P.S., Golding, J.P., Nagata, Y., Hudon, V., Partridge, T.A. and Beauchamp, J.R. (2004) Muscle satellite cells adopt divergent fates: a mechanism for self-renewal? *J Cell Biol* 166, 347–357.
10. Collins, C.A., Olsen, I., Zammit, P.S., Heslop, L., Petrie, A., Partridge, T.A. and Morgan, J.E. (2005) Stem cell function, self-renewal, and behavioral heterogeneity

of cells from the adult muscle satellite cell niche. *Cell* 122, 289–301.
11. Shinin, V., Gayraud-Morel. B., Gomes, D., and Tajbakhsh, S. (2006) Asymmetric division and cosegregation of template DNA strands in adult muscle satellite cells. *Nat Cell Biol* 8, 677–687.
12. Asakura, A., Seale, P., Girgis-Gabardo, A., Rudnicki, M.A. (2002) Myogenic specification of side population cells in skeletal muscle. *J Cell Biol* 159, 123–134.
13. Sampaolesi, M., Torrente, Y., Innocenzi, A., Tonlorenzi, R., D'Antona, G., Pellegrino, M. A., Barresi, R., Bresolin, N., De Angelis, M.G., Campbell, K.P., Bottinelli, R. and Cossu, G. (2003) Cell therapy of alpha-sarcoglycan null dystrophic mice through intra-arterial delivery of mesoangioblasts. *Science* 301, 487–492.
14. Kuang, S., Chargé, S.B., Seale, P., Huh, M. and Rudnicki, M. (2006) Distinct roles for Pax7 and Pax3 in adult regenerative myogenesis. *J Cell Biol* 172, 103–113.
15. Montarras, D., Morgan, J., Collins, C., Relaix, F., Zaffran, S., Cumano, A., Partridge, T. and Buckingham, M. (2005) Direct isolation of satellite cells for skeletal muscle regeneration. *Science* 309, 2064–2067.
16. Bischoff, R. (1975) Regeneration of single skeletal muscle fibers in vitro. *Anat Rec* 18, 215–235.
17. Konigsberg, U.R., Lipton, B.H. and Konigsberg I.R. (1975) The regenerative response of single mature muscle fibers isolated in vitro. *Dev Biol* 45, 260–275.
18. Rosenblatt, J.D., Lunt, A.I., Parry, D.J. and Partridge, T.A. (1995) Culturing satellite cells from living single muscle fibre explants. *In Vitro Cell Dev Biol* 31A, 773–779.
19. Nagata, Y., Partridge, T.A., Matsuda, R. and Zammit, P.S. (2006) Entry of muscle satellite cells into the cell cycle requires sphingolipid signaling. *J Cell Biol* 174, 245–253.
20. Cornelison, D.D., Wold, B.J. (1997) Single-cell analysis of regulatory gene expression in quiescent and activated mouse skeletal muscle satellite cells. *Dev Biol* 191, 270–283.
21. Collins, C.A., Zammit, P.S., Ruiz, A.P., Morgan, J.E., Partridge, T.A. (2007) A population of myogenic stem cells that survives skeletal muscle aging. *Stem Cells* 25, 885–894.
22. Partridge, T.A., Morgan, J.E., Coulton, G.R., Hoffman, E.P. and Kunkel, L.M. (1989) Conversion of mdx myofibres from dystrophin-negative to -positive by injection of normal myoblasts. *Nature* 337, 176–179.
23. Kelly, R., Alonso, S., Tajbakhsh, S., Cossu, G. and Buckingham, M. (1995) Myosin light chain 3F regulatory sequences confer regionalized cardiac and skeletal muscle expression in transgenic mice. *J Cell Biol* 129, 383–396.
24. Tajbakhsh, S., Rocancourt, D., Buckingham, M. (1996) Muscle progenitor cells failing to respond to positional cues adopt non-myogenic fates in myf-5 null mice. *Nature* 384, 266–270.
25. Gross, J.G., Bou-Gharios, G., Morgan, J.E. (1999) Potentiation of myoblast transplantation by host muscle irradiation is dependent on the rate of radiation delivery. *Cell Tissue Res* 298, 371–375.

Part VI

Regeneration of the Vascular System

Chapter 21

Differentiation and Dynamic Analysis of Primitive Vessels from Embryonic Stem Cells

Gefei Zeng and Victoria L. Bautch

Abstract

Embryonic stem (ES) cells, which are derived from developing mouse blastocysts, have the ability to differentiate into various cell types in vitro. When placed in basal medium with added serum, mouse ES cells undergo a programed differentiation favoring formation of cell types that are found in the embryonic yolk sac, including vascular endothelial cells. These in vitro differentiated endothelial cells form primitive blood vessels, analogous to the first vessels that form in the embryo and the yolk sac. This differentiation model is ideal for both genetic and pharmacological manipulation of early vascular development. We have made mouse ES cell lines that express endothelial-specific GFP or H2B-GFP and used these lines to study the processes of mammalian vessel development by real-time imaging. Here we describe protocols for making transgenic ES cells and imaging the processes of blood vessel development. We also provide methods for ES cell maintenance and differentiation, and methods for analysis of vascular marker expression.

Key words: Angiogenesis, in vitro differentiation, murine embryonic stem cell, Histone-2B, time-lapse imaging, cell division orientation, immunofluorescence, GFP.

1. Introduction

Blood vessels in the embryo develop by two processes: vasculogenesis, the de novo formation of vessels from in situ differentiating endothelial cells, and angiogenesis, the generation of new blood vessels from endothelial cells of preexisting vessels (1, 2). Embryonic stem (ES) cells derived from developing mouse blastocysts are pluripotent, or have the ability to differentiate into various cell types in vitro and in vivo. In vitro differentiation of ES cells has been used to study vasculogenesis since the mid-1980s (3–5). When placed in basal medium with added serum, mouse ES cells undergo a programed differentiation favoring formation of cell types that are found in the embryonic yolk sac, including vascular endothelial cells *(5–8)*. These in vitro differentiated endothelial cells form primitive blood vessels, analogous to the first vessels that form in the embryo and yolk sac. In this model, vessels develop in the context of other cell types that are normally associated with vascular development in the embryo. Thus, this model is ideal for studying cellular and molecular aspects of mammalian vessel development. Specifically, ES cell clumps are dissociated and differentiated into embryoid bodies (EBs). The EBs are then attached to a permissive surface, where they adhere and spread while continuing their differentiation program *(8, 9)*. The EBs differentiate into lumenized primitive vessels in three dimensions and maintain a valid biological context, but the "pancake" feature of the flattened EBs allows for high-resolution imaging *(10)*. This feature of the ES cell model, combined with fluorescent protein labeling and confocal microscopy, allows us to actually record the process of vessel development. Although some success in the culture and imaging of whole mouse embryos has been reported, this approach requires a more sophisticated image acquisition and data analysis system *(11)*.

We have made mouse ES cell lines with vascular specific expression of GFP or H2B-GFP and used these cell lines to analyze the dynamic processes of vessel development by real-time imaging *(12, 13)*. ES cells have the potential to be used for reconstituting vessels in a clinical setting, so studies using this model may eventually provide new insights into therapeutic vessel regeneration *(14–16)*. Here we describe protocols for making transgenic ES cell lines and for time-lapse imaging of the primitive vessels. We also provide methods for maintaining and differentiating mouse embryonic stem cells and methods for analyzing the vasculature via immunofluorescence.

2. Materials

2.1. Mouse ES Cell Maintenance

1. Collection of 5637 human bladder carcinoma cell line (ATCC #HTB9) conditioned media: 5637 cells are grown to confluence in Dulbecco's modified Eagle's medium (DMEM) (Gibco/BRL, Bethesda, MD) (*see* **Note 1**) supplemented with 10% lot selected FBS (*see* **Note 2**). The cells are then incubated in collection media (DMEM + 5% FBS), and supernatant is collected every 48 h until the cells look suboptimal (usually after the fifth collection). Spin the supernatant for 10 min at 2,000 g, and filter through 0.22 µm cellulose acetate filter units (Nalgene). Store filtered medium at 4°C or −20°C (if not used within 2 weeks). Combine collections and test the pooled conditioned medium prior to use.

2. 7.5 mM monothioglycerol (Sigma) prepared with PBS, store at 4°C for up to 2 weeks.
3. ES cell culture media consists of 65.9%, 5637 conditioned medium, 17.13% lot selected FBS (see **Note 2**), 82.5 µM (final concentration) monothioglycerol (MTG), and 15.8% DMEM-H.
4. Gelatin (Type A porcine gelatin, Bloom Factor 200, Difco) is dissolved in tissue-culture water at 0.1%, filtered and stored in aliquots at 4°C. Tissue culture dishes are coated with 0.1% gelatin for at least 1 h and up to 1 week at 37°C.
5. Solution of trypsin (0.5 g/L) and EDTA (0.2 g/L) (Gibco/BRL, Cat. #25300) is aliquoted and stored at −20°C. Working solution (0.25×) is prepared by dilution in PBS.
6. Trypsin stop solution: 50% newborn bovine serum (NBS) in PBS.

2.2. In vitro Differentiation of Mouse ES Cells

1. ES cells cultured for 5–7 days after passage in ES cell medium.
2. Differentiation medium: DMEM-H supplemented with 20% lot selected fetal bovine serum (FBS) (see **Note 3**) and 150 µM monothioglycerol (MTG).
3. Dispase Grade II stock (2.4 U/ml) (Boehringer–Mannheim) is aliquoted and stored at −20°C. Working solution is prepared freshly by dilution in PBS (1:1, V/V).
4. 10 cm Kord-Valmark bacteriological Petri dishes (Wilkem Scientific Co.)
5. Autoclaved medidroppers (Fisher Scientific, Cat. #13-711).

2.3. DNA Preparation and Electroporation

1. Mouse ES cells cultured for 3–4 days in ES cell medium.
2. 10 mg/ml Protease K (Invitrogen, Cat. #25530-015).
3. $T_{10}E_{0.1}$ solution: 10 mM Tris–HCl, 0.1 mM EDTA (pH 8.0).
4. 100 and 70% ethanol (EtOH).
5. 3 M NaAc (pH 5.2).
6. Cold and warm PBS.
7. 0.4 cm electroporation cuvettes (BioRad, Cat. #165-2088).
8. BioRad GenePulser II with capacitance extender (BioRad, Hercules, CA).
9. Geneticin (50 mg/ml) and hygromycin (50 mg/ml) from Roche.

2.4. Time-Lapse Imaging

1. Slide flasks (Nunc).
2. Nikon TE300 inverted microscope (Melville, NY) with a Perkin Elmer spinning disk confocal head (Shelton, CT) and a heated stage.

2.5. Platelet Endothelial Cell Adhesion Molecule (PECAM) Immunofluorescent Staining

1. Fixatives: 4% paraformaldehyde (Polysciences Inc.), or freshly mixed cold acetone/methanol (MeOH) (1:1, V/V).
2. Staining media: PBS with 5% FBS (Hyclone) and 0.1% sodium azide (Fisher Scientific).
3. Primary antibody: Rat anti-mouse CD31 (PECAM-1) (Mec 13.3; BD Pharmingen).
4. Secondary antibody: FITC-AffinPure F(ab')$_2$ fragment of donkey anti-rat IgG (Jackson Immunoresearch, Cat. #712-096-150).
5. Fluorescent microscope. We have an Olympus IX-50 inverted microscope (Melville, NY) outfitted with epifluorescence.

3. Methods

3.1. Culturing Mouse ES Cells in the Absence of Feeder Cells

There are numerous published protocols describing mouse ES cell maintenance (5, 7–10, 17). ES cells are traditionally maintained on a feeder layer of mouse embryo fibroblasts or STO cells, which provide LIF (leukemia inhibitory factor) to prevent differentiation. Here we describe the maintenance of mouse ES cells in the absence of feeder cells. In this situation, LIF can be provided by several other sources: (1) commercially available LIF; (2) the medium of COS cells transiently transfected with LIF-expressing plasmids; (3) harvesting medium conditioned by the 5637 cell line. We prefer the 5637 cell conditioned medium because in our hands it preserves the undifferentiated morphology of the ES cells better than the other two options.

All volumes are given assuming that one 6 cm dish of ES cell colonies is being used.

1. Aspirate off medium. Wash two times with 5 ml pre-warmed PBS.
2. Add 1 mL of 0.25× trypsin–EDTA solution to dish (this volume should just cover the bottom of the dish). Place dish in 37°C incubator until a majority of ES cell clumps dissociate upon gentle agitation (1–3 min).
3. Stop trypsinization reaction by adding 4 ml of trypsin stop solution to dish. Gently pipette the ES cells/trypsin stop solution up and down a few times to break up the cell clumps.
4. Remove a gelatin-coated 6 cm dish (*see* **Section 2.1**) from incubator. Aspirate off gelatin solution, and add 5 ml pre-warmed ES medium. Add two to three drops of the cell suspension into the dish. ES cells should be in a single cell suspension, or in clumps of no more than four to six cells/

clump. If cell clumps are significantly larger, then pipette solution to further break apart cell clumps.

5. Place dish in 37°C incubator with 5% CO_2 and gently move dish in a "back and forth" motion in order to evenly disperse ES cells throughout the dish.

3.2. In vitro Differentiation of Mouse Embryonic Stem Cells

The ES cells are cultured for 5–7 days in ES cell medium after normal passage, and then digested with dispase. The ES clumps generated by dispase treatment are resuspended in differentiation medium in bacteriological dishes for 3 days to form EBs that contain endoderm and mesoderm, as well as hemangioblasts, which are the precursors of both vascular and hematopoietic cells. The EBs are then reattached to tissue culture treated dishes to spread and continue the process of differentiation. The process of differentiation is shown in **Fig. 21.1**.

1. Choose the dish that has the best ES cell clumps for differentiation. ES cell clumps should be round and differentiated on the very edge and tight, shiny, and undifferentiated in the middle. ES cell clumps are collected from dishes after incubation at 37°C in ES cell medium for 5–7 days without feeding after normal passage.

2. Aspirate off media from ES cell dish(es). Wash two times with 5 ml of cold 1× PBS. Aspirate PBS.

3. Add 1 ml of cold dispase (diluted 1:1 with cold PBS just before use), and let dish sit at room temperature for 1–2 min. Check to see if the ES cell clumps have detached from dish bottom by shaking the dish. If the majority of cells clumps have not detached, let the solution sit longer.

4. When a majority of the cell clumps have detached from the dish, use a 10 ml pipette to gently transfer the cells into a 50 ml tube containing 35 ml of room temperature 1× PBS. Rinse the dish with 5 ml of 1× PBS and add the rinse to the 50 ml tube. Cap tube, and invert the tube once gently to mix.

5. Let the tube sit until the cell clumps have settled to the bottom of the tube (around 10 min).

6. Aspirate all but 4–5 ml of PBS, carefully avoiding the ES cell clumps.

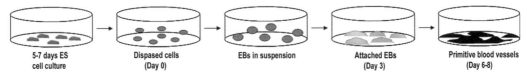

Fig. 21.1. Scheme for programed differentiation of mouse embryonic stem cells. ES cells of 5–7 days old are dispersed in suspension by dispase digestion (day 0). They are allowed to grow in suspension for 3 days, and then allowed to attach to tissue culture plastic dishes. At days 6–8 primitive blood vessels are well established.

7. Add another 35 ml room temperature 1× PBS gently down the side of the 50 ml tube. Gently swirl the tube to redistribute the cell clumps. Cap tube, and invert gently to mix.

8. Let the tube sit until the cell clumps have settled to the bottom of the tube.

9. Aspirate the PBS (leaving 2–3 ml PBS/ES cell clump solution at the bottom), and gently add 5 ml pre-warmed (37°C) Differentiation medium down the side of the tube.

10. Pipette 10 ml of pre-warmed differentiation medium into a Kord Scientific 10 cm Petri dish (see **Note 4**). Using a 25 ml pipette, transfer the contents of the 50 ml tube (cell clumps/PBS/media) to the 10 cm dish.

11. Check the density of cell clumps in each dish. We attempt to achieve approximately 100 clumps/dish. Incubate at 37°C in a humidified incubator with 5% CO_2. The day of the dispase treatment is day 0.

12. Transfer the EBs to a new Petri dish with fresh media on day 2. Gently swirl the old dish in a circular manner so that the EBs migrate to the center of the dish. Transfer the EBs to a new dish with a sterile medidropper.

13. Set up reattachment cultures on day 3 after the dispase treatment. Add 4 ml of pre-warmed differentiation media to each slide flask that is to be seeded with EBs.

14. Use a sterile medidropper to transfer EBs from the dish to slide flasks. Generally, dispense between 30 and 40 EBs per slide flask. Holding the dish up and looking at it from underneath is helpful when determining the number of EBs in a slide flask.

15. Keep plate level and gently move the slide flask when placed in incubator to ensure that the EBs spread evenly in the slide flask. Incubate at 37°C in a humidified incubator with 5% CO_2. Attachment generally occurs within a few hours.

16. Feed the attached cultures every other day until days 6–8 (see **Note 5**). To feed, aspirate off media, then slowly add 4 ml fresh pre-warmed differentiation medium down the top cover of the slide flask so as not to disturb the attached cultures.

3.3. DNA Preparation and Electroporation

1. Linearizing DNA for electroporation: For a 20 kb construct use 20 µg DNA. Linearize plasmid according to standard procedure. Check plasmid for complete digestion on agarose gel.

2. Add SDS to 0.5% final concentration, add Proteinase K to 0.2 µg/µl final concentration, bring up to 500 µl volume with $T_{10}E_{0.1}$. Incubate at 37°C for 30 min.

3. Heat inactivate the restriction enzyme according to the demands of that specific enzyme (see **Note 6**).

4. Add 3 M NaAc to 0.3 M final concentration and two volumes of 100% EtOH. Precipitate on ice for 30 min or store overnight at –20°C.

5. Centrifuge DNA at 12,000 g for 15 min, wash with 1 ml of 70% EtOH, then centrifuge another 10 min. Under the tissue culture hood carefully remove supernatant and let dry for approximately 10 min, then resuspend in sterile PBS to 1 µg/µl.

6. Trypsinize the cells as usual (see **Section 3.1**). Stop trypsin by adding an equal volume of FBS.

7. Spin down the cells in a clinical centrifuge on speed #4 for 2 min. At the same time count the cells using a hemocytometer.

8. Aspirate the supernatant, resuspend the cells in cold PBS to 2×10^7 cells/0.5 ml, and mix with the DNA. We usually split the DNA into two different fractions (such as 17.5 and 2.5 µg) and do two electroporations for each construct.

9. Transfer cell/DNA mixture to chilled cuvettes. Make a control sample of cells without added DNA.

10. Chill cuvettes on ice for 20 min.

11. Electroporate at the following settings: 300 µF, 250 V on high capacitance. Pulse once (see **Note 7**).

12. Incubate cuvettes on ice for 20 min.

13. Transfer each sample to two 10 cm dishes (contain 1/4 or 3/4 cells in 10 ml ES cell medium). Bring the volume up to 10 ml for each dish with ES medium.

14. After 24 h, change to ES medium with the selective drug. We use Geneticin (G418) or Hygromycin at 200 µg/ml.

15. Feed culture every other day with ES medium containing the appropriate drug. Clones should form and expand over the next 12–14 days. There should be no clones on the "no DNA" control plate.

16. When the clones reach approximately 1 mm in diameter, they can be picked and moved to a new dish. Aspirate the media, add 10 ml PBS, aspirate, then add fresh PBS so it only covers the cells (5 ml for one 10 cm plate).

17. Add trypsin–EDTA (1:4 in PBS) to a 48-well plate (20 µl/well), keeping it as a droplet.

18. Pick colonies from the original plate with plugged pipet tips (new tip for each colony) and put one colony in each trypsin droplet. We usually pick up 20–30 colonies.

19. Incubate in 37°C for a few minutes. At the same time, add ES medium (we remove drug from the ES cell medium at this point) to a gelatin-coated 24-well plate.

20. Stop the trypsin–EDTA reaction by adding an equal volume of FBS, and transfer the cells to the wells of the 24-well plate.

21. Expand the colonies (clones) by passing them to six-well plates after 2–3 days, then to individual 6 cm dishes after another 2–3 days.

22. Screen individual clones by in vitro differentiation (*see* **Section 3.2**). Check GFP expressions if the cells are electroporated with a GFP construct. Maintain (*see* **Section 3.1**) and freeze (*see* **Note 8**) the colonies in the meantime. We freeze two to three vials (from one 6 cm dish) for each clone, then freeze more vials for the good clones after screening.

3.4. Time-Lapse Imaging of Blood Vessel Formation

1. Differentiate ES cells using the described protocol (*see* **Section 3.2**). Attach EBs to slide flasks on day 3 (*see* **Note 9**).

2. At days 7–8 of differentiation, take the slide flask out of the incubator and tighten the cap. Using a fluorescent microscope, look for areas with nice vasculature that express GFP or H2B-GFP. Circle the area(s) on the bottom side of the slide flask with a Sharpie marker.

3. Set up the slide flask on a heated stage under an inverted microscope with a spinning disk confocal head, and find the area you wish to image (*see* **Note 10**). The diagram of the system is shown in **Fig. 21.2**.

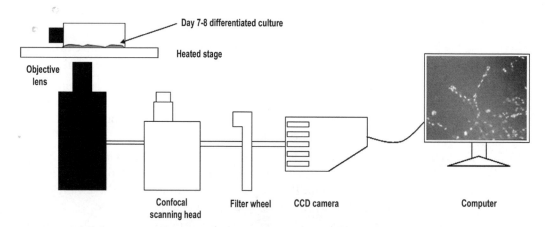

Fig. 21.2. Schematic illustration of a confocal time-lapse imaging system for in vitro differentiated ES cultures. A sealed slide flask containing the culture is placed on a heated stage. The laser light excites the GFP, and the emitted light goes to the objective lens, through the confocal scanning head, the filter wheel, and to the CCD camera. The image is projected onto the computer screen and can be saved as a TIFF file. The compilation of images is played as a time-lapse "movie" after collection.

Imaging ES Cell-Differentiated Vessels   341

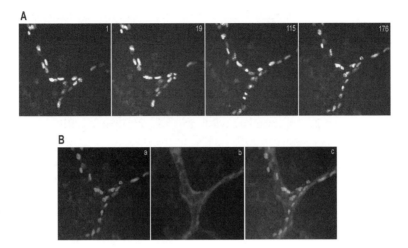

Fig. 21.3. Time-lapse imaging of ES cell differentiation cultures. An ES cell line expressing endothelial cell-specific H2B-GFP was differentiated to day 8. It was imaged on a Perkin Elmer spinning disk confocal microscope. Culture is fixed right after imaging and stained for PECAM-1. (A) Frames of a time-lapse movie. Time is in minutes at the top right corner of each panel. (B) Last frame of the movie in green (**a**), the same field with PECAM-1 stain in red (**b**), and the overlay of the two images (**c**). (*See* Color Plate 21)

4. Acquire time lapse images using acquisition software (*see* **Note 11**). We take images once every minute for 1 s, but other options are possible. We usually image for 2–10 h. Although the latter times sometimes yield some degradation of the sample, we rarely see sample degradation with total times of less than 8 h. Images from a time-lapse movie are shown in **Fig. 21.3(A)** and Color Plate 21.

5. Proceed immediately to fixation and PECAM staining (*see* **Section 3.5.**) after time lapse imaging is done.

3.5. ES Cell Culture Fixation and PECAM Staining

These Procedures Assume the use of One Slide Flask.

1. Aspirate media and wash twice with 4 ml of 1× PBS at room temperature.

2. Add 2 ml of cold fixative and incubate for 5 min at room temperature (1:1 methanol/acetone fixative or 4% PFA fixative) (*see* **Note 12**). Aspirate fixative and wash twice with PBS, incubating 2 min at room temperature for each wash.

3. Gently add 2 ml of staining medium to the slide flask. Incubate for 45 min to 1 h at 37°C.

4. Remove slide flask from incubator, aspirate staining medium, and add 1 ml fresh staining medium containing the properly diluted primary antibody. Incubate for 1–2 h at 37°C.

5. Remove slide flask from incubator, remove primary antibody (*see* **Note 13**), and wash two times (2–3 min/wash) with 3 ml staining medium. Add 1 ml fresh staining medium containing the

properly diluted secondary antibody. Incubate for 1 hr at 37°C, and protect the secondary antibody from light (foil wrapped).

6. Remove slide flask from incubator, remove secondary antibody (*see* **Note 13**), and wash once (2–3 min/wash) with 3 ml staining medium. Aspirate, then wash two times (2 min/wash) using 1× PBS.

7. Aspirate last wash and add 2 ml of 1× PBS to the slide flask, store in PBS at 4°C in the dark.

8. Obtain images of PECAM staining of the imaged area under an inverted microscope outfitted with epifluorescence and a camera. Overlay image with the last time lapse image acquired. Example of the images is shown in **Fig. 21.3 (B)**.

4. Notes

1. The DMEM is supplemented with Gentamicin (0.5 µg/ml final, GIBCO/BRL) before being used for medium preparation.

2. It is important to use serum that has been screened for ES cell maintenance. Prescreened serum is commercially available, but we have had good success screening serum lots in our own laboratory. In general, 50–75% of lots approved for general tissue culture use are suitable for ES cell maintenance.

3. Testing lots of FBS from different manufacturers is critical for optimal ES cell differentiation. We found serum that is not heat inactivated favors differentiation, although the serum for ES cell maintenance should be heat inactivated.

4. Do not use tissue culture treated dishes at this step because the EBs will stick to the plastic. The bacteriological dishes dramatically reduce the attachment of the ES cell clumps to the bottom of the dish during this period, which is crucial in obtaining a good differentiation.

5. Using this protocol, we typically monitor vascular development in cultures that have been differentiated for 7–8 days (day of dispase treatment is day 0). Angioblast formation is generally observed at days 4–6, while vessel formation occurs at days 6–8.

6. If the restriction enzyme cannot be heat inactivated, add an equal volume of tris equilibrated phenol/chloroform. Mix by inverting four to six times, centrifuge at 12,000 g for 10 min. Transfer upper layer to new tube, then continue to step 4 of **Section 3.3**.

7. Electroporation protocol with BioRad GenePulser II can be converted to a compatible one with other electroporator models. For example, we have used an ECM 830 (BTX

Harvard Apparatus), with settings at 250 V, 10 msec, 1 pulse. It worked as efficiently as the BioRad GenePulser II.

8. Generally, freeze two vials (1 ml/vial) of cells per 6 cm dish of ES cells. Prepare a 20% DMSO/80% FBS solution, chill on ice. Detach cells from tissue culture plates per usual. Spin cells down in a clinical centrifuge on speed #4 for 2 min. Resuspend cells in FBS at one-half the final volume. Cool cells on ice for 2–5 min. Dropwise, with gentle swirling of tube, slowly add an equivalent volume of cold 20% DMSO/80% FBS solution. Distribute 1 ml volume of the cells to the cryotube. Freeze cells overnight at –70°C in a Nalgene freezer box for cryo-freezing. Move the cells to liquid nitrogen vessels the next day.

9. Slide flasks are required if you use a simple heated stage as we do, but if you have a more complex (and costly) plexiglass incubator that sits atop the stage for temperature and gas, you can then use whatever fits in the incubator for imaging.

10. We have a Nikon TE300 inverted microscope (Melville, NY) with a Perkin Elmer spinning disc confocal head (Shelton, CT), but there are now several nice options and combinations to choose from.

11. We use Metamorph (Version 6.1, PLACE) software for acquisition and data analysis, but other software packages are available.

12. PECAM-1 and ICAM-2 antibodies work in both PFA and MeOH/acetone fixatives. Both fixatives are listed here since some applications (not described here) favor one fixative over another. We prefer MeOH/acetone fixative since the cell layers stay attached to the dish more tightly, and it yields brighter staining. The PFA fixed cells tend to come up off the dish during the staining procedure. The H2BGFP is stable in both fixatives, while cytoplasmic GFP will diffuse in MeOH/acetone fixation.

13. Primary and secondary antibodies can be saved for subsequent experiments by storing at 4°C (the staining media contains 0.1% sodium azide). We reuse the antibodies up to three times over several months.

Acknowledgements

The authors would like to thank Drs Joseph B. Kearney and Nicholas C. Kappas for setting up the system for imaging differentiated ES cells and Rebecca Rapoport for technical assistance. This work was supported by NIH grants HL 43174 and HL 83262 to V.L.B.

References

1. Poole, T. J., and Coffin, J. D. (1989) Vasculogenesis and angiogenesis: two distinct morphogenetic mechanisms establish embryonic vascular pattern. *J Exp Zool* **251**, 224–231.
2. Risau, W. (1997) Mechanism of angiogenesis. *Nature* **386**, 671–674.
3. Doetschman, T.C., Eistetter, H., Katz, M., Schmidt, W., and Kemler, R. (1985) The in vitro development of blastocyst-derived embryonic stem cell lines: formation of visceral yolk sac, blood islands and myocardium. *J Embryol Exp Morphol* **87**, 27–45.
4. Risau, W., Sariola, H., Zerwes, H.G., Sasse, J., Ekblom, P., Kemler, R., and Doetschman, T. (1988) Vasculogenesis and angiogenesis in embryonic-stem-cell-derived embryoid bodies. *Development* **102**, 471–478.
5. Wang, R., Clark, R., Bautch, V.L. (1992) Embryonic stem cell-derived cystic embryoid bodies from vascular channels: an in vitro model of blood vessel development. *Development* **114**, 303–316.
6. Keller, G.M. (1995) In vitro differentiation of embryonic stem cells. *Curr Opin Cell Biol* **7**, 862–869.
7. Vittet, D., Prandini, M.H., Berthier, R., Schweitzer, A., Martin-Sisteron, H., Uzan, G., and Dejana, E. (1996) Embryonic stem cells differentiate in vitro to endothelial cells through successive maturation steps. *Blood* **88**, 3424–3431.
8. Bautch, V.L. (2002) Embryonic stem cell differentiation and the vascular lineage. *Methods Mol Biol* **185**, 117–125.
9. Bautch, V.L., Stanford, W.L., Rapoport, R., Russell, S., Byrum, R. S., and Futch, T. A. (1996) Blood island formation in attached cultures of murine embryonic stem cells. *Dev Dyn* **205**, 1–12.
10. Kearney, J.B., Bautch, V.L. (2003) In vitro differentiation of mouse ES cells: hematopoietic and vascular development. *Meth Enzymol* **365**, 83–98.
11. Jones, E.A., Baron, M.H., Fraser, S.E., and Dickinson, M. E. (2005) Dynamic in vivo imaging of mammalian hematovascular development using whole embryo culture. *Meth Mol Med* **105**, 381–394.
12. Kearney, J.B., Kappas, N.C., Ellerstrom, C., DiPaola, F. W., and Bautch, V. L. (2004) The VEGF receptor flt-1 (VEGFR-1) is a positive modulator of vasular sprout formation and branching morphogenesis. *Blood* **103**, 4527–4535.
13. Zeng G., Taylor S.M., McColm J.R., Kappas N.C., Kearney J. B., Williams L. H., Hartnett M. E., Bautch V. L. (2007) Orientation of endothelial cell division is regulated by VEGF signaling during blood vessel formation. *Blood* **109**, 1345–1352.
14. Zwaginga, J.J., Doevendans, P. (2003) Stem cell-derived angiogenic/vasculogenic cells: possible therapies for tissue repair and tissue engineering. *Clin Exp Pharmacol Physiol* **30**, 900–908.
15. Alessandri, G., Emanueli, C., Madeddu, P. (2004) Genetically engineered stem cell therapy for tissue regeneration. *Ann N Y Acad Sci* **1015**, 271–284.
16. Madeddu, P. (2005) Therapeutic angiogenesis and vasculogenesis for tissue regeneration. *Exp Physiol* **90**, 315–326.
17. Feraud, O., Prandini, M.H., Vittet, D. (2003) Vasculogenesis and angiogenesis from in vitro differentiation of mouse embryonic stem cells. *Meth Enzymol* **365**, 214–228.

Chapter 22

Derivation of Contractile Smooth Muscle Cells from Embryonic Stem Cells

Sanjay Sinha, Mark H. Hoofnagle and Gary K. Owens

Abstract

Smooth muscle cells (SMCs) play a key role in vascular physiology and pathology. An appreciation of normal SMCs developmental mechanisms will likely lead to a better understanding of disease processes and potentially to novel treatment strategies. We present a method for generating relatively pure populations of SMCs from embryonic stem cells (ESC) which display appropriate excitation and contractile responses to vasoactive agonists. We also present protocols for assessment of SMCs purity and identity by immunofluorescence, quantitative RT-PCR, and FACS. This ESC-based system has tremendous potential for studying developmental regulation of SMC lineage, as well as for possible SMC tissue engineering.

Key words: Stem cell, smooth muscle, embryoid body, contraction.

1. Introduction

Mature smooth muscle cells (SMCs) possess a highly specialized ultrastructure and their principal role in the adult is contraction. However, unlike skeletal or cardiac muscle cells, adult SMCs are able to modulate their phenotype back into an immature form. Such phenotypic plasticity permits necessary vessel remodeling and repair, even in adult organisms. However, this plasticity also makes the cells susceptible to the forms of phenotypic switching that contribute to the pathogenesis of a wide variety of major human diseases, including restenosis, atherosclerosis, vascular aneurysms, and accelerated graft disease (1, 2). An understanding of the normal molecular mechanisms that regulate SMC maturation and gene expression during development is essential for insights into how these mechanisms are dysregulated during disease processes. Moreover, factors that promote the mature contractile state during development may also have the potential to inhibit or reverse the phenotypic modulation of SMC that is a key feature of vascular disease.

A variety of in vitro systems, including 10T1/2 (*3*), A404 (*4*), MONC (*5*), and neural crest cells (*6*), have previously been described in which multipotential cells give rise to SMC-like cells. However, it is unclear whether these systems model the full range of SMC developmental events. Moreover, there is uncertainty as to the developmental origins of these cells and the fact that most of these systems, with the exception of the rat neural crest cells, have not been shown to exhibit the defining feature of mature SMC—contraction. Embryonic stem cells (ESCs), analogous to the pluripotential inner cell mass, recapitulate early embryonic development in vitro when placed in aggregate as embryoid bodies (EBs) (*7*). A major advantage of this system is the development of contractile SMCs (*8–10*), suggesting all required developmental pathways are functional. However, a limitation of the EB system is the heterogeneous mixture of cell types that develop, making specific biochemical studies on SMC difficult. We recently published a method in which contractile SMCs are derived from ESC using SMC-specific promoters to isolate a pure population (*11*) and this method is described in detail below.

2. Materials

2.1. ESC and EB Culture

1. ESC medium: DMEM-H (high glucose) supplemented with 15% FBS (Gibco, Gaithersburg, MD), 1× L-glutamine, 1× non-essential amino acids, pyruvate, β-mercaptoethanol (ME) and leukemia inhibitory factor (LIF, 10^6 U/ml, Chemicon, Temecula, CA).

2. EB medium: as ESC medium but without β-ME and LIF and with penicillin/streptomycin and 20% FBS.

3. All-*trans* retinoic acid (atRA, Sigma, St Louis, MO) stock at 2 mM in ethanol (so need to dilute 200,000×).

4. Sterile bacterial Petri dishes—used to generate EBs up to day 6. EBs need to be in suspension culture from days 3 to 6 and do not adhere readily to Petri dish plastic since it is not tissue culture treated.

5. Collagenase/trypsin mix: 2× collagenase (Collagenase Type IV, Invitrogen, Carlsbad, CA) = 2 mg/ml in DMEM F12, add

equal volume of 1× trypsin, filter with 0.22 μm filter, warm to 37°C.

6. SMC selection media: 5 ml EB media + 2.5 μl puromycin (1 mg/ml stock; final concentration 0.5 μg/ml).

7. 0.22 μm filters, 70 μm nylon cell strainer.

8. Cells: D3 ESC (kindly provided by the Gene Targeting and Transgenic Facility [GTTF], University of Virginia) and mouse embryonic fibroblasts (CF-1) rendered mitotically inactive by gamma irradiation (4,000 Rad).

9. 0.1% (w/v) porcine gelatin (Sigma) autoclaved to dissolve and sterilize.

2.2. Generation of Transgenic ESC Lines

1. Plasmids: smooth muscle α-actin (SMαA)-puromycin N-acetyl transferase (PAC), (p453) and/or smooth muscle-myosin heavy chain (SMMHC)-PAC (p434) constructs (*4*), and pNeo 4.3 (Clontech, Mountain View, CA).

2. DNA purification: phenol:chloroform, chloroform and ethanol.

3. Sterile electroporation cuvettes and electroporator (Gene Pulser II, Bio-Rad, Hercules, CA).

4. G418 (neomycin) stock 50 mg/ml in 100 mM Hepes (pH 7.4).

5. Neomycin-resistant mouse embryonic fibroblasts (neoMEFs, supplied kindly by the GTTF at the University of Virginia) mitotically inactivated by gamma irradiation.

6. Dissecting microscope to pick surviving colonies.

7. 2× freezing medium: ESC medium with 20% DMSO and 40% FBS—add equal amount of ESC suspension, mix gently, and place in controlled freezing chamber.

8. PCR buffer, dNTPs, and Taq polymerase.

2.3. Immunofluorescence

1. Fixation: 4% (w/v) PFA dissolved in phosphate-buffered saline (PBS).

2. Permeabilization solution: 0.5% (v/v) Triton X-100 in PBS.

3. Blocking solution: 3% (w/v) bovine serum albumin (BSA) and 3% (v/v) donkey serum in PBS.

4. Primary antibodies: anti-SMαA (Sigma, clone 1A4); anti-SM-MHC (Biomedical Technologies, Stoughton, MA)—both diluted in blocking solution.

5. Secondary antibodies: Donkey anti-mouse- or anti-rabbit-Cy2 (Jackson, West Grove, PA)—diluted in PBS/0.1% BSA.

6. Nuclear stain: 1 μM DAPI (4,6-diamidino-2-phenylindole) in dH$_2$O.

2.4. Real-Time RT-PCR

1. Standard reagents for reverse transcription and real-time PCR (Applied Biosystems) as described by Bustin (*12*).
2. Primers and probes for real-time reactions as in **Table 22.1**.

2.5. FACS

1. Fixation solution: 4% (w/v) paraformaldehyde (PFA) in PBS.
2. Permeabilization agent: 100% methanol
3. Resuspension buffer: 2% (w/v) BSA in PBS with 0.1% (w/v) sodium azide.
4. Staining antibodies: anti-SMαA-FITC (Sigma 1A4 clone) or anti-SMαA-Cy3 (Sigma). The use of a secondary antibody, in our hands, led to unacceptable levels of nonspecific background staining as well as excessive loss of cells. Consequently, we use primary antibodies to which the fluorescent marker is directly conjugated. The 1A4 clone for anti-SMαA is available in conjugated form. For SM-MHC staining, we have used an antibody from Biomedical Technologies (BT-562) to stain SM-MHC and conjugated this to Alexa Fluor 647 using an antibody labeling kit (Invitrogen). We should point out that we have noticed a reduction in specificity with more recent lots of this antibody.
5. DNA staining: Draq5 (Biostatus, Shepsed, UK)

3. Methods

3.1. ESC Culture

Mouse embryonic fibroblasts are used as feeder layers for the ESCs. The MEFs assist in the maintenance of the ESC pluripotential state by supplying LIF and possibly other essential unknown factors. MEFs need to be mitotically inactivated, by γ-irradiation in our case (although treatment with mitomycin C is an alternative), so that they form a non-proliferative layer of feeder cells to support the growth of ESCs. We use D3 ESC which have their medium replaced daily and are split 1:4–1:10 depending on colony size every 2–3 days. Detailed protocols for the generation and irradiation of MEFs as well as culture of mouse ESC are published elsewhere (*13*). Consequently, these will not be described further here.

3.2. Generation of Transgenic ESC Lines

We have used promoters from SMC-specific genes that we have previously shown to recapitulate endogenous gene expression in vivo (*14; 15*) to drive expression of a puromycin resistance cassette. Promoters from both the SMαA and the SM-MHC genes have been used to create transgenic ESC lines that can subsequently be used to generate the pure contractile SMC populations—APSCs and MPSCs, respectively. The generation of these promoter-selection marker constructs has been previously described (*4*) and these

Table 22.1
Primers and probes used for real-time PCRs

	Probe	Forward primer	Reverse primer
SMαA	CAGCACAGCCCTG GTGTGCGAC	CGCTGTCAGGA ACCCTGAGA	CGAAGCCG GCCTTACAGA
SMMHC	AGAACACTAAACGACA GCAGAGCCCAGC	TGGACCATGT CAGGGAAA	ATGGACACAAGT GCTAAGCAGTCT
SM22α	TGCCCATAGCCT GTCATGCCAGC	AAGCACGTCAT TGGCCTTCA	GCCTTCCCTTC TAACTGATGATC
Calponin	CTGGCTGCTGCT CTTGGCTGGACC	CTACAACTCTGC CTAGGGGC	CCTCATCTCCCAA ACCGTAA
Smoothelin A+B	CCCAACCCTTCTCA GCACCAGCA	AGAGAGCAGAATTG ACTTTGGGAT	GCAGCCTCCTC GGTTACCA
Smoothelin B	TGCTGCCTGCCT CCTGCTGTCCT	TGACCACACTGC TTCGGAGT	GGCACCTTACCA GGGTCCAA
Myocardin	ACTCTGACACCTTGAGATCATC CAGGTTTGG	AAACCAGGCC CCCTCCC	CGGATTGGAAGCT GTTGTCTT
Smemb	TTGGCTGGCTGGTTC ACCGCATCA	GCATTGGCAAAAGCTA CCTATGA	GCTTCTCGTTGGT GTAGTTGATG
Cardiac α-actin	CCAGAGGAACACCCA ACCCTGCT	GGCACCATACATTCTAC AATGAGC	CATCGCCAGAATCCAG AACAATG
Cardiacα-MHC	SybrGreen	AGGGAACAGGTGGAG AACTATTAC	CTGGACACGGAGCTT TTATCTGC

(continued)

Table 22.1 (continued)

	Probe	Forward primer	Reverse primer
Myf5	SybrGreen	GGGAAGGGGGC AAAGTCAC	GCAGAAGAGGCC CGAGTAGG
NeuroD	TGTCTGCCTCGTGT TCCTCGTCCT	TGAGATCGTCACTAT TCAGAACCT	CATGGCTTCA AGCTCGTCCT
18S	TGCTGGCACCAG ACTTGCCCTC	CGGCTACCACATC CAAGGAA	AGCTGGAATTACC GCGGC
Rapsyn	None: conventional PCR	AGGACTGGGTGGCTTCCAACT CCCAGACAC	AGCTTCTCATTGCTGCGC GCCAGGTTCAGG
Neomycin	None: conventional PCR	AGGATCTCCTGTCATCTCA CCTTGCTCCTG	AAGAACTCGTCAAGAAG GCGATAGAAGGCG
Puromycin	None: conventional PCR	GTCACCGAGCTGC AAGAACT	CAGGAGGCCTTC CATCTGT

Reproduced by permission from: Sinha et al. Stem Cells 2006;24:1678–88, © AlphaMed Press.

constructs may be obtained on request from Prof. GK Owens. It is likely that other SMC-selective promoter segments, such as for SM22α, could be used in an analogous manner.

3.2.1. Prior to Transfection

1. The SMαA-PAC (p453) and/or SM-MHC-PAC (p434) constructs (20 µg of each) and pNeo4.3 are linearized and purified by phenol–chloroform extraction.
2. The linear DNA is resuspended with filter sterilized TE and handled with sterile technique in a tissue culture hood. Recovered DNA concentration is quantified on a spectrophotometer or on a gel.
3. The amounts of DNA required for transfection are calculated on the basis of a 5:1 molar ratio between promoter construct:pNeo4.3 (e.g., for SM-MHC-PAC lines, SM-MHC-PAC = 23 kb, pNeo4.3 = 4.3 kb which is a 23/4.3 or approximately six-fold difference in molecular weight, so 18 µg SM-MHC-PAC and 3 µg pNeo4.3 is a 1:1 molar ratio. Therefore, we used 18 µg SM-MHC-PAC and 0.6 mg pNeo4.3 for the final transfection).
4. Irradiated neoMEFs are plated onto 12 × 60 mm dishes (coated overnight with gelatin) and onto 2 × 24 well plates. Will probably need to plate at approximately $1-1.5 \times 10^5$ MEFs/cm^2.
5. Standard (CF-1) MEFs are irradiated and plated onto gelatin-coated T25 and T75 flasks.

3.2.2. Transfection—Day 0

6. Early passage ESCs are used to generate transgenic lines and the medium is changed 2–3 h prior to the procedure.
7. ESCs from a T75 flask, with moderately dense and medium–large colonies, are trypsinized and cell numbers counted. At least two aliquots are required with $2-10 \times 10^6$ cells per aliquot for electroporation.
8. Cells are centrifuged and washed with 10 ml PBS.
9. Cells are centrifuged again and each aliquot resuspended in 0.8 ml cold PBS and placed on ice for 10 min.
10. The 0.8 ml ESC aliquot is placed in a sterile electroporation cuvette. Linearized DNA is added to the test cells and no DNA to controls (~10–20 µg of promoter-PAC DNA and 1/5 molar amount of pNeo4.3).
11. Electroporation is carried out according to the specific voltage and capacitance recommended by the manufacturer (we obtain good results using 280 V and 250 µF with the Gene Pulser II; BioRad). *See* **Note 1**.
12. Cells are rested for 10 min at room temperature.

13. Cells are resuspended in ESC media (*with penicillin/streptomycin*). Viable cell numbers are counted and plated onto γ-irradiated neomycin-resistant MEFs at ~0.5×10^6 cells/plate. Dishes of size 60 mm are used. Only two or three dishes are required for the non-DNA control sample and the rest may be used for the transfected DNA sample.

3.2.3. Day 1

14. ESC medium (including penicillin/streptomycin) is changed. You may need to consider adding G418 (400 μg/ml) if small colonies are visible already.

3.2.4. Day 2

15. ESC medium (including penicillin/streptomycin) is changed. It will probably be necessary to add G418 today. It is important not to let colonies get larger than small in size or G418 will be ineffective (let one control plate grow without G418 to ensure it is effective). If colonies continue to grow on the control plate, then G418 may be used at 800–1,000 μg/ml for a day or two. *See* **Note 2**.

3.2.5. Days 3+

16. Medium is changed daily and supplemented with G418 at 200–400 μg/ml depending on colony growth in control plates. Significant amounts of cell death should be evident with only a few surviving colonies by day 7 in the test plates.

17. Once all the control colonies are dead, the surviving colonies in the DNA group are picked either by eye or using a dissecting microscope in a sterile tissue culture hood.

18. Each colony is placed in 100 μl trypsin in individual wells of a 96-well plate and dispersed using a pipetman. It is not necessary to reduce the colony to single cells, five to six cell pieces will suffice. An equal volume of ESC medium is added and the suspension plated onto neoMEF seeded 24-well plates.

19. Once they are growing well, the colonies are trypsinized with 200 μl trypsin. The cells are dispersed and 800 μl medium added to stop the trypsin.

20. 0.5 ml of this ESC suspension is seeded onto a new 24-well neo-MEF covered plate (which will be harvested for DNA in due course).

21. The remaining 0.5 ml is added to 0.5 ml of 2× freezing medium and frozen to –80°C.

22. The growing colonies are left for several days until the ESCs are overgrown and then harvested for DNA. Two wells of MEFs with no ESCs are also harvested as negative controls.

23. PCR is carried out on the genomic DNA using primers specific for neomycin or puromycin to confirm that the isolated clones have incorporated both transgenes. Each PCR also includes primers for rapsyn, a single copy gene (*16*), which is used as an

internal control to ensure the PCR amplifies genomic DNA successfully. Reaction conditions are as follows: Neomycin – 30 cycles with annealing temperature of 69°C; puromycin – 30 cycles with annealing temperature 65°C. Protocols for setting up and visualizing PCRs are described in detail elsewhere (*17*).

3.3. Generation and Culture of EBs

1. To generate EBs, ESC are trypsinized and resuspended in a small volume (2–3 ml) of EB medium and cell numbers are counted. The number of ESCs (small cells) should greatly outnumber the number of MEFs (large cells) when examined under the microscope.

2. The suspension is diluted to furnish a final cell suspension of 80 cells/μl. Ten microliters of drops (containing 800 cells; *see* **Note 3**) are pipetted onto an inverted 100 mm Petri dish lid using sterile filter tips. The lid is then replaced on the base, which contains 5–8 ml of PBS and placed in the incubator. The inverted drops allow the cells to settle at the drop apex and form an ESC aggregate or EB (*see* **Note 4**).

3. EBs are left undisturbed as hanging drops for 2–3 days and then washed off the Petri dish lids using EB medium. Embryoid bodies may be seen to be adhering to the lid and may need to be washed off vigorously with a pipette (*see* **Note 5**). EBs from two lids (approx. 100–150) may be combined at this stage and placed in suspension in a Petri dish in a final volume of 10 ml EB medium.

4. EB medium is changed every 1–2 days depending on the color of the medium. During this process, the medium is gently agitated using a pipette to ensure the EBs are not adherent to the Petri dish. If the dish is tilted, the EBs, which can be seen by the naked eye, will settle at one end and the old medium may be carefully aspirated without aspirating the EBs!

5. On day 5, 6-well, 24-well or 48-well tissue culture plates are coated with 0.1% gelatin in preparation for plating. Sufficient gelatin solution is added to cover the base of the wells which are left in the 37°C incubator overnight.

6. The following day (day 6), the gelatin is aspirated to dryness. EBs are then transferred to the gelatin coated plates using a P100. Two to four EBs can be placed in each well of a 48-well dish and the numbers can be scaled up with larger plates.

7. Approximately 200 μl of EB medium is then added to each well of a 48-well plate and the plate is swirled if necessary to center the EBs.

8. On each of the following 3 days (days 7, 8, and 9), the medium is replaced with fresh EB medium + 10 nM atRA. Care must be taken not to dislodge or aspirate EBs when removing old medium! atRA stock is at 2 mM in ethanol (so dilute 200,000×).

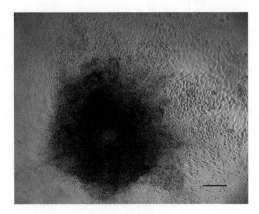

Fig. 22.1. Phase contrast image of differentiating EB. An EB was plated at day 6 and treated with at RA from day 7 to day 10. Phase contrast imaging at day 18 revealed a central mass of differentiating cells in the center of the picture with radial spread and migration of multiple cell types forming a "fried egg" appearance. SMCs are typically seen in clumps around the central mass and peripherally. Bar = 100 μm.

9. EB medium (alone) is replaced daily from days 10 to 28. **See Fig. 22.1** for an example of the typical appearance of a differentiating EB.

3.4. Isolating SMCs from EB Outgrowths

EBs will attach to and grow across the gelatin covered well and differentiate into a wide variety of differentiated cell types. **Figure 22.1** shows the classic appearance of an attached differentiating EB at day 18. Typically, collections of cardiac myocytes can be seen from approximately day 12 to day 20 and are easily identified by regular rapid contractions or twitching under microscopy. SMCs form as both scattered cells and large aggregates that form slowly contracting areas quite different from the rapid contraction of the cardiac myocytes. SMCs express SMC markers by RT-PCR and immunofluorescence at an early stage (day 10 onwards), but coordinated contractile activity is seen later, at approximately day 20 onwards (see video at online supplementary material for Sinha et al. (9)). Patient observation is required to visualize contraction of SMC aggregates, since the periodicity of this process is long and the aggregates may only contract intermittently. Consequently, mature contractile SMC will be detected in the EB outgrowths. However, the EB outgrowths also contain many other cell types and for many studies it is essential to isolate a "pure" population of SMC. In the system, we have developed, all the cells carry a transgene in which puromycin resistance is expressed under the control of the SMC-specific promoters, SMαA or SM-MHC. Pure SMC populations are therefore isolated by enzymatic dispersion of the mature EBs and selection by puromycin to eliminate all non-SMC types.

1. The EBs are harvested at 14 days (for immature SMC) or 28 days (to obtain mature SMC). The EBs are washed with PBS twice.

2. 0.5 ml of collagenase/trypsin mixture is added to each well of a 48-well plate or 1 ml to each well of a 24-well plate and the plates are incubated for 10 min at 37°C.

3. To start dissociating the EBs into a single cell suspension, a P1000 is used to triturate the enzyme mixture. It is important to avoid bubbles!

4. Since the cells in an EB outgrowth are embedded in a well-developed extracellular matrix, the dissociation may take some time. The cell–enzyme mix is incubated for a further 5–10 min and trituration repeated aggressively so that the mass of EB is broken down.

5. All samples for each group are pooled into a 50 ml sterile tube and trituration repeated with a P1000 tip (or a 10 ml pipette if volume too large). There should be a large number of single cells and small cell clusters by this stage. If not, the cells may be incubated further and re-triturated.

6. The dispersed EB mix is squirted forcefully through a 70 μm nylon filter, placed over a 50 ml tube, and washed through with 10 ml EB medium, to collect any remaining single cells and to inactivate the enzymes.

7. Cells are centrifuged for 5 min to pellet and then resuspended with 3 ml SMC selection media.

8. Cells are plated into 3× 12-wells (1 ml per well).

9. The next day, dead adherent cells and debris are washed off the surviving SMCs two to three times with PBS.

10. The cells are now ready for immunostaining, RNA or protein harvesting or may be trypsinized for FACS studies or other experiments. Alternatively, puromycin selection may be left to proceed for up to 3 days under low dose puromycin (0.05 μg/ml) to ensure no undifferentiated pluripotential cells remain in the cell population (*see* **Note 6**).

3.5. Immunofluorescence

Indirect immunofluorescence is used to assess SMC localization within the intact EB as well as to obtain a preliminary assessment of purity of SMC following puromycin selection. We commonly use antibodies to SMαA and SM-MHC.

1. EBs or puromycin selected cells are washed briefly with PBS.

2. Cells are then fixed with 4% PFA in PBS or 10% buffered formalin for 5 min.

3. Three times 3 min PBS washes are carried out.

4. Cell membranes are permeabilized with 0.5% triton in PBS for 10 min.

5. Three times 3 min PBS washes.

6. Non-specific-binding sites are blocked with 3% BSA/3% donkey serum in PBS for 30–45 min (*see* **Note 7**).

7. The primary antibody is applied diluted in block for 1 hr at room temperature (or overnight at RT or 4°C). Dilution for SMA is 1:4,000 and use mouse IgG as a negative control (2 mg/ml). For SM-MHC, the BTI antibody is diluted at 1:500 and normal rabbit serum used as negative control.
8. Three times 3 min PBS/0.1% BSA washes.
9. The fluorophore-labeled secondary antibody (donkey anti-mouse or anti-rabbit-Cy2 at 1:400) in PBS/0.1% BSA is applied for 30–45 min.
10. Three times 3 min PBS washes.
11. DAPI 1 μM is used to stain cell nuclei for 3 min.
12. PBS wash for 10 min.
13. EB/cells are left in PBS and visualized under fluorescence microscopy. **Figure 22.2** and Color Plate 22, shows examples

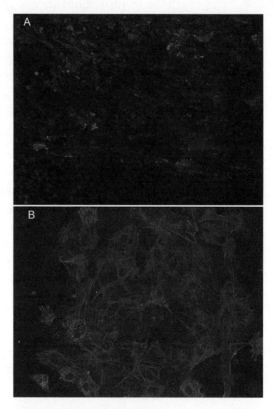

Fig. 22.2. Immunofluorescence for SMαA in intact EBs and puromycin purified SMCs. Immunostaining was carried out as described in 3.5 on intact day 28 EBs (**A**) and on cells following enzyme dispersal and overnight selection (**B**). A "sheet" of SMCs may be seen in an intact day 28 EB with adjacent non-staining non-SMC at the bottom of the image (**A**). Such groups of SMC may be seen to contract spontaneously under the microscope. Following puromycin selection, relatively pure populations of cells with a typical SMC morphology were seen (**B**). Red staining: SMαA immunofluorescence, Blue: DAPI nuclear staining. (*See* Color Plate 22)

of SMαA immunofluorescence of SMC within intact EBs and following dispersion and selection.

3.6. RT-PCR

Expression of multiple genes in either EBs or puromycin selected cells may be assessed quantitatively using real-time RT-PCR. RNA is harvested using Trizol and following the manufacturer's instructions.

3.6.1. Reverse Transcription and Quantitative PCR

Reverse transcription (RT) is carried out on 0.2 or 1 μg RNA, depending on yield from sample, according to standard protocols (*17*). Samples for –RT controls are also included. We prefer a two step RT-PCR procedure, particularly if RNA yield is limited, since there is sufficient cDNA product for the analysis of at least 30–40 different PCRs. Target gene expression is measured in parallel with 18S or GAPDH expression to normalize target gene levels for variations in RNA extraction efficiency and RT variations between samples. 0.5 μl of the RT reaction mix is used in each PCR (*see* **Note 8**) and standard methods are used for real-time PCR using either Taqman chemistry probes or SYBR green (*12*). Expression of a variety of SMC-specific and other cell type-specific genes is quantified in **Fig. 22.3**.

3.7. Flow Cytometry

We have used flow cytometry, based on the expression of the specific markers SMαA and SM-MHC, to quantify the degree of purification of SMC in this system. Since there are no well-characterized extracellular markers that delineate the SMC lineage, it is necessary to fix and permeabilize the cells before sorting. See **Fig. 22.4** for an example of the results obtained from this system. Below is the protocol for flow assisted sorting on the basis of SMαA staining. We have also sorted cells on the basis of SM-MHC staining using this protocol. Other SMC specific markers could potentially be used if a suitable antibody is available.

3.7.1. Fixation

1. EB derived SMCs after puromycin selection are trypsinized and then spun down in a 50 ml tube and the supernatant removed.
2. Cells are chilled on ice and then resuspended in 1 mL ice-cold PBS. Use a 1 ml pipette if necessary to break up the pellet. Any aggregates formed at this point will be permanently fixed together.
3. Cold 1 ml PFA is added dropwise while vortexing gently (lowest possible speed to generate vortex, usually a setting of two to three out of ten) to a final concentration of 2% formaldehyde. Adding cold PFA and warming to room temperature while shaking will decrease cellular aggregation.
4. Cells are allowed to fix for 15 min at room temperature with rapid orbital shaking and then centrifuged at 1,000 *g*.

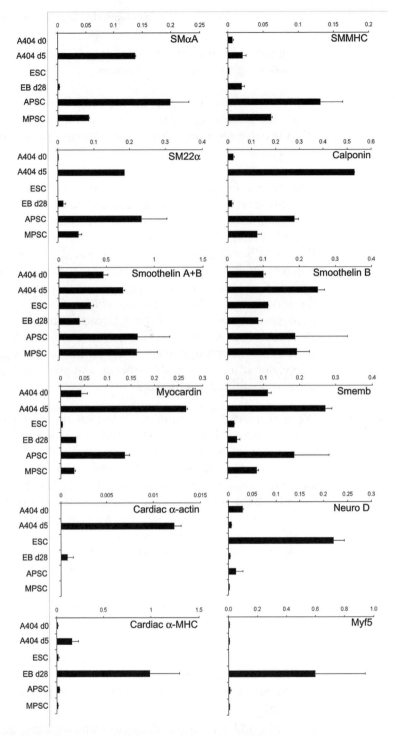

Fig. 22.3. mRNA expression profile of ESC-isolated cells compared to precursor ESC, EB, and a P19-derived A404 cell line. Quantitative real-time RT-PCR was used to measure the expression of a wide variety of SMC-specific markers in APSC, MPSC, their precursor EB and ESC. A P19-derived A404 cell line that expresses SMC-specific markers on stimulation with retinoic acid was used as a positive control. Control genes specific to cardiomyocytes (cardiac α-actin and cardiac α-MHC), skeletal myocytes (Myf5), and neurons (neuro D) were also measured. The data suggest a SMC identity for APSC and MPSC and confirm a high degree of specificity within these populations. APSC (SMαA puromycin-selected cells), MPSC (SMMHC puromycin-selected cells). Reproduced by permission from: Sinha et al. Stem Cells 2006;24:1678-88, © AlphaMed Press.

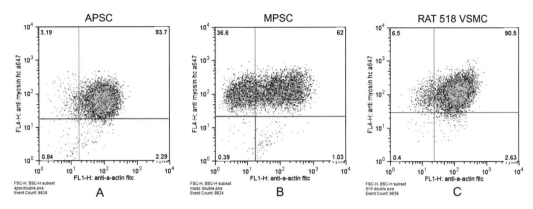

Fig. 22.4. FACS analysis of ESC-derived SMC and a primary SMC line for levels of SMC markers. APSC (**A**), MPSC (**B**) or rat aortic SMC (**C**) were stained with FITC-conjugated SMαA and Alexa 647-conjugated SMMHC antibody and analyzed by flow cytometry. All three cell types had similar homogenous staining for the highly specific SMC marker, SMMHC. Conversely, while the rat aortic SMC and APSC also demonstrated similar staining for SMαA, the MPSC had a bimodal distribution of SMαA with approximately 40% of cells staining weakly for SMαA, while the remainder exhibited strong staining. Thus APSC and MPSC both expressed SMC-specific proteins although the MPSC comprised at least two distinct populations: APSC (SMαA puromycin-selected cells) and MPSC (SMMHC puromycin-selected cells). Reproduced by permission from: Sinha et al. Stem Cells 2006;24:1678-88, © AlphaMed Press.

3.7.2. Permeabilization

5. The supernatant is poured off and the fixed cells are resuspended in the remaining volume. If necessary 100 μl PBS may be added to get cells back into solution.

6. Cells are permeabilized by adding ice-cold 100% methanol (*see* **Note 9**) dropwise to prechilled cells while gently vortexing so that final concentration is at least 90% methanol (~1.8 ml).

7. Incubate on ice for 30 min, then centrifuge at 500–600 g (EBs −1,000 g).

8. Cells are resuspended in PBS–BSA-2% and transferred to a 2 ml clear tube (*see* **Note 10**).

9. Cells are counted using a coulter counter or stained with DAPI and examined under a fluorescence microscope with a hemocytometer (*see* **Note 11**).

10. Spin and remove supernatant.

11. Cells are resuspended at a concentration of approximately 1–5 million cells/ml.

12. Keep at 4°C overnight.

3.7.3. Staining (start staining 2 h before flow sorting)

Cells are set aside from each group for no primary controls and backup cells. Depending on the application different concentrations of antibodies and different controls will be used (*see* **Note 12**). Before performing the experiment for the first time, each antibody is titrated to determine the ideal concentration for your application. For determining total SMαA protein content of a cell (to quantitate SMC differentiation or changes with drug or treatment for instance), a saturating concentration of antibody is required so that all epitopes are occupied. For determining total number of SMαA positive cells, a lower concentration of antibody may be ideal.

13. To about 10^5 (20–100 μl) cells in each tube is added anti-SMαA-FITC (Sigma 1A4 clone) or anti-SMαA-Cy3 in concentrations ranging from 1:50 to 1:5,000 (*see* **Note 13**). In particular, the range of 1:100–1:750 will be saturating for SMC, lower concentrations for non-SMC or cells with a smaller proportion of SMC present.

14. A similar amount is added to a non-staining cell line (we use ESCs) to quantify nonspecific staining at each concentration (concentrations >1:750–1:1,000 will lead to significant staining).

15. Incubation is carried out for 15 min − 1 h with agitation (shorter times may decrease nonspecific staining).

16. 2 ml PBS–BSA is added and the cells are rotated or agitated for 5 min to wash off primary (more washes will decrease nonspecific binding without altering specific binding much – but yield decreases rapidly).

17. Cells are spun down and resuspended in 500 μl.

18. Saturating concentration is determined, and an antibody concentration is chosen that is at or near saturation without causing too much nonspecific staining of the ESCs.

19. As concentration of antibody decreases, nonspecific staining should decrease more rapidly than specific staining. But to quantify protein in the cells, the concentration must be as near to saturation as possible, otherwise subtle shifts in protein content will be missed.

20. For mere identification of SMαA-positive cells, lower concentrations (1:2,000) are adequate, and will eliminate nonspecific staining. Non-staining cells should still be included to determine gates (unless there is a inherent non-staining population in your cells that can be reliably identified and used for gating).

21. For APSCs/MPSCs, we found that about 1:500 anti-SMαA-FITC is an ideal concentration for saturation of signal at 5 million cells/ml. For rat primary SMC lines, we have found between 1:250 and 1:500 is ideal. For EBs, 1:1,000–1:2,000 may saturate the positive cells, but it is difficult to be certain due to variability in the system.

22. Once each antibody to be tested has been titrated, perform the experiment the same, 1 h incubation (shorter times may decrease nonspecific binding) on each cell type or treatment groups to be tested (*see* **Note 14**).

23. DNA staining is included—ideally Draq5 or something in the far red region to prevent spectral overlap to identify nucleated cells for gating. At concentrations of 1–5 million cells/ml, Draq5 may be used at concentrations of 1:500–1:1,000. Avoid PI, it will overlap in many channels, 7-AAD is a little better.

3.7.4. Controls

24. Unstained cells—to determine FSC, SSC, and baseline signal in FL1, FL2, FL3, FL4, etc., keep enough to tune the machine.
25. Compensation controls—must always be included. Consists of each fluorophore to be tested used singly on positive cells (also called single-stain controls).
26. Non-staining cells—our recommended method for setting gates/boundaries. It should be combined with FMO (fluorescence-minus-one) controls ideally. ES cells make a good negative control for many proteins.
27. Isotype controls—not a very good control. Depending on labeling, purity, or tendency of the isotype to stick to some epitope in a specific fashion, these controls do not say anything meaningful for intracellular flow. Fine for extracellular, but FMO and non-staining cells we believe are more stringent and more accurate in most cases.

3.8. Assessment of Physiological Responses

A major strength of this system is that the SMC generated in this way display appropriate physiological responses including agonist induced Ca^{2+} transients and a contractile response. However, as the focus of this chapter is on the generation of SMCs from ESCs, there is insufficient space to fully describe the methods we used to assess these physiological responses. Since these methods have been described in detail elsewhere (*18;19*), we present only a brief overview of our approach.

3.8.1. Assessment of Electrophysiological Activity of Puromycin-Selected SMCs

APSCs or MPSCs are loaded with the Ca^{2+}-sensitive fluorophore, Fluo-4 AM and mounted into a constant-flow superfusion chamber, as previously described (*18*). Cell fluorescence is assessed in response to endothelin-1, angiotensin II, high potassium or ionomycin exposure. In addition, whole cell Ca^{2+} currents are determined with a standard whole cell voltage-clamp technique as previously described (*19*). These studies showed that depolarization with 80 mM KCl resulted in an increase in intracellular Ca^{2+} levels as did treatment with the potent vasoconstrictors angiotensin II and endothelin-1 (*see* **Fig. 22.4** in Sinha et al. (*11*)). Moreover, we demonstrated that typical L-type voltage-gated Ca^{2+} channel current traces and current–voltage (I–V) relationships were present consistent with a SMC identity for our puromycin purified cells. Interestingly, Ca^{2+} influx was not consistent between different cells within APSC and MPSC populations and some cells displayed no response to the agonists suggesting that the purified cells, whilst all expressing SMC markers, were nevertheless heterogeneous in nature.

3.8.2. Smooth Muscle Fiber Preparation and Contractility Assessment

Contractile function is assessed by generating reconstituted muscle fibers as described by Oishi and colleagues (20). Briefly, cultured D3 ES cells, APSC or MPSC, are seeded at a density of $8–10 \times 10^6$ cells/ml in a collagen I solution and dispensed into a rectangular Sylgard 184 mold. Once gelled, the collagen–cell suspensions are covered with and maintained in EB media. Incubation for 7 days leads to the formation of a "dumbell"-shaped muscle fiber between the two poles. Small longitudinal strips are cut from reconstituted muscle fibers on day 7, and connected to a force transducer. These reconstituted muscle fibers generated from either APSC or MPSC exhibit a K^+-induced contraction similar to intact smooth muscle tissue (**Fig. 22.5**). Moreover, endothelin-1 or sphingosine-1-phosphate stimulation leads to force generation which can be inhibited by Y27632 (21) suggestive of a Rho kinase-dependent pathway characteristic of mature SMCs.

3.9. Conclusion

We describe a method for purifying SMC populations from ESC-derived embryoid bodies. We also present protocols for assessments of SMC purity and identity by immunofluorescence, quantitative RT-PCR, and FACS (**Figs. 22.2–22.4**). A major advantage of this system is that it appears to provide all necessary environmental cues required for programing totipotent ESC into fully functional SMC, and permits purification of the resulting SMC using a stably incorporated negative selection system. Since ESCs are relatively amenable to genetic modification, this system has great potential for studying developmental regulation of SMC lineage using heterozygous or homozygous knockout of genes of interest. Moreover, the ESC–EB system is uniquely placed to take advantage of several large multinational consortia such as the International Gene Trap Consortium (http://www.igtc.ca/) that are systematically generating null ESC lines to cover the entire mouse genome. In addition, this system offers a more rigorous assessment of cell autonomous versus non-cell autonomous gene functions as compared to conventional knockout mice, since cell lineage programing in the ESC-EB is independent of the requirement for a fully functioning cardiovascular system. The puromycin-derived SMCs may potentially be used in tissue engineering applications. However, a major concern in this field when using ESC-derived cells is the risk of teratoma formation. Our studies have shown that this risk is eliminated following 72 h of negative selection and emphasizes the critical need for a selection

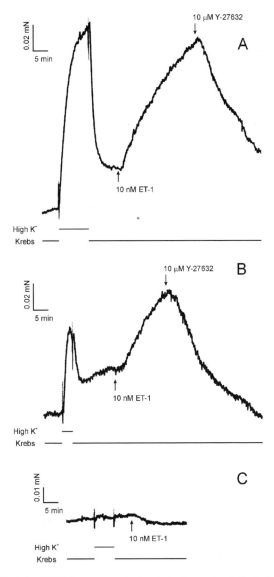

Fig. 22.5. Contraction of ESC-derived SMC in response to depolarization and endothelin-1 in reconstituted fiber preps. Artificial muscle fibers were prepared by seeding collagen gels with APSC (**A**), MPSC (**B**), or D3 progenitor cells (**C**) as described in 3.8.2 and in REF. After 7 days, small strips (1.5 × 0.75–0.80 mm) were cut from the reconstituted fiber and the ends tied to wire hooks of a force transducer. Contraction was measured in response to 80 mM KCl and endothelin-1. APSC and MPSC, but not D3 progenitor cells, displayed typical SMC contractile responses in which the endothelin-1 induced contraction was inhibited by the Rho kinase inhibitor Y-27632. APSC (SMαA puromycin-selected cells) and MPSC (SMMHC puromycin-selected cells). Reproduced by permission from: Sinha et al. Stem Cells 2006;24:1678–88, © AlphaMed Press.

strategy to eliminate residual pluripotent cells when using ESC-derived cells for tissue engineering.

It is important to note, however, that the precise identities of the SMC isolated using this system are not fully defined. The FACS data suggest that the puromycin purified cells are composed of more than one sub-population, with a bimodal distribution of SMαA expression in the MPSCs as a clear example of heterogeneity. Indeed, since the EB recapitulates embryonic development, such heterogeneity may be expected and it is likely that the population of cells isolated includes vascular, visceral, and airway SMC. Although methods to isolate each sub-type may be envisaged, by modification of the SMC promoter segments, for example, the current system is ideally suited to studying developmental mechanisms that are common to all SMC sub-types. In any case, given the remarkable phenotypic plasticity of SMC, it is possible that this heterogeneity may not be a significant limitation in terms of prospective tissue engineering applications using these isolated cells.

4. Notes

1. An alternative to electroporation is nucleofection. Recently we have been using the nucleofector from Amaxa, with the ES cell nucleofector kit, with good results.

2. If growth of false positive clones is a problem with poor antibiotic selection efficiency, then consider adding cells to resistance/selection media by splitting. In other words, 24 h after electroporation and allowing cells to grow and express the resistance gene, split the cells again into G418. Adding antibiotic to cells that are already plated may contribute to selection of false-positive colonies.

3. The exact number of cells, and thus different sized EBs, to generate optimal SMC may depend on the starting line of ESCs. Also, larger EBs are easier to visualize and work with.

4. 70–80 drops can be comfortably placed on each 100 mm lid and the process is considerably easier if a repeating pipette is used.

5. This is best accomplished using 3–5 ml medium and holding the inverted dish lid at an angle to allow the fluid to flow from one side of the dish to the other. Care must be taken at this stage to ensure the operator's finger tips do not come into contact with the washing medium and that foaming does not occur.

6. We investigated the behavior of puromycin selected SMC in vivo (*11*). Purified cells obtained by an 18 h selection from d28 ESC–EBs were labeled with di-I, resuspended in matrigel and injected subcutaneously into mice genetically matched to the D3 ESC line. Large subcutaneous tumors developed in these mice at the injection sites within 2–3 weeks (*11*) and

histological appearance was consistent with the formation of teratomas. However, teratoma growth was eliminated by maintaining puromycin selection for 72 h before injecting the cells in vivo.

7. Block with serum from the species that the secondary antibody was raised in (donkey in this case).

8. To ensure equal amounts of cDNA are distributed in each well, the cDNA sample is diluted 1:20 in dH$_2$O and sufficient amounts are made to allow 10 μl for each test gene + 18S and/or GAPDH internal controls + 1.

9. Methanol is a denaturing agent and may not be optimal for antibodies that require an intact epitope. Alternatively, one may use 0.1–0.3% tween 20, or triton, but these severely affect scatter patterns. Usually 15 min of incubation in the presence of one of these detergents is adequate for permeabilization. Different permeabilization techniques should be attempted to find the ideal for each cell and antibody. Note that methanol is incompatible with GFP—use Tween instead.

10. After methanol treatment, the cells will stick to Falcon tubes, and regular Eppendorfs perform terribly for viewing and removing supernatant from small pellets.

11. Any count done with light microscopy, even phase, will likely be wrong, especially with EBs. Cells in aggregates should also be counted as they will also be absorbing stain.

12. We recommend against using an isotype control, because unless it is labeled with an equal quantity of dye per antibody it does not give a valid comparison of nonspecific staining. Rabbit IgG in particular is poor for intracellular flow, due to excessive nonspecific staining. Better controls are the fluorescence minus one technique, and inclusion of a non-staining control population of cells (such as undifferentiated ES cells). It is unwise to attempt secondary staining in intracellular flow, too many cells are lost after permeabilization with each spin, and too much nonspecific staining will occur. It is best to either label the primary antibody with Rhodamine (Pierce) or with FITC or one of the Alexafluors (Molecular probes). Alternatively, Zenon kits (Molecular probes) are a fast and easy way to label antibodies but do not always work well. Note that the storage solution for the antibody may need to be exchanged before labeling, and after labeling aliquots should be frozen back for single use to prevent degradation of the antibody.

13. FITC labeling gives a brighter signal, while Cy3-tends to have less spectral overlap. In cells with large amounts of SMαA, Cy3 will prevent detector overload since it is not detected as well on FL2 as FITC is on FL1.

14. If you have multiple antibodies with the same fluorophore, you only need one single stain for each fluorophore, not each antibody. Always use the brightest cell population for compensation. FMO controls require that for every antibody–fluorophore added, every combination of those must be tested to measure spectral overlap. For two fluorophores: FITC and Draq5, for instance, the compensation controls are adequate. For three fluorophores (e.g., FITC, Cy3, and Draq5), you have to assemble every possible combination in addition to single-stained controls (FITC+Cy3, FITC+Draq5, and Cy3+Draq5). This only needs to be done with one population of cells, and it is to show where the different fluorophores spectrally overlap. In other words, the controls set the gates or quadrants determining positive and negative populations. For instance, the FITC + Cy3 group will indicate where to set the line indicating Draq5 positivity by showing the contribution to the Draq5 signal from the other fluorochromes. FITC+Draq5 shows where Cy3 signal is positive/negative and Cy3+Draq5 will show where the FITC signal is positive/negative.

Acknowledgements

This work was supported by NIH HL R01 HL57353 and R21 HL071976 to GKO and AHA Post-Doctoral Fellowship grant 0120501U to SS.

Reference

1. Owens, G.K., Kumar, M.S., Wamhoff, B.R. (2004). Molecular regulation of vascular smooth muscle cell differentiation in development and disease. *Physiol Rev* **84**, 767–801.
2. Schwartz, S.M., deBlois, D., O'Brien, E.R. (1995). The intima. Soil for atherosclerosis and restenosis. *Circ Res* **77**, 445–465.
3. Hirschi, K.K., Rohovsky, S.A., D'Amore, P.A. (1998). PDGF, TGF-beta, and heterotypic cell-cell interactions mediate endothelial cell-induced recruitment of 10T1/2 cells and their differentiation to a smooth muscle fate. *J. Cell Biol* **141**, 805–814.
4. Manabe, I., Owens, G.K. (2001). Recruitment of serum response factor and hyperacetylation of histones at smooth muscle-specific regulatory regions during differentiation of a novel P19-derived in vitro smooth muscle differentiation system. *Circ Res* **88**, 1127–1134.
5. Jain, M.K., Layne, M.D., Watanabe, M., et al. (1998). In vitro system for differentiating pluripotent neural crest cells into smooth muscle cells. *J Biol Chem* **273**, 5993–5996.
6. Oishi, K., Ogawa, Y., Gamoh, S., et al. (2002). Contractile responses of smooth muscle cells differentiated from rat neural stem cells. *J Physiol* **540**, 139–152.
7. Desbaillets, I., Ziegler, U., Groscurth, P., et al. (2000). Embryoid bodies: an in vitro model of mouse embryogenesis. *Exp Physiol* **85**, 645–651.
8. Drab, M., Haller, H., Bychkov, R., et al. (1997). From totipotent embryonic stem cells to spontaneously contracting smooth muscle cells: a retinoic acid and db-cAMP in

vitro differentiation model. *FASEB J* **11**, 905–915.
9. Sinha, S., Hoofnagle, M.H., Kingston, P.A., et al. (2004). Transforming growth factor-beta1 signaling contributes to development of smooth muscle cells from embryonic stem cells. *Am J Physiol Cell Physiol* **287**, C1560–C1568.
10. Yamada, T., Yoshikawa, M., Takaki, M., et al. (2002). In vitro functional gut-like organ formation from mouse embryonic stem cells. *Stem Cells* **20**, 41–49.
11. Sinha, S., Wamhoff, B.R., Hoofnagle, M.H., et al. (2006). Assessment of contractility of purified smooth muscle cells derived from embryonic stem cells. *Stem Cells* **24**, 1678–1688.
12. Bustin, S.A. (2000). Absolute quantification of mRNA using real-time reverse transcription polymerase chain reaction assays. *J Mol Endocrinol* **25**, 169–193.
13. Hogan, B., Beddington, R., Constantini, F., et al. (1994). *Manipulating the Mouse Embryo: A Laboratory Manual* Cold Spring Harbor Press.
14. Mack, C.P., Owens, G. K. (1999). Regulation of smooth muscle alpha-actin expression in vivo is dependent on CArG elements within the 5' and first intron promoter regions. *Circ Res* **84**, 852–861.
15. Madsen, C.S., Regan, C.P., Hungerford, J.E., et al. (1998). Smooth muscle-specific expression of the smooth muscle myosin heavy chain gene in transgenic mice requires 5'-flanking and first intronic DNA sequence. *Circ Res* **82**, 908–917.
16. Frail, D.E., McLaughlin, L.L., Mudd, J., et al. (1988). Identification of the mouse muscle 43,000-dalton acetylcholine receptor-associated protein (RAPsyn) by cDNA cloning. *J Biol Chem* **263**, 15602–15607.
17. Sambrook, J., Russell, D.W. (2001). *Molecular cloning: a laboratory manual* Cold Spring Harbour Laboratory Press.
18. Wamhoff, B.R., Dixon, J.L., Sturek, M. (2002). Atorvastatin treatment prevents alterations in coronary smooth muscle nuclear $Ca2+$ signaling in diabetic dyslipidemia. *J Vasc Res* **39**, 208–220.
19. Wamhoff, B.R., Bowles, D.K., McDonald, O.G., et al. (2004). L-type voltage-gated $Ca2+$ channels modulate expression of smooth muscle differentiation marker genes via a rho kinase/myocardin/SRF-dependent mechanism. *Circ Res* **95**, 406–414.
20. Oishi, K., Itoh, Y., Isshiki, Y., et al. (2000). Agonist-induced isometric contraction of smooth muscle cell-populated collagen gel fiber. *Am J Physiol Cell Physiol* **279**, C1432–C1442.
21. Uehata, M., Ishizaki, T., Satoh, H., et al. (1997). Calcium sensitization of smooth muscle mediated by a Rho-associated protein kinase in hypertension. *Nature* **389**, 990–994.

Part VII

Regeneration of the Pancreas and Liver

Part VII

Regeneration of the Pancreas and Liver

Chapter 23

Islet-Derived Progenitors as a Source of In Vitro Islet Regeneration

Stephen Hanley and Lawrence Rosenberg

Abstract

Current therapies do not prevent the complications of diabetes. Furthermore, these therapies do not address the underlying pathology; the lack of functional β-cell mass that occurs in both types 1 and 2 diabetes. While pancreas and islet transplantation do serve to increase β-cell mass, a lack of donor organs limits the therapeutic potential of these treatments. As such, expansion of β-cell mass from endogenous sources, either in vivo or in vitro, represents an area of increasing interest. One potential source of islet progenitors is the islet proper, via the dedifferentiation, proliferation, and redifferentiation of facultative progenitors residing within the islet. We have developed a tissue culture platform whereby isolated adult human pancreatic islets form proliferative duct-like structures expressing ductal and progenitor markers. Short-term treatment with a peptide fragment of islet neogenesis-associated protein (INGAP) induces these structures to reform islet-like structures that resemble freshly isolated islets with respect to the frequency and distribution of the four endocrine cell types, islet gene expression and hormone production, insulin content, and glucose-responsive insulin secretion. As such, the plasticity of adult human islets has significant implications for islet regeneration.

Key words: Dedifferentiation, islet, islet neogenesis-associated protein (INGAP), progenitor, redifferentiation.

Abbreviations: DLS: duct-like structure, DMEM: Dulbecco's modified Eagle's medium, EGF: epidermal growth factor, FBS: fetal bovine serum, GLP-1: glucagon-like peptide-1, HBSS: Hank's-buffered salt solution, HRP: horseradish peroxidase, IE: islet equivalent, ILS: islet-like structure, INGAP: islet neogenesis-associated protein, PBS: phosphate-buffered saline.

1. Introduction

Diabetes can be defined as a deficiency of insulin-producing β-cell mass (1, 2). While transplantation-based therapeutic approaches aimed at restoring the lost β-cell mass have yielded mixed results

(3), the most recent area of research interest is the endogenous expansion of β-cell mass within the native pancreas. The obvious benefits of this approach include the potential to reduce or remove the need for immunosuppression, as well as obviating the need for donated organs and the inherent associated shortages. There exists, however, a debate as to the actual mechanisms of β-cell mass expansion, with some suggesting that replication of pre-existing β-cells is essentially the only mechanism at work *(4)*, while others argue that islet neogenesis—the *de novo* formation of islets from non-endocrine progenitor cells—plays an important role in overall β-cell mass dynamics *(5, 6)*.

The identification of an islet cell progenitor that can be expanded and differentiated as part of the process of islet neogenesis remains to be determined. Pancreatic ductal cells are commonly cited as the most likely cell candidate *(5–9)*, but acinar *(10, 11)* and even dedifferentiated islet cells *(12–14)* have been proposed as well. In the latter case, however, there also exists a semantic debate as to whether an islet begetting an islet, via a dedifferentiated intermediate cell type(s), constitutes replication or islet neogenesis *(15–17)*.

Furthermore, several agents have been proposed as potential inducers of endogenous β-cell mass, including but not limited to glucagon-like peptide-1 (GLP-1) and related compounds *(18–20)*, combination treatment with gastrin and epidermal growth factor (EGF) *(10, 21, 22)*, and islet neogenesis-associated protein (INGAP) *(23, 24)*, each acting through replication, neogenesis, or a combination thereof.

We have developed a culture platform whereby isolated adult human pancreatic islets can be induced to form highly proliferative duct-like structures (DLS) that express putative progenitor cell markers *(25)*. By treating these dedifferentiated structures with a peptide-derivative of INGAP, the DLS can be induced to regenerate islet-like structures (ILS) that resemble freshly isolated islets with respect to the frequency and distribution of endocrine cell types. Moreover, these ILS express endocrine hormones and relevant transcription factors at levels matching those of freshly isolated islets. Finally, these structures also contain equivalent amounts of insulin, and secrete insulin in a glucose-responsive manner *(16)*.

2. Materials

2.1. Embedding and Cell Culture

1. Dulbecco's modified Eagle's medium: Nutrient mix F12 (DMEM/F12) (Invitrogen, Carlsbad, CA) contains L-glutamine and sodium bicarbonate.

2. Heat-inactivated fetal bovine serum (FBS; Montreal Biotech Inc, Montreal, Qc) is stored in aliquots at −20°C, and added to culture medium as needed.

3. Dexamethasone (Sigma-Aldrich, Oakville, ON) is dissolved in 95% ethanol at 1 mM, stored at −20°C, and added to culture medium as needed.

4. 1,000× penicillin+streptomycin+fungizone (Invitrogen) is stored in aliquots at −20°C and added to culture medium as needed.

5. EGF (Sigma-Aldrich) is dissolved in Hank's-buffered salt solution (HBSS; Invitrogen) at 10 μg/mL, stored in aliquots at −20°C, and added to culture medium as needed.

6. Humulin R insulin (Eli Lilly, Toronto, Ont.) is dissolved in HBSS at 16 U/mL, stored at 4°C, and added to culture medium as needed.

7. Cholera toxin (Sigma-Aldrich) is dissolved in culture-grade water at 500 μg/mL, stored at 4°C and added to culture medium as needed.

8. Type I rat tail collagen is prepared fresh from frozen rat tails *(26)* or obtained from Becton-Dickinson (Franklin Lakes, NJ), stored at 4°C, and used for embedding as needed.

9. Neutralization solution is prepared by mixing 10× Waymouth buffer (Sigma-Aldrich) and 0.34 N NaOH in a ratio of 6:4. This solution is filtered, stored at 4°C, and used for embedding as needed.

10. INGAP peptide (NH$_2$-IGLHDPSHGTLPNGS-COOH; Sheldon Biotechnology Centre, Montreal, Qc) is dissolved in culture-grade water at 250 μg/mL, aliquoted, and stored at −80°C, and added to culture medium as needed.

2.2. Immunocytochemistry

1. Type XI collagenase (Sigma-Aldrich) is dissolved in DMEM/F12 at 2 mg/mL, stored at 4°C, and added to cultures as needed.

2. Agarose is prepared by dissolving 1 g agarose (Invitrogen) in 50 mL culture-grade water and heating until clear. The agarose gel can be reheated as needed.

3. 10% phosphate-buffered formalin, xylene, petroleum ether, 30% H$_2$O$_2$, methanol, ethanol and Permount are from Fisher Scientific (Nepean, Ont.).

4. Endogenous peroxidase quenching solution is prepared fresh as a 9:1 methanol:30% H$_2$O$_2$ solution.

5. Citrate buffer is prepared by dissolving citric acid to 0.1 M in distilled water, and adjusting to pH 6 with NaOH.

6. Blocking buffer, secondary antibody reagent, and horseradish peroxidase (HRP) enzyme conjugate are components of the Histostain Plus Horseradish Peroxidase broad spectrum kit (Zymed, San Francisco, CA), and are all ready-to-use solutions.

7. Primary antibodies are diluted in blocking buffer prior to use. The following antibodies are used: guinea pig anti-insulin (1:500), rabbit anti-glucagon (1:500), rabbit anti-somatostatin (1:500), rabbit anti-pancreatic polypeptide (1:500), mouse anti-cytokeratin AE1/AE3 (1:200) (all from Dako Cytomation, Carpinteria, CA).

8. DAB substrate (Sigma-Aldrich) is prepared fresh by vortexing 1 gold-packaged tablet and 1 silver-packaged in 5 mL distilled water.

9. Hematoxylin is from Sigma-Aldrich.

10. Sodium borate solution is prepared by dissolving Na_3BO_3 to 0.05 M in distilled water, storing at room temperature.

2.3. Real-Time PCR

1. Type XI collagenase (Sigma-Aldrich) is dissolved in DMEM/F12 at 2 mg/mL, stored at 4°C, and added to cultures as needed.

2. Buffer RLT, buffer RW1, RPE concentrate, RNase-free water, and RNeasy columns and collection tubes are components of the RNeasy mini kit (Qiagen, Mississauga, Ont.).

3. Buffer RLT+βME is prepared fresh by adding 10 μL β-mercaptoethanol (Sigma-Aldrich) to 1 mL buffer RLT.

4. Buffer RPE+EtOH is prepared fresh by diluting RPE concentrate 1:4 in anhydrous ethanol (Fisher).

5. Buffer RDD and DNase I are components of the RNase-free DNase set (Qiagen).

6. DNase I is dissolved in 550 μL supplied RNase-free water, aliquoted, and stored at −20°C until needed.

7. Buffer RDD+DNase is prepared by diluting DNase I 1:7 in buffer RDD.

8. RT buffer, dNTP mix, and reverse transcriptase are components of the Omniscript RT kit (Qiagen).

9. Oligo-dT is from Invitrogen and is stored at −20°C until use.

10. RNase inhibitor is from Promega (Madison, WI) and prepared by diluting 1:3 in RNase-free water, aliquoted, and stored at −20°C until use.

11. PCR master mix and RNase/DNase-free water are components of the QuantiTect SYBR green PCR kit (Qiagen) and are supplied in a ready-to-use form.

12. Custom primers are synthesized by Invitrogen and dissolved in RNase/DNase-free water to 100 μM. Working dilutions of 1 μM are prepared in RNase/DNase-free water and aliquoted and stored at −20°C until use. Primers used, with annealing temperatures, are listed in **Table 23.1**.

Table 23.1
Primer sequences, suggested annealing temperatures and amplicon lengths for real-time PCR-based analysis of islets, DLS, and ILS

Gene	Primer	$T_{annealing}$ (°C)	Amplicon (bp)
Insulin	GCTGGTAGAGGGAGCAGATG AGCCTTTGTGAACCAACACC	58	243
Glucagon	ACAAGGCAGCTGGCAACGTTCCCT CCTTCCTCCGCCTTTCACCAGCCA	58	343
Somatostatin	CCAGACTCCGTCAGTTTCTG GCCATAGCCGGGTTTGAGTTA	58	204
Pancreatic polypeptide	CCACCTGCGTGGCTCTGTTA AGAAGGCCAGCGTGTCCTC	58	189
pdx1	CCTTTCCCATGGATGAAGTC TTCAACATGACAGCCAGCTC	54	199
nkx2.2	TTGTCATTGTCCGGTGACTC TCTACGACAGCAGCGACAAC	58	154
isl1	TTCCCACTTTCTCCAACAGG GTTACCAGCCACCTTGGAAA	58	240
ck19	CTGGAGATGCAGATCGAAGG CGGTTCAATTCTTCAGTCCG	58	241
caII	CAGCAGCGAGCAGGTGTTGA CCAAGGTAGATCTGTGCTTAGC	58	205
ngn3	CGGTAGAAAGGATGACGCCT TGCTTGCTCAGTGCCAACTC	58	249
nestin	CAGGAGAAACAGGGCCTACA GTGTCTCAAGGGTAGCAGGC	58	294
β-actin	AGAGCTACGAGCTGCCTGAC AGCACTGTGTTGGCGTACAG	58	181

3. Methods

Several tissue culture platforms have been developed with a view to inducing adult human islets to transform into a heterogeneous population containing facultative islet progenitor cells. While other methods involve the formation of an islet-derived monolayer (17, 27, 28), we describe a method wherein islets are embedded within a matrix of type 1 rat tail collagen, as a required microenvironment to effect transformation into DLS (25).

Initial post-isolation recovery of the islets is completed in DMEM/F12 medium containing 10% FBS, 1 µM dexamethasone, penicillin+streptomycin+fungizone, 10 ng/mL epidermal growth factor, and 24 mU/mL insulin (= recovery medium).

Culture of embedded islets to induce DLS formation is completed in DMEM/F12 medium containing 1 µM dexamethasone, penicillin+streptomycin+fungizone, 10 ng/mL epidermal growth factor, 24 mU/mL insulin, and 200 ng/mL cholera toxin, but without serum (= dedifferentiation medium).

To induce islet regeneration, DLS are cultured in DMEM/F12 medium containing 10% FBS, 1 µM dexamethasone, penicillin+streptomycin+fungizone, 10 ng/mL epidermal growth factor, 24 mU/mL insulin, and 250 ng/mL (167 nM) INGAP peptide (= redifferentiation medium). Control samples are cultured in the above medium without INGAP peptide.

3.1. Embedding and Cell Culture

1. After isolation, islets are cultured in suspension in recovery medium, overnight at 37°C in 100 mm non-tissue culture-treated dishes, at a density of 1,000 IE/mL and 10,000 IE/dish (10 mL/dish).

2. On the day of embedding, islets are collected into a 50 mL tube, and washed (2 min at 900 rpm) twice in recovery medium. Samples of 6,000 IE are then aliquoted into 15 mL tubes, pelleted, and resuspended in 500 µL recovery medium. Six milliliters of cold collagen is added to each 15 mL tube, and tubes are stored on ice.

3. Required materials are made ready, including three 35 mm tissue culture-treated dishes per 15 mL tube, 1.5 mL neutralization solution per 15 mL tube, vortex, and 10 mL pipette.

4. While vortexing gently, neutralization solution is added dropwise until a bubblegum pink color is attained. Neutralization of 6 mL collagen (the contents of one 15 mL tube) typically requires 1 mL neutralization solution (see **Note 1**).

5. Quickly, the neutralized solution is mixed gently by pipetting, and transferred to three 35 mm dishes, with each dish receiving approximately 2 mL neutralized collagen

containing the islets (~2,000 islet equivalents [IE]). Dishes are gently shaken horizontally to spread the collagen solution across the entirety of the dish's surface area (*see* **Note 2**).

6. Dishes are allowed to stand for 15 min at room temperature to allow the collagen to solidify completely. At this point, islets should be evenly dispersed throughout the entire thickness of the gel (~2 mm) and across the entire surface area of the dish.

7. Dishes receive 2 mL dedifferentiation medium, pre-warmed to 37°C. Cultures are placed in an incubator at 37°C with 5% CO_2/95% air.

8. To change medium every other day, dishes can be tipped gently to allow the replacement of nearly all liquid medium. Care should be taken not to disrupt the collagen gel, which can become fragile.

9. After 14 days of culture, virtually all islets should be transformed into DLS, as assessed morphologically. Representative photomicrographs of the various stages of the transformation from islet into DLS are shown in **Fig. 23.1** and Color Plate 23.

10. To induce the regeneration of ILS from DLS, dedifferentiation medium is replaced with redifferentiation medium. Control cultures receive redifferentiation medium without INGAP peptide supplementation (*see* **Note 3**). Cultures are maintained in an incubator at 37°C with 5% CO_2/95% air, with medium changed every other day as above. Cultures should be maintained for 4–8 days before islet regeneration is assessed.

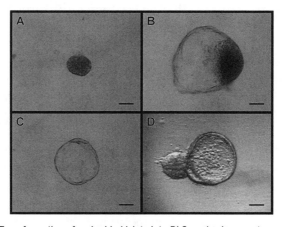

Fig. 23.1. Transformation of embedded islets into DLS, and subsequent regeneration of ILS. **(A)** Immediately after embedding, islets are characterized by a solid-spheroid shape. **(B)** DLS formation appears to initiate in specific foci, **(C)** until a DLS replaces the islet. **(D)** Treatment with INGAP induces the budding of regenerating ILS from DLS (bar = 100 μm). (*See* Color Plate 23)

3.2. Immunocytochemistry

1. To remove samples from the collagen gel, type XI collagenase is added to the culture medium to a final concentration of 250 μg/mL. Cultures are returned to the incubator and allowed to digest for 45 min at 37°C. The contents of the dish are collected into a 50 mL tube and washed twice in 1× PBS (*see* **Note 4**).

2. Pellets are transferred to 1.5 mL Eppendorf tubes, pelleted, resuspended in 1 mL 10% phosphate-buffered formalin, and stored at 4°C for several days.

3. To allow for routine immunocytochemical processing, pellets must be embedded in agarose. Samples are pelleted and resuspended in 400 μL molten agarose. The contents are mixed quickly by pipetting and centrifuged (2 min at 10,000 rpm) to ensure that the sample forms a pellet that is fully entrapped within the polymerizing agarose.

4. The Eppendorf containing the sample+agarose plug is stored on ice for 15 min to ensure completed polymerization, and is then removed, wrapped in lens paper, and placed in a pathology cassette for routine immunocytochemical processing and paraffin embedding (*see* **Note 5**).

5. Sections 4 μm in thickness are cut and applied to standard microscope slides.

6. Samples are dewaxed by two 5 min xylene incubations, followed by 2 min in petroleum ether.

7. Endogenous peroxidase activity is quenched by incubating the slides in quenching solution for 25 min at room temperature.

8. Slides are washed briefly in running water.

9. Antigen retrieval is carried out by placing the samples in citrate buffer and boiling for 10 min, followed by cooling for another 10 min (*see* **Note 6**).

10. Slide areas containing sample are circled using an ImmunoEdge pen (Vector Laboratories, Burlingame, CA), and the samples are incubated with blocking buffer for 15 min at room temperature.

11. Blocking buffer is replaced with a solution containing the primary antibody diluted in blocking buffer. Slides are incubated with this solution overnight at 4°C.

12. Slides are washed thrice with 1× PBS for 5 min, and exposed to the secondary antibody reagent for 15 min at room temperature.

13. Following another wash step, as in Step 12, slides are exposed to the HRP enzyme conjugate solution for 15 min at room temperature.

14. Following another wash step, as in Step 12, slides are incubated with the DAB substrate for 10 min (*see* **Note 7**).

15. If a second antibody is to be used, then Steps 11–14 are repeated, taking care to use a different enzyme conjugate and substrate.

16. Slides are rinsed in running water for 5 min.

17. To counterstain, slides are placed, sequentially, in hematoxylin for 1 min, running water for 1 min, sodium borate solution for 1 min, and running water for 1 min.

18. To dehydrate the slides and prepare for the xylene-based coverslipping solution, slides pass through solutions with increasing ethanol concentrations; that is 1 min each in 70, 80, 90 and 100% ethanol. Slides then return to xylene for two 5 min incubations before finally being coverslipped with Permount.

19. This immunocytochemical protocol has been optimized for small samples of cell and tissue pellets as are produced from this tissue culture platform. Representative images are shown in **Fig. 23.2** and Color Plate 24.

3.3. Real-Time PCR

1. To remove samples from the collagen gel, type XI collagenase is added to the culture medium to a final concentration of 250 μg/mL. Cultures are returned to the incubator and allowed to digest for 45 min at 37°C. The contents of the dish are collected into a 50 mL tube and washed twice in 1× PBS (*see* **Note 8**).

2. Pellets are transferred to 1.5 mL Eppendorf tubes, pelleted, resuspended in 700 μL buffer RLT+βME, and vortexed. At this point, RNA isolation can proceed immediately or samples can be stored at −80°C for later use.

3. An equal volume (700 μL) 70% ethanol is added and the sample is mixed by pipetting. Half of the sample is applied to an RNeasy column fitted with a 2 mL collection tube and centrifuged for 15 sec at 10,000 rpm, with the flow-through discarded. The second half of the sample is applied to the column and centrifuged again for 15 s at 10,000 rpm, with the flow-through again being discarded.

4. 350 μL buffer RW1 is applied to the column, centrifuged for 15 s at 10,000 rpm and the flow-through discarded.

5. 80 μL buffer RDD+DNase is applied directly to the column membrane. The column is incubated at room temperature for 15 min, then 350 μL buffer RW1 is applied to the column, centrifuged for 15 sec at 10,000 rpm and the 2 mL collection tube containing the flow-through is discarded.

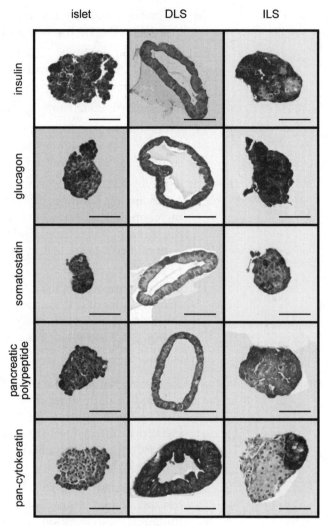

Fig. 23.2. Immunocytochemistry. Analysis of islets, DLS, and ILS demonstrates that while islets and ILS express endocrine hormones (insulin, glucagon, somatostatin, and pancreatic polypeptide) in approximately the same proportions, these hormone$^+$ cells are absent in DLS. Conversely, staining for a ductal cell marker (pan-cytokeratin) is observed primarily in DLS. Furthermore, while few or no ductal cells are observed in islets, ILS can be observed to be "budding" from ductal structures (bar = 100 μm). (*See* Color Plate 24)

6. The column is fitted with a fresh 2 mL collection tube. 500 μL buffer RPE+EtOH is applied to the column, centrifuged for 15 sec at 10,000 rpm and the flow-through discarded.

7. Another 500 μL buffer RPE+EtOH is applied to the column, this time centrifuging for 2 min at 10,000 rpm, with the second 2 mL collection tube containing the flow-through discarded.

8. The column is fitted with a 1.5 mL collection tube. 30 µL RNase-free water is applied to the column and centrifuged for 1 min at 10,000 rpm.

9. The flow-through, containing the purified mRNA, is stored in the 1.5 mL collection tube. RNA quantification and reverse transcription can proceed immediately or the sample can be stored at −80°C for later use.

10. To assess the purity and quantity of the RNA sample, absorbance at 260 and 280 nm is measured; 2 µL RNA sample is diluted to 100 µL in RNase-free water and read in a spectrophotometer at 260 and 280 nm, using 100 µL RNase-free water as a blank. To estimate purity, the A_{260}/A_{280} ratio is calculated, and should be close to 1.8. The RNA concentration is estimated as follows: [RNA] (ng/µL) = A_{260} × 44 ng/µL × dilution factor. In this case, the dilution factor is 50 (100 µL/2 µL = 50), while it has been established that $1A_{260}$ = 44 ng/µL RNA.

11. RNA samples must be reverse transcribed into cDNA before real-time PCR can be undertaken. 2.0 µL RT buffer + 2.0 µL dNTP mix + 0.5 µL oligo-DT + 1.0 µL RNase inhibitor + 1.0 µL reverse transcriptase per reaction are combined.

12. A volume of sample containing 1 µg RNA is diluted to 13.5 µL and added to the above reaction mix, giving a total reaction volume of 20 µL.

13. The reverse transcription reaction is allowed to proceed as follows: 1 h at 37°C, 5 min at 90°C and hold at 4°C. Again, real-time PCR analysis can proceed immediately or samples can be stored at −80°C for later use.

14. Samples for real-time PCR are prepared as follows: 10 µL PCR master mix + 5 µL primers + 4 µL RNase/DNase-free water per reaction.

15. Reactions are prepared in triplicate, with 1 µL cDNA sample being added to each reaction mixture. In addition to the genes of interest, a housekeeping gene and control reactions with sample omitted should also be included.

16. The real-time PCR is allowed to proceed as follows: 15 min at 95°C, then 40 cycles of 30 sec at 94°C, 30 sec at 58°C, 30 sec at 72°C, and a plate reading (*see* **Note 9**).

17. Relative transcript enrichment is calculated using the $2^{-\Delta\Delta}$ method *(29)*. Briefly, a fluorescence threshold in the linear portion of the reaction curve is selected. Cut-off values (C_t) are determined for all samples, and these values are used to calculate the average C_t for the housekeeping gene and genes of interest for all samples. The difference between housekeeping gene C_t and the gene of interest C_t represents the

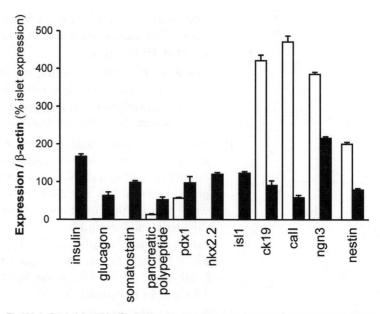

Fig. 23.3. Real-time PCR. Confirming the previous immunocytochemical results, endocrine hormone (insulin, glucagon, somatostatin, and pancreatic polypeptide) expression levels in ILS virtually match those in islets, while said hormones are almost undetectable in DLS. Ductal cell markers (cytokeratin-19 [ck19] and carbonic anhydrase-II [caII]) and putative progenitor markers (neurogenin-3 [ngn3] and nestin) are overexpressed in DLS. While most endocrine transcription factors assessed (NK homeobox gene-2.2 [nkx2.2] and islet-1 [isl1]) are found primarily in islets and ILS, the expression of one such factor that has also been proposed as a progenitor cell marker (pancreatoduodenal homeobox gene-1 [pdx1]), was somewhat maintained in DLS (□ = DLS expression level relative to islets; ■ = ILS expression level relative to islets).

first Δ. Using one sample as a reference, in this case a sample of islets taken at the time of embedding, the difference in Δ values between samples and the reference sample represents the second Δ. To calculate actual changes in expression levels, the $2^{-\Delta\Delta}$ formula is applied. Examples of real-time PCR data obtained from this tissue culture platform are displayed in **Fig. 23.3**.

4. Notes

1. Neutralization solution need not be added dropwise initially, as the buffering capacity of the neutralization solution is focused near the desired neutralization point. Alternatively, the volume of solution required to neutralize a specific volume of collagen can be pre-determined and this specific volume added to avoid problems with the titration step.

2. It is imperative to establish a timeline from neutralization to plating and shaking that is fast enough that the collagen does not polymerize in the tube or the pipette, but slow enough that islets do not sediment and reach the surface of the dish prior to complete collagen polymerization (islets that attach to the plastic surface of the dish will not form DLS). Trial and error using small numbers of islets is strongly suggested.

3. Care should be taken to ensure that no residual islets remain in the DLS cultures, as these islets may be mistaken for regenerated islet-like structures (ILS) at a later point. If significant islet contamination exists, DLS can be removed from culture by collagenase digestion as described, handpicked, and re-embedded to establish a pure population.

4. Samples containing less than 5,000 IE or equivalent numbers of DLS or ILS should be avoided as the pellet formed during the agarose embedding step will be small and hard to find during subsequent slide cutting and staining.

5. After processing but before paraffin embedding, sample+agarose plugs must be removed from the lens paper. It is suggested that the plug be oriented in the paraffin wax to allow a maximal cross-sectional area during slide cutting and staining.

6. Boiling in citrate buffer can be accomplished by heating in a microwave on medium-low power (setting 3) for 10 min.

7. Care should be taken not to touch the DAB substrate and to dispose properly with all DAB waste as this chemical is a potent toxin.

8. Samples containing less than 2,000 IE or equivalent numbers of DLS or ILS should be avoided as RNA recovery and sample concentration will be adversely affected.

9. Annealing temperature is specific to each primer pair and should optimized prior to experimental use. Values in **Table 23.1** are suggestions for the primer sets listed.

Acknowledgements

We thank Emily Austin, Mauro Castellarin, Deborah Driver, Al-Maleek Jamal, Xinfang Li, Mark Lipsett, Julia Makhlin, and Ryan Scott for technical assistance, as well as Québec-Transplant for coordination of organ availability. This work was supported in part by the Canadian Institutes of Health Research (CIHR) and the Stem Cell Network of Canada. S. Hanley is supported by fellowships from the Canadian Diabetes Association (CDA)/CIHR and Fonds de la Recherche en Santé Québec (FRSQ). L. Rosenberg is a chercheur national (national scientist) of the FRSQ.

References

1. Atkinson, M. A. and Eisenbarth, G. S. (2001) Type 1 diabetes: new perspectives on disease pathogenesis and treatment. *Lancet* **358**, 221–229.
2. Butler, A. E., Janson, J., Bonner-Weir, S., Ritzel, R., Rizza, R. A., and Butler, P. C. (2003) Beta-cell deficit and increased beta-cell apoptosis in humans with type 2 diabetes. *Diabetes* **52**, 102–110.
3. Shapiro, A. M., Lakey, J. R., Ryan, E. A., Korbutt, G. S., Toth, E., Warnock, G. L., Kneteman, N. M., and Rajotte, R. V. (2000) Islet transplantation in seven patients with type 1 diabetes mellitus using a glucocorticoid-free immunosuppressive regimen. *N Eng J Med* **343**, 230–238.
4. Dor, Y., Brown, J., Martinez, O. I., and Melton, D. A. (2004) Adult pancreatic beta-cells are formed by self-duplication rather than stem-cell differentiation. *Nature* **429**, 41–46.
5. Bonner-Weir, S., Baxter, L. A., Schuppin, G. T., and Smith, F. E. (1993) A second pathway for regeneration of adult exocrine and endocrine pancreas. A possible recapitulation of embryonic development. *Diabetes* **42**, 1715–1720.
6. Rosenberg, L., Brown, R. A., and Duguid, W. P. (1983) A new approach to the induction of duct epithelial hyperplasia and nesidioblastosis by cellophane wrapping of the hamster pancreas. *J Surg Res* **35**, 63–72.
7. Gu, G., Dubauskaite, J., and Melton, D. A. (2002) Direct evidence for the pancreatic lineage: NGN3+ cells are islet progenitors and are distinct from duct progenitors. *Development* **129**, 2447–2457.
8. Heremans, Y., Van De Casteele, M., in't Veld, P., Gradwohl, G., Serup, P., Madsen, O., Pipeleers, D., and Heimberg, H. (2002) Recapitulation of embryonic neuroendocrine differentiation in adult human pancreatic duct cells expressing neurogenin 3. *J Cell Biol* **159**, 303–312.
9. Wang, R. N., Kloppel, G., and Bouwens, L. (1995) Duct- to islet-cell differentiation and islet growth in the pancreas of duct-ligated adult rats. *Diabetologia* **38**, 1405–1411.
10. Rooman, I., Lardon, J., and Bouwens, L. (2002) Gastrin stimulates beta-cell neogenesis and increases islet mass from transdifferentiated but not from normal exocrine pancreas tissue. *Diabetes* **51**, 686–690.
11. Lipsett, M. and Finegood, D. T. (2002) Beta-cell neogenesis during prolonged hyperglycemia in rats. *Diabetes* **51**, 1834–1841.
12. Fernandes, A., King, L. C., Guz, Y., Stein, R., Wright, C. V., and Teitelman, G. (1997) Differentiation of new insulin-producing cells is induced by injury in adult pancreatic islets. *Endocrinology* **138**, 1750–1762.
13. Guz, Y., Nasir, I., and Teitelman, G. (2001) Regeneration of pancreatic beta cells from intra-islet precursor cells in an experimental model of diabetes. *Endocrinology* **142**, 4956–4968.
14. Zulewski, H., Abraham, E. J., Gerlach, M. J., Daniel, P. B., Moritz, W., Muller, B., Vallejo, M., Thomas, M. K., and Habener, J. F. (2001) Multipotential nestin-positive stem cells isolated from adult pancreatic islets differentiate ex vivo into pancreatic endocrine, exocrine, and hepatic phenotypes. *Diabetes* **50**, 521–533.
15. Dor, Y. (2006) Beta-cell proliferation is the major source of new pancreatic beta cells. *Nat Clin Pract Endocrinol Metab* **2**, 242–243.
16. Jamal, A. M., Lipsett, M., Sladek, R., Laganiere, S., Hanley, S., and Rosenberg, L. (2005) Morphogenetic plasticity of adult human pancreatic islets of Langerhans. *Cell Death Differ* **12**, 702–712.
17. Gao, R., Ustinov, J., Korsgren, O., and Otonkoski, T. (2005) In vitro neogenesis of human islets reflects the plasticity of differentiated human pancreatic cells. *Diabetologia* **48**, 2296–2304.
18. Xu, G., Stoffers, D. A., Habener, J. F., and Bonner-Weir, S. (1999) Exendin-4 stimulates both beta-cell replication and neogenesis, resulting in increased beta-cell mass and improved glucose tolerance in diabetic rats. *Diabetes* **48**, 2270–2276.
19. Tourrel, C., Bailbe, D., Meile, M. J., Kergoat, M., and Portha, B. (2001) Glucagon-like peptide-1 and exendin-4 stimulate beta-cell neogenesis in streptozotocin-treated newborn rats resulting in persistently improved glucose homeostasis at adult age. *Diabetes* **50**, 1562–1570.
20. Xu, G., Kaneto, H., Lopez-Avalos, M. D., Weir, G. C., and Bonner-Weir, S. (2006) GLP-1/exendin-4 facilitates beta-cell

neogenesis in rat and human pancreatic ducts. *Diabetes Res Clin Pract* in press.
21. Brand, S. J., Tagerud, S., Lambert, P., Magil, S. G., Tatarkiewicz, K., Doiron, K., and Yan, Y. (2002) Pharmacological treatment of chronic diabetes by stimulating pancreatic beta-cell regeneration with systemic co-administration of EGF and gastrin. *Pharmacol Toxicol* **91**, 414–420.
22. Suarez-Pinzon, W. L., Yan, Y., Power, R., Brand, S. J., and Rabinovitch, A. (2005) Combination therapy with epidermal growth factor and gastrin increases beta-cell mass and reverses hyperglycemia in diabetic NOD mice. *Diabetes* **54**, 2596–2601.
23. Rosenberg, L., Lipsett, M., Yoon, J. W., Prentki, M., Wang, R., Jun, H. S., Pittenger, G. L., Taylor-Fishwick, D., and Vinik, A. I. (2004) A pentadecapeptide fragment of islet neogenesis-associated protein increases beta-cell mass and reverses diabetes in C57BL/6 J mice. *Ann Surg* **240**, 875–884.
24. Lipsett, M., Hanley, S., Castellarin, M., Austin, E., Suarez-Pinzon, W. L., Rabinovitch, A., and Rosenberg, L. (submitted) The role of islet neogenesis-associated protein (INGAP) in islet neogenesis. *Cell Biochem Biophys*.
25. Jamal, A. M., Lipsett, M., Hazrati, A., Paraskevas, S., Agapitos, D., Maysinger, D., and Rosenberg, L. (2003) Signals for death and differentiation: a two-step mechanism for in vitro transformation of adult islets of Langerhans to duct epithelial structures. *Cell Death Differ* **10**, 987–996.
26. Richards, J., Larson, L., Yang, J., Guzman, R., Tomooka, Y., Osborn, R., Imagawa, W., and Nandi, S. (1983) Method for culturing mammary epithelial cells in a rat tail collagen gel matrix. *Journal of Tissue Culture Methods* **8**, 31–36.
27. Gershengorn, M. C., Hardikar, A. A., Wei, C., Geras-Raaka, E., Marcus-Samuels, B., and Raaka, B. M. (2004) Epithelial-to-mesenchymal transition generates proliferative human islet precursor cells. *Science* **306**, 2261–2264.
28. Bonner-Weir, S., Taneja, M., Weir, G. C., Tatarkiewicz, K., Song, K. H., Sharma, A., and O'Neil, J. J. (2000) In vitro cultivation of human islets from expanded ductal tissue. *Proc Natl Acad Sci* **97**, 7999–8004.
29. Livak, K. J. and Schmittgen, T. D. (2001) Analysis of relative gene expression data using real-time quantitative PCR and the 2(-Delta Delta C(T)) Method. *Methods* **25**, 402–408.

Chapter 24

Isolation and Characterization of Hepatic Stem Cells, or "Oval Cells," from Rat Livers

Thomas D. Shupe, Anna C. Piscaglia, Seh-Hoon Oh, Antonio Gasbarrini, and Bryon E. Petersen

Abstract

The pace of research on the potential therapeutic uses of liver stem cells or "oval cells" has accelerated significantly in recent years. Concurrent advancements in techniques for the isolation and characterization of these cells have helped fuel this research. Several models now exist for the induction of oval cell proliferation in rodents. Protocols for the isolation and culture of these cells have evolved to the point that they may be set up in any laboratory equipped for cell culture. The advent of magnetic cell sorting has eliminated reliance on expensive flow cytometric sorting equipment to generate highly enriched populations of oval cells. Our laboratory has had much success in using the oval cell surface marker Thy-1 in combination with magnetic sorting to produce material suitable for testing the influence of a myriad of chemical signaling molecules on the oval cell phenotype. This chapter will describe our basic strategy for oval cell induction and isolation. Additionally, two in vitro procedures are described which the reader may find useful in the early stages of developing an oval cell research project.

Key words: Liver stem cells, oval cells, 2-acetylaminofluorene (2AAF), partial hepatectomy (PH), collagenase perfusion, simple gravity enrichment, immunomagnetic sorting (IMS), flow cytometric sorting (FACS), proliferation assay, migration assay.

1. Introduction

The past decade has shown great strides in our understanding of the potential therapeutic use of adult, tissue-committed stem cells. The extensively characterized rat hepatic stem/progenitor cell, or oval cell, has been demonstrated by our laboratory, as well as several others, to show particular promise as a tool for the

---

T.D. Shupe and A.C. Piscaglia contributed equally

treatment of genetic liver disorders and as a bridging therapy or alternative to liver transplant *(1–4)*.

The term "oval cells" (OC) is currently used to define the population of small proliferating cells with oval-shaped nuclei which arise within the liver following certain types of liver injury *(1–7)*. OC proliferation is activated when the proliferative capacity of hepatocytes is impaired, as well as in certain models of carcinogenesis. In OC-mediated liver regeneration, OC first appear in the portal tract periphery, and eventually migrate deep into the lobular parenchyma. Due to their bipotency and high clonogenicity, OC have been considered putative liver stem cells. Ongoing research is beginning to elucidate the signaling molecules involved in controlling the activation, proliferation. and differentiation of this unique cell population *(5–7)*.

2. Materials

1. 2AAF (2-acetylaminofluorene). 70 mg 2AAF pellets, 28 day time release (Innovative Research of America; Sarasota, FL).
2. Isoflurane (Webster Veterinary; Sterling, MA).
3. Iodine solution.
4. 70% Ethanol (EtOH).
5. 2–0 silk suture (Ethicon Inc.; Somerville, NJ).
6. S&M solution (solution A): 500 mg KCl, 8.3 g NaCl, 2.4 g HEPES, 190 mg NaOH. Bring to 1 L with $_dH_2O$. Adjust pH to 7.4 (*see* **Note 6**).
7. $CaCl_2$ solution (solution B): 636 mg $CaCl_2 \cdot 2H_2O$. Bring to 1 L with S&M solution (*see* **Note 1**).
8. Collagenase H (Roche; Germany), at a working concentration of 320 μg/ml (*see* **Note 2**).
9. 20-gauge catheter, 1–1/4" (Medex; Carlsbad, CA).
10. Nylon mesh (100–150 μm pore diameter).
11. MACS® cell separator, columns, and anti-FITC microbeads (Miltenyi Biotec Inc., Auburn, CA).
12. FITC-conjugated Thy-1/mouse CD90.1 (Thy-1.1) monoclonal antibody. (BD Biosciences, Pharmingen; San Diego, CA).
13. FITC-mouse IgG_1 isotype control (Vector, Burlingame, CA).
14. PBS buffer (pH 7.4).
15. IMDM supplemented with 10% FBS, 1% insulin, and antibiotics (GIBCO; Grand Island, NY).
16. Bovine serum albumin, BSA, 0.5% (Roche; Germany).
17. Transwell culture dishes with 5 μm pore filters, 6.5 mm diameter, 24-well cell clusters (Coring Inc.; NY, USA).

3. Methods

3.1. Oval Cell Activation: 2AAF/PH Protocol

While several methods are available for the activation of oval cells, they all involve the same basic principal: inhibition of the proliferative potential of mature hepatocytes followed by liver injury *(1–7)*. Chemical inhibition of hepatocyte proliferation is made possible by the unique expression of mixed function oxidases (P450s) within the parenchymal cells of the liver. Popular OC induction models in the rat include choline-deficient diet followed by ethionine exposure, galactosamine, 2-acetylaminofluorene (2AAF)/CCl₄, 2AAF/partial hepatectomy (PH), and allyl alcohol. In our experience, the 2AAF/PH protocol has generated the most reproducible OC response in rats and has become the gold standard method within this field *(8, 9)*.

3.1.1. 2AAF Administration

The 2AAF/PH protocol is a well-established model of OC activation in rats *(8, 9)*. 2AAF is metabolized by hepatocytes to an N-hydroxyl derivative, which interferes with the cyclin D1 pathway. Biliary epithelial cells and oval cells lack the ability to convert the 2AAF to its toxic metabolite. Therefore, the administration of 2AAF prior to PH inhibits hepatocyte proliferation and force oval cell recruitment to mediate liver regeneration. This procedure results in a robust oval cell response following PH (peaking between day 9 and 11 post-surgery) and within 14 days, OC start to differentiate into hepatocytes *(10)*.

Rats must be continuously exposed to 2AAF for 7 days prior to liver injury. Two strategies for treatment with 2AAF have been successfully employed by our laboratory. The first is administration by gastric gavage. 2AAF is dissolved in corn oil and administered daily by gavage at a dosage of 10 mg/kg for 14 days, with PH being performed after 1 week *(11)* (*see* **Note 3**). Our laboratory currently commissions the production of time release 2AAF pellets (Innovative Research of America; Sarasota, FL). These pellets can be custom made for different release rates and durations. In our experience, 70 mg 2AAF, 28-day pellets result in almost complete inhibition of hepatocyte proliferation and little toxicity in F344 rats implanted subsequent to 10 weeks of age (*see* **Note 4**). We insert these pellets intraperitoneally through a small (∼5 mm) incision in the right lower quadrant of the abdomen.

3.1.2. Partial Hepatectomy

Following 7 days exposure to 2AAF, a partial hepatectomy is performed. The surgical technique we employ is a modification, optimized by Piscaglia, of the method previously described by Higgins and Anderson *(12)*.

The rat is maintained under anesthesia with isoflurane, and standard aseptic technique should be used. After shaving the abdomen, disinfect with iodine solution, followed by 70% EtOH. A 2.5–3.0 cm median-line longitudinal incision is made through the skin, above the xiphoid process of the sternum. The incision is continued through the abdominal muscles (<1 cm), just to the right side of the linea alba (*see* **Note 5**). The upper edge of the incision should approach the diaphragm and extreme care must be

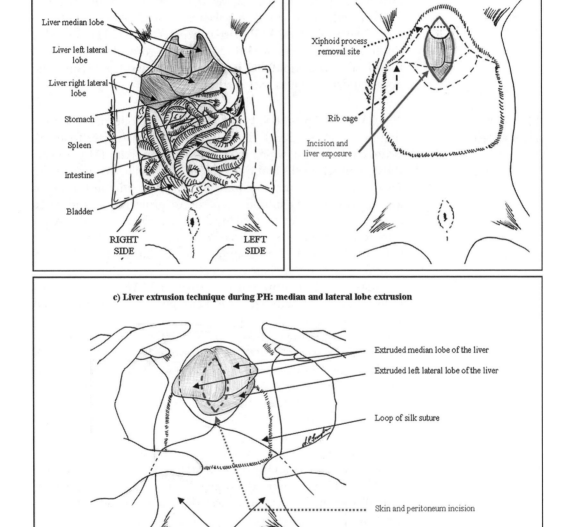

Fig. 24.1. Partial hepatectomy: (**a**) Normal anatomy of the rat abdominal viscera. (**b**) Following incision, three lobes should be clearly visible. (**c**) The median and left lateral lobes of the liver should be externalized, through the loop of silk suture, by pressure from the thumbs. The index fingers brace the lower portion of the rib cage.

used so as to not puncture the thoracic cavity, because a pneumothorax is rapidly fatal to the animal (*see* **Fig. 24.1a** for a diagram of the rat abdominal viscera). At this point the xiphoid process may be removed with dissecting scissors. The laparotomy may be enlarged (up to 1.5–2.0 cm), until the median lobe and the left lateral lobe are clearly visible (**Fig. 24.1b**) (*see* **Note 6**). A loop of 2–0 silk suture is placed loosely over the incision. The medial and left lobes of the liver are gently extruded through the incision, using the index fingers of both hands to brace the lower rib cage while using the thumbs to push in and up below the liver. The right central and left central portions of the medial lobe will come out first, followed by the left lateral lobe (**Fig. 24.1c**). This will result in 65–75% of the entire liver mass protruding through the suture loop. The suture is carefully constricted around the base of the exposed liver lobes and tightened. Care must be given not to capture skin or muscle while tightening the loop. The exposed lobes may now be cut off distal to the ligature. The remaining liver stump should be examined for active bleeding before suturing the muscle and skin. No special post-operative care is needed, complications are extremely rare, and post-operative mortality is uncommon (<3%).

3.2. Two-Step Collagenase Perfusion

The two-step collagenase perfusion is a common technique, employed to obtain viable cells from a solid organ. This procedure uses two liver perfusates. The first consists of an isotonic buffer to flush blood cells from the vasculature. The second perfusate contains collagenase which digests the extracellular matrix, allowing for the collection of a single cell suspension suitable for isolation of discrete cell populations *(13)*.

Regarding OC isolation from liver perfusion following 2AAF/PH in rats, the peak of the OC response will occur between days 9 and 11 following partial hepatectomy *(8–10)*. It is during this time frame that the following procedures will yield the greatest number of OC (*see* **Note 7**).

1. For each rat to be perfused, aliquot 250 ml of solutions A and B. Each of these should be pre-warmed to 40°C in a circulating water bath. Ideally, the solution should cool to 37°C as it is pumped from the water bath to the animal. The containers should be weighted, so they do not float and tip over during the perfusion. An amount of collagenase appropriate for 250 ml should be aliquoted and stored on ice.

2. Set up a peristaltic pump directly next to the water bath. A section of appropriate diameter tubing should be used to transfer solutions from the water bath, through the pump, to the animal. Use a sufficient length to allow free manipulation of the tubing during catheterization, but be aware that the solution should not cool below 37°C before it enters the animal.

3. Anesthetize the rat for a terminal procedure (*see* **Note 8**).

4. Secure the rat to a dissection board by taping or pinning the extremities. The abdomen may be shaved prior to disinfecting if sterility of the product is of concern. Rinse the abdomen and lower thorax with 70% EtOH. Cut a large mid-line incision through the skin and peritoneum. Make a subsequent cut perpendicular to the laparotomy incision on both sides below the subcostal margin (**Fig. 24.2a**). Retract the wound edges and expose the abdominal viscera (**Fig. 24.2b**).

5. Displace the intestines toward the lower left quadrant and expose the IVC. Use a gauze pad in each hand to tear the fascia covering the IVC. Identify the renal veins which branch from the IVC. Also, identify the portal vein which will be the large vessel entering the liver within the lesser omentum (**Fig. 24.2c**).

6. Run a 2-0 silk suture under the IVC immediately above to the renal vein branches. Tie but do not tighten the suture leaving a 1 cm diameter loop above the IVC. Run a second suture under the IVC distal to the renal branches. Do not tie this suture (**Fig. 24.2d**).

7. Turn on the pump and recirculate solution A through the tubing back into the flask containing the S&M solution. Gently tap the tubing to dislodge as many air bubbles as possible. Once the tubing is completely full of solution with no air bubbles, turn off the pump.

8. Cannulate the IVC using a 20-gauge catheter, between the two sutures that were put in place in step 6. Insert the catheter very carefully, keeping it just a few degrees from parallel to the IVC (*see* **Note 9**). Gently slide the catheter into the IVC until the end is past the most proximal loop of suture (**Fig. 2d**). Blood should begin to ooze back into the catheter. If this does not happen, use a small syringe to apply negative pressure to the catheter.

9. Without moving the catheter, carefully tighten the suture loop around the IVC and catheter. Tie a second knot on top of the first to secure. Bring the ends of the second suture over the top of the portion of catheter which is external to the IVC and secure with a double knot. At this point the catheter should be locked securely within the IVC with no blood flow through the IVC due to constriction of the vessel by the sutures. (It is helpful to have a second set of hands while performing this step.)

10. Attach the outflow end of the tubing to the catheter and turn on the pump with a flow rate of approximately 15 ml/min. Attach the tube to the catheter and secure the tube with tape so that the catheter is not pulled out of position. Blanching of the liver indicates successful perfusion. Immediately cut the

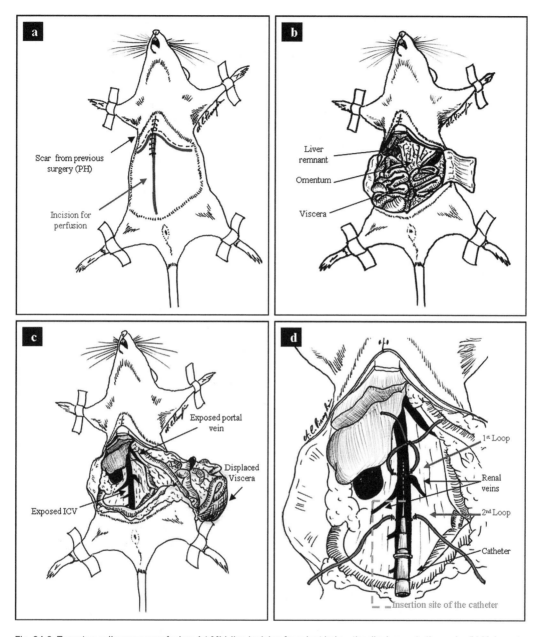

Fig. 24.2. Two step collagenase perfusion: (a) Mid-line incision from just below the diaphragm to the groin. (b) Make cuts perpendicular to the main incision to expose viscera. (c) Identify the IVC and portal vein. (d) Diagram showing the correct placement of the silk sutures and catheter.

portal vein between the liver and the stomach with dissecting scissors. A large volume of blood followed by solution A will flow from the severed vein.

11. Open the chest cavity, cutting through the diaphragm, above the liver. Insert a curved hemostat clamp through this hole and clamp off the superior vena cava.

12. The liver may have patches that do not perfuse well. Gently massage the liver lobes with a cotton swab, to assist in moving the blood from the sinusoids, until the effluent is clear. At this point, add the collagenase to solution B, mix well and keep the solution with collagenase at 40°C. Allow solution A to run until no blood is visible within the liver and the effluent is clear (up to 15 min, *see* **Note 10**).

13. Once the liver is completely blanched, turn off the pump and move the in-flow end of the tubing to the solution B with collagenase. Turn the pump on and confirm liver perfusion as described in step 12. Decrease the flow rate to about 12 ml/min and let it run for 10–15 min (the duration of this step is largely dependent on the activity of the collagenase used, *see* **Note 11**).

14. When adequate digestion of the liver has been achieved, shut off the pump and disconnect the tubing from the catheter. The liver must now be carefully dissected away from tissues which connect to it. Be aware that the liver is extremely fragile following digestion.

15. Transfer the liver into a beaker containing about 50 ml of ice-cold solution A. Keep the top of the beaker covered at all times to limit contamination.

16. From this point onwards, every step must be performed under a laminar flow hood if cells are to be cultured. Grasp the tissue with sterile pliers and shake/agitate so that it breaks apart, dispersing cells in the buffer, which will turn brown (the darker the color, the better the perfusion).

17. Carefully pass the cell suspension through a nylon mesh (100–150 μm pore diameter) placed over the top of a 250 ml beaker. Wash the mesh with an adequate volume of S&M buffer to collect all of the cells. Aliquot the supernatant into 50 ml conical tubes.

18. Proceed to simple gravity enrichment, to separate hepatocytes from non-parenchymal cells.

3.3. Simple Gravity Enrichment

A crude separation of hepatocytes and non-parenchymal cells (NPC) may be accomplished by centrifugation alone. This procedure will result in a very pure NPC population. Analysis of this population by immunostaining for P450 (CYP3A2) demonstrates less than 2% of the fractionated NPC express this hepatocyte marker. Oval cells will partition to the NPC fraction. Purity of the parenchymal cell fraction will vary depending on the number of washing steps, though it is difficult to achieve greater than 75% enrichment by this method.

1. Centrifuge the cell suspension at 55 g for 5 min at 4°C. Transfer the supernatant (enriched for NPC) to another

container and save it on ice. Resuspend the pellet (enriched for parenchymal cells) in an additional 50 ml ice-cold S&M buffer and repeat centrifugation (*see* **Note 12**). Discard the final pellet, unless you need to collect the hepatocyte fraction.

2. Centrifuge the supernatants at 220 g for 5 min at 4°C. Collect the final pellets, which represent the enriched non-parenchymal cell fraction. Resuspend and pool the NPC, using an adequate volume of filtered phosphate-buffered saline (PBS; pH 7.4).

3. Wash cells twice in PBS (220 g for 5 min at 4°C).

4. Check cell viability by trypan blue exclusion and proceed to OC isolation.

3.4. Oval Cell Isolation: Immunomagnetic and Flow Cytometric Cell Sorting

Several methods have been employed for the isolation of OC. These techniques range in complexity and include Nicodenze gradient separation, and sorting by flow cytometric (FACS) or immunomagnetic sorting (IMS). FACS had been the gold standard for isolating OC prior to the introduction of IMS (13). IMS has become a widely used method for separating different cell types. It has numerous advantages compared to other methods. It is a convenient, easy to use system which can yield enrichment of up to 99% depending on the specificity of the cell surface marker used. Stem cell biology has benefited significantly from the introduction of magnetic cell sorting (14, 15).

3.4.1. Thy-1+ Cell Sorting

Thy-1 (CD90) is a GPI-anchored membrane glycoprotein of the Ig superfamily which is involved in signal transduction (16–18), and it is expressed by a wide spectrum of hematopoietic stem and progenitor cells (16–19), as well as non-hematopoietic cells, including neurons (16), edothelium at inflammatory sites (20), and hepatic OC (21, 22). High purity OC enrichment is achieved through IMS using the OC surface marker Thy-1. We demonstrated that, using Thy-1 in conjunction with FACS sorting, a 95–97% enriched population of OC may be obtained (21).

Prior to sorting, the NPC population must be isolated by the simple gravity enrichment method described above. Because the collagenase perfusion protocol eliminates virtually all blood cells from the liver, the Thy 1+ cell population is comprised almost entirely of oval cells. Thy-1+ cells from rat liver can be isolated using indirect labeling strategies. We currently employ fluorescein isothiocyanate (FITC)-conjugated mouse anti-rat CD90 (Thy-1.1) to positively select OC.

3.4.2. Immunomagnetic Sorting

IMS technology is based on the use of microbeads, columns, and a magnetic field. The microbeads are super-paramagnetic particles (approximately 50 nm in size, biodegradable, and non-toxic to cells), which are coupled, directly or indirectly, to specific monoclonal antibodies. These beads magnetically label the target cell population. The patented column technology (MACS®, Miltenyi Biotec Inc., Auburn, CA) is specifically designed to generate the high-strength magnetic field that is required to retain the labeled cells, while maintaining optimal cell viability and function. By placing the column in a permanent magnet (called the "separator"), the magnetic force is sufficient to retain the target cells. By simply rinsing the column with buffer, all the unlabeled cells are washed away. Once the column is detached from the magnet, the labeled fraction can be eluted (**Fig. 24.3**). *See* **Note 13**.

1. Separate the NPC fraction through simple gravity enrichment and wash it three times, as described above.
2. Resuspend the pellet in 10 ml of ice-cold filtered PBS and transfer the cell suspension to a 15 ml tube.
3. Add 100 µl FITC-conjugated mouse anti-rat CD90 and incubate for 1 h at 4°C. Cover the tube with sterile aluminum foil, and leave on ice in a shaker (gentle agitation, ~1 oscillation/sec).
4. Centrifuge the cell suspension at 220 g for 5 min at 4°C, discard the supernatant and resuspend the pellet in PBS. Repeat the washing step.
5. Resuspend the pellet in 900 µl of PBS and add anti-rat FITC microbeads (100 µl), mix well and incubate for 30 min at 4°C. Cover the tube with sterile aluminum foil, and leave on ice in a shaker as described above.

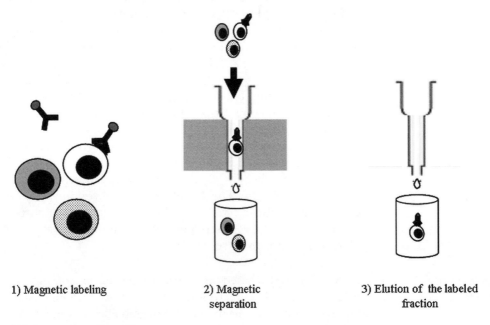

Fig. 24.3. Immunomagnetic sorting.

6. Add PBS (up to 15 ml) and centrifuge at 220 g for 5 min at 4°C. Remove the supernatant, and resuspend the pellet in 3 ml of PBS. At this point, a small sample may be removed for flow cytometric analysis.

7. Place the column (combined with the appropriate column adapter) in the magnetic field of the MACS Separator. Place an empty tube below the column. Equilibrate the column by washing with 1 ml of ice-cold PBS.

8. When the buffer has passed through, apply the cell suspension to the top of the column. Let the unlabeled cells pass through. Rinse twice with 3 ml of PBS.

9. Remove the column from the separator, and place it over a suitable collection tube. Pipette 1 ml of PBS onto the column and flush out magnetically labeled fraction using the plunger supplied with the column. Repeat two more times, using 1 ml PBS each time (final volume 3 ml). Push hard and fast, in order to get complete cell elution.

10. To achieve higher purity, apply the magnetically labeled fraction onto a new, freshly prepared column. Let the unlabeled cells pass through. Rinse with 2×3 ml of PBS. Elute the magnetically labeled fraction as described above.

11. Check cell viability by trypan blue exclusion, and count the cells on a hemocytometer. From one rat, 5×10^7 Thy-1$^+$ cells with 90–95% viability can be achieved on a regular basis.

3.4.3. Flow Cytometric Sorting

1. Separate the NPC fraction through simple gravity enrichment and wash it three times, as described above.

2. Resuspend the pellet and incubate cell suspension with FITC-conjugated mouse anti-rat CD90 (1 μl/10^6 cells), for 20 min at 4°C. Cover the tube with sterile aluminum foil, and leave on ice in a shaker (gentle agitation, ∼1 oscillation/sec).

3. Wash the cells twice in PBS supplemented with 1% fetal bovine serum. Store at 4°C in the dark until sorting.

4. Sort cells into Thy-1.1$^+$ and Thy-1.1$^-$ fractions, using the FACS flow cytometer.

5. Check cell viability by trypan blue exclusion, and count the cells on a hemocytometer. From one rat, 5×10^7 Thy-1$^+$ cells with 90–95% viability can be achieved on a regular basis.

3.5. Oval Cell Characterization: In Vitro Assays

Several laboratory techniques may be used to characterize OC. Cytospins from OC suspensions can be employed for immunohistochemistry or immunofluorescence (**Fig. 24.4**). RNA, DNA, or protein can be extracted and processed for molecular biology. Additionally, OC may be cultured for in

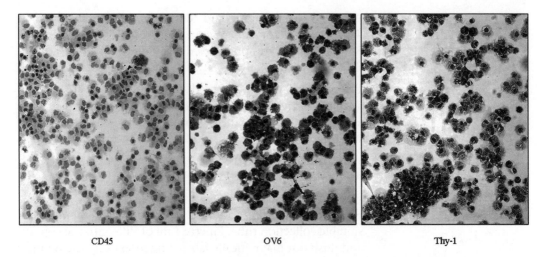

Fig. 24.4. Oval cell cytospin: immunomagnetic sorting of Thy-1+ oval cells yields a cell population, which is almost entirely Thy-1+/OV-6+ (both are oval cell markers) while almost completely devoid of CD45+ leucocytes.

vitro assays. Several media have been proposed to culture OC, with varying degrees of success. It is very difficult to maintain and expand OC in culture without spontaneous differentiation. We have been able to culture OC for up to 5 months without signs of senescence or spontaneous differentiation (Piscaglia et al., unpublished data) using a relatively minimal medium (see **Note 14**).

In vitro assays are very useful in analyzing the properties of OC and their response to exogenous factors. In particular, two properties are critical to the OC phenotype: proliferation following certain types of liver injury, and migration from the periportal space to the liver parenchyma *(1–7)*. Trafficking, mobilization, and homing of OC are multifactorial processes that are regulated by several factors, including adhesion molecules, cytokines, and chemotactic molecules *(6)*. Therefore, proliferation and migration assays may be useful in clarifying the molecular mechanisms underlying OC activation. In our hands, in vitro assays on OC isolated through the previously described techniques have contributed to the identification of several factors, such as SDF-1, SST, and G-CSF, which are involved in OC activation *(23–25)*.

3.5.1. Proliferation Assay

Proliferation assays may be used to assess the effects of a specific factor on oval cell proliferation. Briefly, the factor is added at different dosages to the basic OC medium (IMDM). Growth kinetics of the OC are evaluated at different time points (usually at 1, 3, 5, and 7 days). A positive control (FBS 10%) and a negative control (bovine serum albumin,

BSA, 0.5%) are required. Every determination should be performed at least in triplicate.

1. Seed cells in six-well plates (10^5 cells/well) in IMDM supplemented with 10% FBS, 1% insulin, and antibiotics, and incubate at 37°C, 5% CO_2 overnight.
2. The following day, prepare proliferation buffers for each experimental group (*see* **Note 15**). Count the attached cells in three wells to establish initial cell number.
3. Begin the proliferation assay by replacing medium with the various experimental buffers.
4. Count cells at the pre-determined time points (*see* **Note 16**).

3.5.2. Migration Assay

Migration assays may be used to assess the effects of a specific factor on oval cell motility. Briefly, OC are seeded on the microporous membrane of a transwell (**Fig. 24.5**). The factor is added at different dosages to the basic OC medium (*see* **Note 17**) in the lower chamber. This will result in the migration of OC through the membrane, following the chemoattractive factor gradient. OC motility can be assessed at different time points (*see* **Note 18**). A negative control and a chemokinetic control are required (*see* **Note 19**). **Fig. 24.6** shows an example of migration assay.

1. Prepare a 24-well plate by pre-coating wells with 0.006% RTC (rat tail collagen in 0.1% acetic acid) for 3 h at 37°C, 5% CO_2 (store the plates in a clean bag at RT for up to several hours until used). Add 0.6 ml of migration buffer (IMDM + insulin 1% + FBS 10% + antibiotics)/well. Insert a transwell into each well. Incubate for a few minutes at 37°C, 5% CO_2.
2. Seed cell suspension on the top of each transwell (10^5 cells in 100 µl of migration buffer).
3. Allow the cells to attach overnight at 37°C, 5% CO_2.

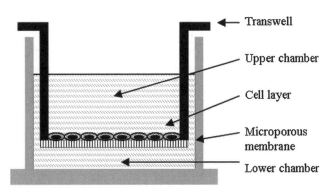

Fig. 24.5. Schematic representation of transwell in chamber for migration assay.

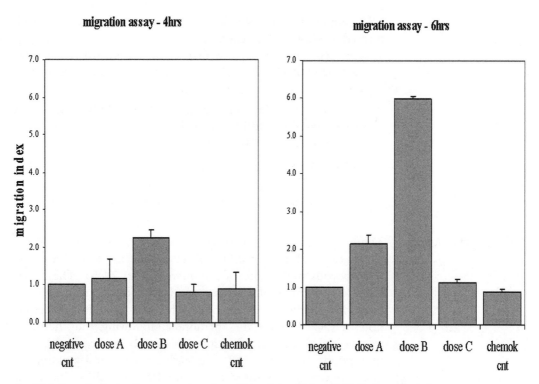

Fig. 24.6. Example of migration assay, where three different doses of a specific factor are tested.

4. Before beginning the test, prepare the various migration buffers by adding the proper dose of the experimental factor. Add each migration buffer to a new cluster plate.

5. Carefully remove non-adherent cells prior to the assay with a 100 μl pipette. Begin the migration assay by transferring the transwell chambers to the cluster plate containing the various buffers.

6. Incubate at 37°C, 5% CO_2 for either 4 or 6 h (time 1 and 2, respectively).

7. At the end of the test, fix and stain the cells as described by Stolz et al. *(26)* (*see* **Note 20**). Remove the collagen and stationary cells from the top of the transwell filter by rubbing with a cotton-tipped applicator. Transfer the transswells to a new, clean 24-well plate, and dry overnight.

8. Enumerate the cells that have migrated to the lower chamber, by counting each transwell chamber at ×10 magnification.

3.6. Applications of Isolated Liver Stem Cells

During the last 20 years, the development of new biological technologies has set the stage for a plethora of novel therapies. In particular, the field of stem cell biology has demonstrated enormous potential in the prevention and treatment of several human diseases *(27, 28)*. Stem cells play a fundamental role in the maintenance of tissue homeostasis. Improving our knowledge of stem cell biology is the *conditio sine qua non* to regenerative medicine. Advantages of the use of adult stem cells are many, and include the availability of cells and the lack of any ethical concerns *(29)*.

Acute or chronic liver diseases may benefit greatly from stem cell therapies *(30)*. Inborn errors of metabolism, such as Crigler-Najjar syndrome, LDL-receptor defect, hereditary hemochromatosis, Wilson's disease, etc., may, someday, be treated with genetically modified, Thy-1-sorted cells *(31)*. Finally, identifying the complex mechanisms involved in maintenance of the hepatic stem cell phenotype may lead to the development of novel therapies for the treatment of liver cancers *(32, 33)*.

Continued improvement of techniques for the identification, isolation, and in vitro expansion of the hepatic oval cells will contribute to the development of stem cell-based approaches for treating genetic liver disorders, preventing hepatic cancers, and providing bridging therapies or alternatives to liver transplant *(1–4)*.

4. Notes

1. If cells are to be used for culture or transplantation, the solutions S&M (solution A) and CaCl$_2$ (solution B) should be filtered through a 0.2 µm filter, under a laminar flow hood. Antibiotics must be added to each buffer.

2. Many collagenase preparations are commercially available. The activity of collagenase varies among different preparations, and it is important to determine an appropriate concentration for the particular collagenase which will be used. Too little activity will result in incomplete disassociation of cells within the liver, whereas too much activity will quickly compromise the integrity of the vasculature resulting in non-uniform perfusion. Also, high collagenase activity can be toxic, decreasing the viability of harvested cells. We have had good success with collagenase H from Roche, Germany, at a working concentration of about 320 µg/ml. The following protocol has been optimized for this particular collagenase preparation.

3. Gastric gavage can be fraught with complications and is time consuming. Most often, a person with limited experience in gavage will occasionally insert the gavage needle into the trachea, resulting in almost certain death of the animal. Time released pellets are expensive, but well worth the price.

4. Younger animals seem to be more susceptible to the toxicity of 2AAF (most notably necrosis within the liver and intestines). A lower dosage may be required for younger animals.

5. Compared to the classical mid-line incision *(12)*, our technique is more effective in exposing the liver, but it also increases the risk of cutting the right epigastric vessels. Therefore, attention must be paid in identifying and sparing these vascular structures.

6. Avoid making the incision longer than necessary, as this can make it difficult to extrude the liver without also externalizing portions of intestine.

7. With slight modification, this protocol may be adapted to mice. A smaller catheter must be used for cannulation. Also, we suggest using collagenase A (Boehringer Mannheim) at a working concentration of 1 mg/ml.

8. It is critical that the heart does not stop beating before the inferior vena cava (IVC) is cannulated. The resulting drop in blood pressure will make insertion of the catheter very difficult.

9. The best way to insert the catheter is to carefully "catch" the surface of the vessel on the bevel of the catheter needle, and lift it up slightly (less than 1 mm) before pushing it in all the way. This will eliminate the possibility of puncturing the back wall of the vessel.

10. Closely monitor the surface of the liver during perfusion. A moist shiny appearance indicates continued perfusion of the organ. If the surface of the liver begins to dry, this may indicate that the perfusion has been interrupted. Also, a gentle heaving of the liver in synchrony with the rollers of the peristaltic pump should be apparent. It is often difficult to determine whether or not the clear perfusate is flowing from the severed portal vein. Use a cotton swab to soak up some blood and squeeze it out above the portal vein. The blood should allow you to visualize flow from the severed vein. If perfusion is interrupted, gently reposition the catheter until flow from the portal vein is re-established.

11. Pay attention to the texture of the liver surface, as this will be the main indicator of the degree of collagenase digestion. Towards the end of the perfusion, channels will begin to form under the hepatic capsule. Trabeculation and fragmentation indicate a breakdown of the extracellular matrix of the liver. These channels will progressively enlarge and connect up with each other, giving the organ a "brain-like" appearance when digestion is sufficient.

12. This step may be repeated several times to increase purity of the parenchymal cell fraction; we suggest at least three centrifugations before proceeding to the next step.

13. For cell culture, all procedures must be performed in a laminar flow hood under aseptic conditions, and all buffers and solutions must be filtered and supplemented with antibiotics.
14. Iscove's modified Dulbecco's medium (IMDM) supplemented with insulin 1%, fetal bovine serum (FBS) 10%, and antibiotics, at 37°C, 5% CO_2.
15. Negative control: IMDM + 0.5% BSA; positive control: IMDM + 10% FBS, treatment groups: IMDM + BSA 0.5% + experimental factor at various doses.
16. Usually 1, 3, 5, and 7 days.
17. OC basic medium: IMDM + insulin 1% + FBS 10%.
18. Usually at 4 and 6 h.
19. Negative control: add the experimental factor to neither the lower chamber nor the upper chamber. Chemokinetic control: add the factor to both the chambers. Every determination should be performed at least in triplicate.
20. Briefly, remove the medium from the upper chamber of each transwell and transfer the transwells to a new 24-well plate containing 0.6 ml of PBS/well. Add 0.5 ml PBS to the upper chamber of each transwell to wash the membrane. Transfer the transwells into a new 24-well plate containing 0.6 ml of 4% paraformaldehyde (PFA) in PBS/well; add 0.5 ml of PFA to the upper chamber of each transwell. Incubate for 30 min at RT. Remove PFA from the chamber of each transwell by transferring the transwells to new 24-well plates containing 0.6 ml PBS/well with 0.1% Coomassie blue (CB), 10% MeOH, and 10% acetic acid. Add 0.5 ml of this same solution to the upper chamber of each transwell. Incubate for 1–2 h at RT. Wash with PBS at least twice by transferring the transwells into new 24-well plates, containing 0.6 ml of PBS/well.

References

1. Newsome, P.N., Hussain, M.A., Theise, N.D. (2004) Hepatic oval cells: helping redefine a paradigm in stem cell biology. *Curr Top Dev Biol* **61**, 1–28.
2. Oh, S.H., Hatch, H.M., Petersen, B.E. (2002) Hepatic oval 'stem' cell in liver regeneration. *Semin Cell Dev Biol* **13**, 405–409.
3. Petersen, B.E. (2001) Hepatic "stem" cells: coming full circle. *Blood Cells Mol Dis* **27**, 590–600.
4. Piscaglia, A.C., Di Campli, C., Gasbarrini, G., Gasbarrini, A. (2003) Stem cells: new tools in gastroenterology and hepatology. *Dig Liver Dis* **35**, 507–514.
5. Lowes, K.N., Croager, E.J., Olynyk, J.K., Abraham, L.J., Yeoh, G.C. (2003) Oval cell-mediated liver regeneration: Role of cytokines and growth factors. *J Gastroenterol Hepatol* **18**, 4–12.
6. Libbrecht, L., Desmet, V., Van Damme, B., Roskams, T. (2000) Deep intralobular extension of human hepatic 'progenitor cells 'correlates with parenchymal inflammation in chronic viral hepatitis: can 'progenitor cells' migrate? *J Pathol* **192**, 373–378.
7. Mohle, R., Bautz, F., Denzlinger, C., Kanz, L. (2001) Transendothelial migration of hematopoietic progenitor cells. Role of

chemotactic factors. *Ann N Y Acad Sci.* **938**, 26–34; discussion 34–35.
8. Petersen, B.E., Zajac, V.F., Michalopoulos, G.K. (1997) Bile ductular damage induced by methylene dianiline inhibits oval cell activation. *Am J Pathol* **151**, 905–909.
9. Petersen, B.E., Zajac, V.F., Michalopoulos, G.K. (1998) Hepatic oval cell activation in response to injury following chemically induced periportal or pericentral damage in rats. *Hepatology* **27**, 1030–1038.
10. Alison, M.R., Golding, M., Sarraf, C.E., Edwards, R.J., Lalani, E.N. (1996) Liver damage in the rat induces hepatocyte stem cells from biliary epithelial cells. *Gastroenterology* **110**, 1182–1190.
11. Evarts, R.P., Hu, Z., Omori, N., Omori, M., Marsden, E.R., Thorgeirsson, S.S. (1996) Precursor-product relationship between oval cells and hepatocytes: comparison between tritiated thymidine and bromodeoxyuridine as tracers. *Carcinogenesis* **17**, 2143–2151.
12. Higgins, G. M., and Anderson, R. M. (1931) Experimental pathology of the liver. *Arch Pathol* **12**, 186–201.
13. Petersen, B.E., and Hatch, H.M. (2002) Stem Cell Culture: Liver Stem Cells. *Methods of Tissue Engineering*: 429–437. Academic Press. 2002
14. Schmitz, B., et al. (1994) Magnetic activated cell sorting (MACS)–a new immunomagnetic method for megakaryocytic cell isolation: comparison of different separation techniques. *Eur J Haematol* **52**, 267–275.
15. Von Schönfeldt, V., Krishnamurthy, H., Foppiani, L., Schlatt, S. (1999) Magnetic Cell Sorting Is a Fast and Effective Method of Enriching Viable Spermatogonia from Djungarian Hamster, Mouse, and Marmoset Monkey Testes. *Biology Reprod* **61**, 582–589.
16. Campbell, D.G., Gagnon, J., Reid, K.B.M., Williams, A.F. (1981) Rat brain Thy-1 glycoprotein. The amino acid sequence, disulphide bonds and an unusual hydrophobic region. *Biochem J* **195**, 15–30.
17. Hermans, M.H.A., and Opstelten, D. (1991) In situ visualization of hemotopoietic cell subsets and stromal elements in rat and mouse bone marrow by immunostaining of frozen sections. *J Histochem Cytochem* **39**, 1627–1634.
18. Garnett, D., Barclay, A.N., Carmo, A.M., Beyers, A.D. (1993) The association of the protein tyrosine kinases p56*lck* and p60*fyn* with the glycosyl phosphatidylinositol anchored proteins Thy-1 and CD48 in rat thymocytes is dependent on the state of cellular activation. *Eur J Immunol* **23**, 2540–2544.
19. Crook, K., and Hunt, S.V. (1996) Enrichment of early fetal-liver hemopoietic stem cells of the rat using monoclonal antibodies against the transferrin receptor, Thy-1, and MRC-OX82. *Dev Immunol* **4**, 235–246.
20. Ishizu, A., et al. (1995) Thy-1 induced on rat endothelium regulates vascular permeability at sites of inflammation. *Int Immunol* **7**, 1939–1947.
21. Petersen, B.E., Goff, J.P., Greenberger, J.S., Michalopoulos, G.K. (1998) Hepatic oval cells express the hematopoietic stem cell marker Thy-1 in the rat. *Hepatology* **27**, 433–445.
22. Mason, D.W., and Williams, A.M. (1980) The kinetics of antibody binding to membrane antigens in solution and at the cell surface. *Biochem J* **187**, 1–20.
23. Hatch, H.M., Zheng, D., Jorgensen, M.L., Petersen, B.E. (2002) SDF-1alpha/CXCR4: a mechanism for hepatic oval cell activation and bone marrow stem cell recruitment to the injured liver of rats. *Cloning Stem Cells* **4**, 339–51.
24. Jung, Y., Oh, S.H., Zheng, D., Shupe, T.D., Witek, R.P., Petersen, B.E. (2006) A potential role of somatostatin and its receptor SSTR4 in the migration of hepatic oval cells. *Lab Invest* **86**, 477–489.
25. Piscaglia, A.C., Shupe, T.D., Oh, S., Gasbasrini, A., Petersen, B.E. (2007) Granulocyte-Colony Stimulating Factor Promotes Liver Repair and Induces Oval Cell Migration and Proliferation in Rats. *Gastroenterology* **133**, 619–631.
26. Stolz, D.B., and Michalopoulos, G.K. (1997) Synergistic enhancement of EGF, but not HGF, stimulated hepatocyte motility by TGF-beta 1 in vitro. *J Cell Physiol* **170**, 57–68.
27. Weissman, I.L. (2002) Stem cells - scientific, medical, and political issues. *New Eng J Med* **346**, 1576–1579.
28. Marshall, E. (2000) The business of stem cells. *Science* **287**, 1419.
29. Winstor, R. (2001) Embryonic stem cell research. The case for... *Nat Med* **7**, 396–397.

30. Ferber, S. (2000) Can we create new organs from our own tissues? *Isr Med Assoc J* **2**(suppl 2), 32–36.
31. Grompe, M. (2001) Liver repopulation for the treatment of metabolic diseases. *J Inherit Metab Dis* **24**, 231–244.
32. Paul, S., and Regulier, E. (2000) Molecular basis of oncogenesis. *Ann Biol Clin* **59**, 393–402.
33. Bach, S.P., Renehan, A.G., Potten, C.S. (2000) Stem cells: the intestinal cell as a paradigm. *Carcinogenesis* **21**, 469–476.

Chapter 25

Reprogramming of Liver to Pancreas

Shifaan Thowfeequ, Wan-Chun Li, Jonathan M.W. Slack, and David Tosh

Abstract

Islet grafts have demonstrated that patients with diabetes would benefit greatly by β-cell therapy. However, the paucity of available islets for transplantation as well as the immunological barriers faced in allogeneic transplantation represent a tremendous barrier to regenerative approaches to the treatment of diabetes. Here, we present a strategy and protocols to transdifferentiate developmentally related hepatocytes into β-cells by the ectopic expression of critical β-cell transcription factors.

Key words: Two-step collagenase liver perfusion, hepatocyte culture, pancreatic transcription factors, Pdx1, adenovirus, transdifferentiation, immunofluorescence.

1. Introduction

Cell therapy means treating diseases with cells, ideally those of the same individual to preserve full immunological compatibility. The ability to produce differentiated cell types at will offers a compelling new approach to cell therapy and therefore for the treatment and cure of degenerative disorders such as diabetes. Potential sources of cells for cellular therapy include embryonic or adult stem cells. However, so far there is a limited progress towards being able to drive differentiation of these types to a β-cell phenotype. An alternative approach is to produce differentiated cells by a process called transdifferentiation. Transdifferentiation is the irreversible conversion of one differentiated cell into another differentiated cell type. Transdifferentiation may on occasion occur naturally between embryologically related tissues and at the molecular level is brought about by the altered expression of a master switch gene whose function is to distinguish the two tissues during development. The expression of the master switch gene can, under suitable circumstances, trigger an alternate programme of gene expression that demarcates the new cell type *(1)*.

A type of transdifferentiation currently being sought by several labs is the conversion of hepatic cells to pancreatic endocrine cell types. This is because reprogramming of hepatocytes could act as an innovatory method of generating transplantable insulin-secreting β-cells for patients with type-1 diabetes, who lack functional β-cells of their own. The liver offers a number of advantages over other endoderm-derived tissues as a source of cells for making pancreatic cells. The first is related to the normal developmental biology of the liver. During development, the liver and pancreas originate from a common progenitor cell population in the foregut endoderm *(2)*. If the rules for transdifferentiation hold true, then overexpression of a pancreatic transcription factor(s) in hepatocytes might cause the induction of a pancreatic phenotype. The second advantage concerns the large size and easy accessibility of the liver. The third is related to the regenerative potential of the liver. It would be possible to surgically remove a lobe (or part of a lobe) of liver, dissociate the cells, induce transdifferentiation to β-cells in vitro, and then return them to the patient. In the meantime, the liver can regenerate the cells lost during the removal of the lobe *(3)*. In this chapter, we describe progress on reprogramming of liver to pancreas based on the overexpression of pancreatic transcription factors in two different liver cell types. The first is the human hepatoma cell line HepG2 and the second is adult rat hepatocytes.

HepG2 cells were chosen as a source for generating pancreatic cells for two reasons. The first is related to their ability for long-term cell subculture, and the second is because they retain some properties of a well-differentiated hepatic phenotype. For example, HepG2 cells secrete serum proteins (e.g. albumin) and express liver-enriched transcription factors (e.g. hepatocyte nuclear factor 4α (HNF4α)). However, HepG2 cells lack some differentiated properties (e.g. urea cycle activity) and they are tumorigenic. Therefore, we have also used primary cultures of isolated rodent or human hepatocytes as a source of cells to induce transdifferentiation towards a pancreatic phenotype.

Based on recent investigations, cultured human hepatoma cells and adult rat hepatocytes maintained in culture can be reprogrammed into pancreatic-like cells expressing β-cell-specific makers by the expression of transcription factors that are normally required for β-cell development. These include *Pdx1*, essential for initial pancreatic specification, and/or *Neurogenin3*, *NeuroD*, *Nkx6.1*, *Nkx2.2*, *Pax4*, *Pax6* and *Islet1*.

The protocols below describe the overexpression of pancreatic transcription factors by transfection (HepG2 cells) or by adenoviral infection (adult rat hepatocytes). Adenoviral vectors are an excellent vehicle for gene delivery as they are capable of transducing non-dividing cells such as hepatocytes. Furthermore, genes delivered by adenoviral vectors do not integrate into the host genome, thus their expression is short-lived (4, 5). This is advantageous from a safety point of view since there is little risk of insertional mutagenesis. Under suitable circumstances, we believe that a transient overexpression of regulatory transcription factors can cause a permanent transdifferentiation by activation of the cell's endogenous machinery to maintain the newly acquired phenotype (hit and run hypothesis) (1). It should be noted, however, that the methods described do not produce complete transdifferentiation to β-cells and that further refinement of the conditions will be necessary to achieve this goal.

2. Materials

2.1. HepG2 Cell Culture

1. The HepG2 cell line can be obtained from European Collection of Cell Cultures (ECACC) or the American Type Culture Collection (ATCC).
2. 1× trypsin–EDTA solution (Gibco™/Invitrogen Life Technologies).
3. Culture medium: Dulbecco's modified Eagle's medium (DMEM; Sigma, UK D5546) containing 100 U/ml penicillin, 100 µg/ml streptomycin, 2 mM L-glutamine, 1 X MEM non-essential amino acids (all from Sigma, UK) and 10% fetal bovine serum (FBS; Gibco™/Invitrogen Life Technologies).
4. FBS mixed with 10% (v/v) dimethyl sulfoxide (DMSO; VWR International, UK) for cell storage.

2.2. Rat Hepatocyte Isolation and Culture

1. Perfusion buffer 1: 0.05 (w/v) KCl in 10 mM HEPES buffer, 5 mM D-glucose, 200 µM EDTA and 0.001% (v/v) phenol red (Sigma, UK), all made up in 500 ml of sterile phosphate-buffered saline (PBS; pH 7.4). The buffer can be made fresh or can be stored at 4°C for up to 24 h and should be pre-warmed in a water bath to 39°C before use. All the solutions used to make the perfusion buffer should either be sterilised by autoclaving or filtered through a 0.22 µm filter.
2. Perfusion buffer 2: 0.05 (w/v) KCl in 10 mM HEPES buffer, 20 mM HEPES buffer, 5 mM D-glucose (Sigma, UK), 1 mM $CaCl_2$ (Fisher Scientific), 0.001% (v/v) phenol red (Sigma, UK), all made up in 500 ml of sterile PBS (pH 7.4). The buffer can be made fresh or can be stored at 4°C for up to 24 h. The buffer should be pre-warmed in a water bath to 39°C before use.

3. 50 mg of type 2 collagenase (Worthington, Lakewood, NJ) to be dissolved in 150 ml of perfusion buffer 2 just before use (*see* **Note 1**).

4. Adult male Wistar rat (body weight 280–300 g).

5. Surgical instruments: one pair of sharp round ended Mayo scissors, two pairs of blunt curved forceps, a pair of Spencer-Well's artery forceps, 18GA cannula (BDVenflon), non-absorbable suture (4-0; Ethicon™ Mersilk™), a 2–3 cm long matchstick.

6. Perfusion apparatus comprising a peristaltic pump and tubing containing a bubble trap (a sterile administration 'giving' set can be used: Non-air vented, filter 200 μm, rotary luer lock; for reference Baxter: RMC2071B). Including a bubble trap in the perfusion apparatus minimises the risk of introducing air bubble during the perfusion process.

7. Isoflurane anaesthesia apparatus.

8. 70 μm filter (Gore-Tex, USA).

9. 35 mm vented plastic culture dishes (Nunclon) with square coverslips inserted to set up cultures for immunocytochemistry and 60 mm vented plastic culture dishes to set up cultures for RNA extraction.

10. Attachment medium: William's Medium E (Sigma) supplemented with 10% (v/v) fetal bovine serum (FBS) and containing 100 U/ml penicillin, 100 μg/ml streptomycin and 2 mM L-glutamine. The attachment medium can be stored at 4°C and pre-warmed in a water bath to 37°C before use.

11. Hepatocyte culture medium: Keratinocyte serum-free medium (KSFM; Gibco™) with 100 U/ml penicillin, 100 μg/ml streptomycin, 50 μg/ml gentamycin, 100 ng/ml amphotericin B, 2 mM L-glutamine, 10 nM dexamethasone (*see* **Note 2**) and supplements (which includes 5 ng/ml recombinant human EGF and 50 μg/ml bovine pituitary gland extract; from Gibco™). The KSFM medium is light sensitive and should be stored at 4°C in the dark. The medium should pre-warmed in a water bath to 37°C before use.

2.3. Transfection of HepG2 Cells

1. GeneJuice transfection reagent (Novagene, USA): GeneJuice transfection reagent exhibits the optimal conditions for transfection of plasmid DNA into HepG2 cells and displays low cytotoxicity after transfection. Follow manufacturer's instructions.

2. Serum-free DMEM medium (same as 2.1(2) without containing 10% FBS).

3. Plasmid DNAs-encoding transgene cassette.

2.4. Adenovirus Infection	1. Adenoviruses of known titre, measured as the number of infectious units per ml (I.U./ml), are prepared. The viruses should always be stored at −80°C until required (*see* **Note 3**).
2. Hepatocyte culture medium (same as **2.2.11**).
3. 1% (w/v) Virkon (DuPont).
2.5. Immunofluorescence Detection of Exogenous Gene Expression	1. 4% (v/v) paraformaldehyde (PFA) in PBS.
2. 1% Triton X-100 (Sigma) in PBS (PBST×).
3. 2% (v/v) blocking buffer in PBS. Stocks of 10% w/v blocking buffer (Roche) can be made in 10× MAB (1 M maleic acid, 1.5 M NaCl, pH 7.5), autoclaved and stored at −20°C. The 2% blocking buffer can be stored at 4°C.
4. Primary antibodies (**Table 25.1**) and appropriate fluorescent secondary antibodies. All antibodies can be diluted in the 2% blocking buffer for use.
2.6. Reverse Transcription Polymerase Chain Reaction (RT-PCR)	1. For RNA extraction: TRI REAGENT™ (Sigma, UK), 100% chloroform (should not contain isoamylalcohol or other additives), 100% isopropanol, 75% (v/v) ethanol, RNAse-free water and cell scrapers.
2. Spectrophotometer.
3. For reverse transcription: 10 mM dNTP mix, 0.5 µg/µl Oligo (dT)$_{12-18}$, 10× RT buffer, 0.1 M DTT, 25 mM MgCl$_2$, 40 U/µl RNaseOUT™ Recombinant Ribonuclease Inhibitor, 200 U/µl SuperScript™ II Reverse Transcriptase, 2 U/µl RNase H (all from Invitrogen, UK).
4. For the polymerase chain reaction: PCR ReddyMix™ Master mix (AbGene) is used. The forward and reverse primers for the genes of interest are diluted to 1 µg/µl in distilled water.

Table 25.1
Antibodies for use in detecting pancreatic marker expression

Antigen	Supplier	Dilution	Species
Pdx-1	J.M.W.Slack, Bath, UK	1:100	Rabbit
Neurogenin 3	Michael German, UCSF, CA	1:200	Mouse
Insulin	DAKO	1:200	Guinea pig
c-peptide	Acris antibodies	1:100	Rabbit
Amylase	Sigma	1:100	Rabbit
Glucagon	Sigma	1:100	Mouse

3. Methods

The protocol used for rat hepatocyte isolation is based on a modification of the technique previously described by Tosh and colleagues (6). There are a number of excellent reviews already on the subject of adenovirus preparation (7) hence it will not be discussed here. The success of an adenoviral infection can be followed either by the detection of various reporters incorporated into the transgene construct or the direct immunocytochemical detection of the transgene product itself. The novel pancreatic genes expressed as a consequence of the introduction of the transgene can be detected by reverse transcription polymerase chain reaction (RT-PCR) even at very low levels of expression. Detailed instructions for the fluorescent immunocytochemistry and RT-PCR protocols will be given below.

3.1. Rat Hepatocyte Isolation and Culture

1. Rinse the perfusion apparatus with PBS and then with perfusion buffer 1. Set the flow rate to 30–35 ml/min.
2. Anaesthetise the rat with 4.5% isoflurane in oxygen, flowing at a constant rate of 1,500 ml/min. When the animal is fully anaesthetised (see **Note 4**), it can be maintained for surgery with the aid of a face mask in 3% isoflurane in oxygen, flowing at 1,500 ml/min.
3. Lay the anaesthetised rat dorsal side down on a flat tray and attach the four legs onto the tray with masking tape so that the rat is slightly stretched.
4. Pour 70% ethanol to disinfect the fur over the abdomen and using the scissors make a 'U'-shaped incisions starting from the lower abdomen and up to the lateral aspect of the rib cage. Fold back the skin over the chest, fully exposing the internal abdominal organs.
5. Push the intestines to the left side of the rat exposing the portal vein. If the large lobe of the liver is obstructing the view of the portal vein, the lobe can temporarily be folded back and held in position with a piece of wet tissue paper.
6. Insert a matchstick underneath the portal vein and using the blunt end of the curved forceps, remove any excess adipose tissue around the vein.
7. Using the forceps as a guide, insert 5–8 cm of suture underneath the portal vein and tie it loosely around the portal vein.
8. Run some perfusion buffer 1 through the cannula to remove any air bubbles and then insert the cannula into the portal vein (see **Note 5**). Secure the cannula in position by tightening the suture around it. Remove the inner needle of the cannula.

9. Using sharp scissors cut the abdominal blood vessels and the spinal cord.

10. Attach the cannula to the perfusion pump and start perfusing with perfusion buffer 1 at a constant flow rate of 30–35 ml/min. Cut open the thoracic cavity and cut the thoracic portion of the inferior vena cava. Then clamp the abdominal blood vessels just above the cut using a pair of artery forceps thereby establishing a continuous flow of the buffer from the cannula in the portal vein, through all the lobes of the liver and out through the vena cava into the thoracic cavity.

11. Continue perfusion until 200 ml of the perfusion buffer 1 has run through (*see* **Note 6**). The liver should clear as blood is flushed out.

12. As buffer 1 is perfusing through the liver, dissolve 50 mg of collagenase in 150 ml of perfusion buffer 2. Making sure that no air bubbles are introduced into the perfusion apparatus beyond the bubble trap, replace EDTA-perfusion buffer 1 with perfusion buffer 2 maintaining the flow rate at 25–30 ml/min.

13. Continue perfusion with perfusion buffer 2 until the liver becomes paler and gradually takes on a slightly swollen appearance. This usually takes 10–12 min (*see* **Notes 7 and 8**).

14. Once the liver has been fully perfused, terminate the perfusion. Carefully cut the connective tissue around the liver attaching the tissue to the abdominal wall and organs and remove the liver into 30 ml of perfusion buffer 2.

15. In a class II biosafety cabinet, using two pairs of curved forceps, gently dissociate the liver.

16. Pass the suspension through a 70 μm filter (or cell strainer) and transfer the cell suspension that passes through the filter into two 50 ml sterile tubes. Top up with perfusion buffer 2 and spin at 50 g for 2 min. The liver parenchymal hepatocytes will sediment in the pellet while the non-parenchymal hepatocytes and any dead cells remain in the supernatant.

17. Repeat the washing procedure by discarding the supernatant and suspending the pellet in fresh perfusion buffer 2, until the supernatant appears clear. The last wash could be done in the attachment medium.

18. After the final wash, the pellet can be resuspended in 10 ml of the attachment medium. A single perfusion should yield between 100 and 400 million hepatocytes (*see* **Note 9**).

19. The hepatocytes can be inoculated at a final density of 4×10^5 cells/ml of attachment medium. Add 1.5 and 3 ml of the cell suspension to 35 and 60 mm dishes, respectively, and culture

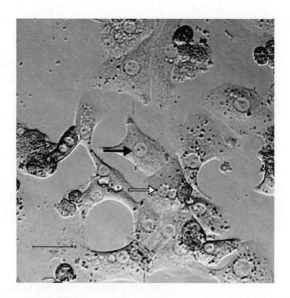

Fig. 25.1. Isolated rat hepatocytes following 3 days of culture. The cells maintain their hepatic phenotype and show characteristic hepatocyte morphology being either binucleated (*white arrowhead*) or with large single nucleus (*black arrow*).

in a 5% CO_2 incubator at 37°C allowing the cells to attach onto the bottom of the dish and flatten out into a monolayer.

20. After 4–5 h, the cells should have attached and their hepatocyte morphology (often binucleated) should be clearly visible (*see* **Fig. 25.1**). Replace the attachment medium with the same amount of hepatocyte culture medium. A quick wash can be performed in PBS if removal of all serum is essential. The cells can be cultured in a 5% CO_2 incubator at 37°C and the medium can be replaced every 48 h.

3.2. Adenovirus Infection

1. Based on the virus titre and the number of cells per dish, calculate the volume of virus required to infect at a total multiplicity of infection (M.O.I) of 10–20 (*see* **Note 10**).

2. In a class II biosafety cabinet, add the appropriate amount of virus made up in hepatocyte culture medium, into each dish and then transfer the dishes to a 37°C incubator for infection to take place.

3. After 1 h, replace the virus-containing culture medium in each dish with fresh hepatocyte culture medium. Discard any virus-contaminated solutions into a bottle of 1% Virkon^{TM} and dispose accordingly.

4. After 24–48 h, the success of infection can be visualised based on reporters included in the gene construct (*see* **Fig 25.2A & B**).

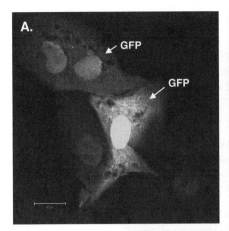
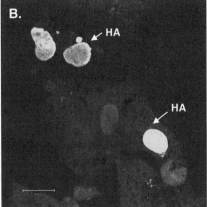
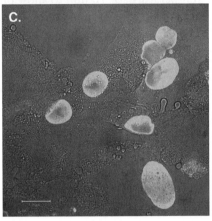

Fig. 25.2. (**A, B**) The success of an infection can be determined by following the expression of reporters such as GFP and the hemagglutinin (HA) tag. In a multiple infection, this enables the number of cells that has been infected with each of the different viruses to be calculated. (**C**) The expression of the transgene itself can be visualised by immunocytochemistry. Hepatocytes were transduced with a CMV-Ngn3 construct, and the nuclear expression of the Ngn3 gene product (seen as fluorescent nuclei) was visualised using an anti-Ngn3 antibody.

3.3. Immunofluorescence Detection of Exogenous Gene Expression

1. Remove the culture medium from the 35 mm dishes containing the cells cultured on the coverslips and wash the cells with PBS.
2. Fix the cells with 1.2 ml of 4% PFA for 20 min. Wash off the fixative with PBS and permeabilise in 1.2 ml of 1% PBSTx for 20 min (*see* **Note 11**).
3. Block in 2% blocking buffer for 1–2 h to minimise any unspecific binding of antibodies.
4. Add the primary antibody diluted in 2% blocking buffer and incubate at 4°C overnight.
5. Wash off the primary antibody (3 × 15 min) with PBS. Add the secondary antibody diluted to 1:100 in 2% blocking buffer and incubate at room temperature in the dark for 1–2 h (*see* **Note 12**).

6. In the dark, wash off the secondary antibody (3 x 15 min) with PBS.

7. Mount the coverslips onto slides using a mounting media containing an anti-fading agent compatible with the immunofluorescence and observe under a fluorescent microscope (*see* **Fig. 25.2C**).

3.4. RT-PCR for the Detection of induced Pancreatic Genes

1. 1 ml of TRI REAGENT™ can be used to directly lyse the cells from two 60 mm dishes. Transfer the homogenous lysate into a tube and to ensure complete dissociation of nucleoprotein complexes, stand the sample at room temperature for 5 min.

2. Add 0.2 ml of chloroform, vortex for 15 s and stand at room temperature for 10 min to allow phase separations. Centrifuge at 12,000 g for 15 min at 4°C. Centrifugation would enhance the phase separation into a bottom red organic phase of protein, an interphase of DNA and a colourless upper aqueous phase containing RNA.

3. Transfer the upper aqueous phase into a fresh tube, add 0.5 ml of isopropanol, vortex, and stand at room temperature for 10 min to allow RNA precipitation. Centrifuge at 12,000 g for 10 min at 4°C.

4. Wash the precipitated RNA pellet with 1 ml of 75% ethanol by vortexing and then centrifuging at 7,500 g for 5 min at 4°C. Decant off the ethanol and air dry the RNA pellet (*see* **Note 13**).

5. Dissolve the RNA pellet in distilled RNase-free water and measure the absorbance on a spectrophotometer. The ratio of absorbance at 260–280 nm should ideally be in the range of 1.8–2.0.

6. For the RT reaction, in a PCR tube mix 3 μg of RNA with 1 μl of 10 mM dNTP, 1 μl Oligo(dT)$_{12-18}$ and make up to 10 μl with RNase-free water. Stand in a 65°C water bath for 5 min and then on ice for 1 min.

7. To the PCR tube add 2 μl of 10× RT buffer, 2 μl of 0.1 M DTT, 4 μl of 25 mM MgCl$_2$, 40 U RNaseOUT™ and 200 U of SuperScript™ II Reverse Transcriptase. Leave in a 42°C water bath for 52 min and then inactivate the reaction on a 70°C heat block for 15 min.

8. mRNA–cDNA hybrids can be degraded by incubating at 37°C for 20 min with 2 U of RNase H in each reaction tube (*see* **Note 14**).

9. Run the PCR with 1 μl of the cDNA and 500 ng of the forward and reverse primers in 20 μl of the PCR mastermix. β-Actin can be used a good loading control for hepatocytes (*see* **Fig. 25.3**).

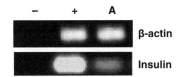

Fig. 25.3. RT-PCR showing β-cell insulin gene expression induced in hepatocytes as a consequence of adenovirus expressing the pancreatic transcription factor Pdx1. Lane 1: negative control (water), lane 2: positive control (RIN cells) and lane 3: hepatocytes infected with Pdx1. β-Actin controls are also shown.

4. Notes

1. This weight of collagenase is only a guide. It is crucial to batch test collagenase lots as well as different suppliers.
2. The 1 mM dexamethasone stock is made up in 100% ethanol and can be stored in –20°C.
3. Thawing and freezing the virus will reduce its titre by a factor of 10 each time.
4. Level of anaesthesia can be determined by testing for the absence of a pedal reflex by pinching the paw pads of the hind legs of the rat.
5. Steps 7–10 should be carried out quickly to prevent clotting of blood.
6. Perfusion buffer 1 facilitates the dissociation of hepatocytes through EDTA-mediated chelation of calcium (which is necessary for cell–cell adhesion).
7. Collagenase added to perfusion buffer 2 digests the mainly collagen extracellular matrix surrounding the hepatocytes. Calcium needs to be included in perfusion buffer 2 as it is required for collagenase activity.
8. It may be necessary to recycle perfusion buffer 2 from the thoracic cavity back into perfusion apparatus using a 10 ml syringe. Insufficient perfusion can result in clumps of cells during isolation and over-perfusion reduces cell viability. If the liver starts to become swollen and leaky, reduce the flow rate.
9. Following isolation and prior to setting up the cultures, the cell viability can be determined using a hemocytometer and 0.4% (w/v) trypan blue. The minimum cell viability should be around 85–90% for successful culture.
10. The M.O.I. for an infection is the theoretical mean number of infectious particles per cell. A low M.O.I will result in inefficient gene delivery and a high M.O.I will have a cytopathic effect. For example, if a 35 mm dish (that takes 1.5 ml of the culture medium) has 5×10^5 cells, infection with 10 μl of a

virus of titre 10^9 I.U./ml would give an M.O.I. of 20. If using more than one virus, it is still advisable not to exceed a total M.O.I of 20. An Ad-Null virus carrying no gene of interest, can be used to maintain the total M.O.I of 20 while keeping the individual M.O.Is constant. For example, if using Ad-A, Ad-B, Ad-C and Ad-D all together, each at M.O.I. of 5 for condition 1, when using Ad-A on its own for condition 2, still use Ad-A at an M.O.I. of 5 and use Ad-Null at M.O.I. of 15, maintaining the total M.O.I at 20.

11. Depending on the antigen you are looking for, post-fixation and antigen retrieval protocols may be adopted prior to blocking. Consult recommendations from the manufacturers for special requirements of antibodies.

12. If performing multiple staining, prevent cross-reactivity by adding the secondary antibodies in the correct sequence taking into account the species it has been raised in.

13. Do not allow the RNA pellet to completely dry out as this would reduce its solubility. Do not use centrifugation under vacuum.

14. The cDNA synthesised in the RT reaction is stable at −20°C for up to 6 months.

References

1. Li, W.-C., Yu, W.-Y., Quinlan, J. M., Burke, Z. D., and Tosh, D. (2005) Molecular basis of transdifferentiation. *Journal of Cellular and Molecular Medicine* **9**, 569–582.
2. Slack, J. M. W. (1995) The developmental biology of the pancreas. *Development* **121**, 1569–1580.
3. Michalopoulos, G. K. (1990) Liver regeneration: molecular mechanisms of growth control. *FASEB J.* **14**, 176–187.
4. Thomas, C. E., Ehrhardt, A., and Kay, M. A. (2003) Progress and problems with the use of viral vectors for gene therapy. *Nature Reviews Genetics* **4**, 346–358.
5. McConnell, M. J., and Imperiale, M. J. (2004) Biology of adenovirus and its use as a vector for gene therapy. *Human Gene Therapy* **15**, 1022–1033.
6. Tosh, D., Alberti, K. G. M. M., and Agius, L. (1988) Glucagon regulation of gluconeogenesis and ketogenesis in periportal and perivenous rat hepatocytes. *Biochemical Journal* **256**, 197–204.
7. Tollefson, A. E., Hermiston, T. W., and Wold, W. S. M. (1998) Preparation and titration of CsCl-banded adenovirus stock, in *Adenovirus Methods and Protocols*, Humana, Totowa, NJ.

INDEX

A

2-Acetylaminofluorene 387–389
Adenovirus 152, 407, 411–412, 414, 417
Adipocyte(s) 68, 128, 159, 276, 281, 282, 304
Adult neurogenesis 143–144
Adult stem cells 21, 45, 72, 296, 401, 407
Angiogenesis 333
Antennapedia 43
Aorta 102, 106, 211
Aorta-gonad-mesonephros 62
Astrocytes 66, 146, 186, 187, 188, 192, 195
Asymmetric cell division 146, 295–297
Axon 172, 176–178, 180, 181

B

B220 .... 44, 95, 260
β-cells 75
Biological transport 43
Blast colony-forming cells (BL-CFC) 64
Blastocyst(s) 3–8, 14, 15, 17, 18, 57, 58, 199
Blood 46, 56, 60, 62–64, 68, 69, 74, 91,
102, 109, 110, 120, 127–129, 132, 133, 135, 144,
146, 162, 270, 271, 275, 281, 282, 286, 391–395,
402, 413, 417
Blood vessels 73, 222, 333, 340, 413
Bone marrow 186, 269
  bone marrow-derived mesenchymal stem cells, isolation
    etc 259, 281–283, 285, 287–289, 291
  human red blood cells, ex vivo
    generation 127–128, 131–132
  interaction between HSCs and their niche 93–94,
    97, 100, 102–104, 106
  measurement of cell-penetrating peptide uptake 44,
    46, 48, 49, 52
Bone morphogenic protein (BMP) 56, 57
Bromodeoxyuridine (BrdU) 201, 206, 209, 227,
230, 275, 296–298, 300–305, 307–311, 315
Bulge stem cells 215–216, 223, 229

C

Cardiomyocytes 58, 70, 71, 72, 281, 282, 289, 358
Cartilage 56, 67, 70, 97, 105, 159,
259, 270, 276, 301, 302
CD2 .... 113, 116
CD3 .... 113, 116, 260
CD5 .... 44, 260
CD11b 44, 260, 282, 284, 288
CD14 .... 113, 116, 282, 284, 288
CD16 .... 113, 116
CD19 .... 113, 116
CD24 .... 113, 116
CD25 .... 271, 275
CD29 .... 68, 282, 284, 288
CD31 .... 260, 261, 263, 264, 265, 266, 267, 282,
284, 288, 336
CD34 .... 68, 109–110, 112–114, 115–117,
127, 131, 137, 216, 217, 223–230, 269–271, 282,
320, 321
CD41a 110, 111, 114, 117–124
CD42b 110, 117–124
CD44 .... 68, 269–271, 282, 284, 288
CD45 .... 68, 96, 100, 103, 106, 117, 260, 261–267,
269–271, 282, 284, 288, 398
CD45R 44
CD56 .... 113, 116
CD61 .... 110, 111, 117
CD62 .... 114, 117, 121, 124
CD66b 113, 116
CD73 .... 68, 269, 270, 271, 282, 284, 288
CD90 .... 68, 269, 270, 271, 282, 395, 396, 397
CD105 .... 68, 269, 270, 271, 282, 284, 288
CD106 .... 68, 269, 270
CD146 269–271
CD166 .... 282, 284, 288
CDX4 64
Cell cycle 144, 215, 227, 296
Cell division orientation 333
Cell-penetrating peptides 43, 64
Cell sorting 40, 63, 95, 100, 101, 110, 151, 215,
216, 259, 267, 281, 387, 395
Cell surface markers 71, 215, 281–282
Cell therapy 55–56, 61–62, 65, 67, 72–74,
281–282, 407
Cell viability 45, 51, 52, 117, 120, 121, 169, 221,
230, 248, 250, 395–397, 417
CFU-F 259–260, 267, 270, 275, 282
Collagen 69, 70, 122, 156, 272, 276, 277, 362, 363,
373, 376–379, 382, 383, 399, 400, 417
Chick neural tube 171–172, 174, 179–180
Collagenase perfusion 387, 391, 393, 395

419

Colony-forming unit fibroblast (CFU-F) 260, 282
Colony-forming unit megakaryocyte (CFU-MK) 123
Compact bone 68, 259, 261–264
Competitive transplantation assay 63
Conditioned medium 151, 166, 281, 285, 288–292, 334, 336
Confocal microscopy 43–46, 49–50, 334
Contraction 345–346, 354, 362–363
Cord blood 63, 109–110, 127–129, 132–133, 270, 281–282
Cornea 233–235, 237–241, 243–247, 249, 251–255
Corneal epithelial cell 233–234, 238–240, 243–246, 251–254
Culture medium design 109
Culture medium optimization 109
Cytokines 16, 21, 56, 109, 112–115, 128–129, 132–133, 281, 283, 398

D

Dedifferentiation 56, 70, 371, 376
Differentiation 186, 234, 282, 359
    embryonic stem cell-derived dopaminergic neurons 199–201, 211–212
    gene knock-downs in human embryonic stem cells 35, 57–60, 64–68, 70–75
    islet-derived progenitors 371, 376–377
    hepatic stem cells 388, 398
    human umbilical cord perivascular cells 269, 271, 275–276, 278
    mammalian neural stem cells 143–146, 152–157
    mouse skin-derived precursors 159–161, 163–169, 172–173, 175–177, 179, 181
    primitive vessels from embryonic stem cells 333–338, 340–342
    patterning embryonic stem cells 21
    reprogramming from liver to pancreas 407–409
    skeletal muscle stem cells 295, 315
    single muscle fibres 319–320
Dopaminergic neuron 65–67, 186, 199
Drug discovery 281

E

Electron microscopy 320
Embryo(s) 3, 6–9, 14–18, 57, 58, 61, 62, 67, 146, 147, 152, 159, 160–163, 171–174, 179–181, 182, 298, 315, 333, 334, 336
Embryoid body 64, 345
Embryonic neurogenesis 143
Embryonic stem cells
    derivation and manipulation of, murine 3
    derivation of contractile smooth muscle cells from 345–346, 348, 351–353, 360–362, 364

differentiation and dynamic analysis of primitive vessels from 333
gene knockdowns using lentiviral-based RNAi, human 35–36, 39–42
patterning using micro-contact printing 21–23, 25–31
regenerative medicine 57–59, 61–67, 69–72, 75–76
xenotransplantation, motor neuron 171
Endothelial cells 58, 62, 64, 71–74, 93, 146, 260, 267, 282, 289, 333, 334, 336, 341
Epidermis 62, 163, 213, 215–216, 221–221, 226, 230, 233–234, 247, 249, 281–282
Epithelial stem cells 215, 217, 219, 221, 223, 225, 227, 229, 231
Erythroid culture 127, 129, 132, 134, 137
ESCs, *see* Embryonic stem cells
Extracellular matrix (ECM) 21, 22, 24–28, 32, 128, 234, 342, 355, 391, 402, 417

F

FACS
    bone marrow-derived mesenchymal stem cells 281, 288
    derivation of contractile smooth muscle cells 345, 348, 355, 359, 362–363
    haemopoietic stem cells and their niche 95, 110, 115, 120–122
    isolation of epithelial stem cells 215–221, 223, 225, 227–229
    isolation of hepatic stem cells 387, 395, 397
    isolation of mesenchymal stem cells 259–262, 263–264, 266–267
    patterning embryonic stem cells 30, 40–41, 45, 47, 48–49
    See also Cell sorting
Feeder(s) 5, 7, 9–12, 16, 17, 29, 38, 181, 200, 205, 210, 218, 220, 227, 228, 231, 234, 267, 336, 348
Fetal brain 65, 199
Fetal liver 62
Fibrin gel 233–234, 254
Fibroblast(s) 9, 16, 61, 73, 128, 181, 200, 233, 234, 238, 243, 245, 249, 250, 251, 253, 260, 282, 291, 304, 336, 347, 348
Fibroblast growth factor 23, 56, 201
Fibronectin 24, 28, 156
Fixation 52, 208, 307, 315, 341, 343, 347, 348, 357, 418
Flow cytometry 68, 182, 231
    analysis of SMC markers 357, 359
    bone marrow-derived mesenchymal stem cells, isolation etc. 281, 287–288, 291

Flow cytometry *(continued)*
    ex vivo expansion, human cord blood
        stem cells 109–112, 114, 116–122, 124
    hemotopoietic stem cells sorting, instrument
        setting 96
    isolation of human umbilical cord perivascular
        cells 274–275
    measurement of cell-penetrating
        peptide 43–48
    patterning ESCs using micro-contact
        printing 21–22, 30–31, 40
Fluorescence 8, 22, 30, 45, 47–49, 51, 52, 95,
    96, 100, 103, 106, 168, 187, 215, 223, 224, 228,
    229, 259, 263, 264, 277, 281, 313, 356, 359, 361,
    365, 381
Foreskin 241
Functional cultured RBC 127

G

Gene expression 61, 152, 345, 348, 357, 371, 408,
    411, 415, 417
Gene delivery 408, 417
Gene therapy 57, 281
Gene transfer 152, 281
Glia 64, 65, 66, 143, 144
Glial cells 66, 146, 154, 159, 201
GR-1 .... 44, 95, 260
Green fluorescent protein (GFP) 8, 74, 365, 415
    asymmetric cell division in skeletal muscle
        stem cells 299–300, 306, 313, 315
    bone marrow-derived mesenchymal stem
        cells 284, 288, 291
    differentiation of primitive vessels
        from ESC 333–334, 340–341, 343
    ESC-derived motor neuron 175–177, 179–182
    gene knockdowns of human ESC 37, 40–41
    isolation of epithelial stem cells 228–230
    neural stem/progenitor cells,
        transplantation 187–188, 195

H

Haematopoietic stem cells, *see* Hematopoietic stem cell
Hair follicles 69, 215–216
Heart 56, 62, 70, 71, 72, 159, 289, 292, 402
Heart valve 73
Hematopoietic stem cells 144, 282
    human cord blood stem cells, ex vivo
        expansion 109
    human red blood cells, ex vivo
        generation 127–128, 132
    measurement of cell penetrating peptide uptake 43
    regenerative medicine 60, 62, 63, 74

Hepatocyte 389, 394, 395, 407, 408, 409
    culture 407, 410–411, 414
Hepatocyte growth factor 67, 73
HepG2 408–410
High throughput screening 21
Histone-2B 333
Hoechst 25, 30, 31, 45, 46, 49, 52, 53, 68, 71,
    161, 167, 168, 296, 310
Homing 63, 69, 93, 96, 103, 398
Homologous recombination 3, 4, 8, 14, 35
Human embryonic stem cell 21, 22–24, 27, 29,
    31, 35–37, 39–42
Human ESC, *see* Human embryonic stem cell

I

Immortal template DNA 295
Immunofluorescence 320, 397
    analysis, dopaminergic neuron 200–204, 208, 210
    asymmetric division, skeletal muscle stem
        cells 298, 308, 316
    detection, exogenous gene
        expression 407, 411, 415–416
    differentiation & dynamic, primitive vessels
        from ESCs 333–334
    purity & identity, smooth muscle cells 345, 347,
        354–357, 362
Immunomagnetic sorting (IMS) 387, 395–396, 398
Integrin(s) 71, 216, 217, 225–227, 229, 230
In vitro differentiation 68, 200, 333, 335, 337, 340
In vitro red blood cell (RBC) production 127
INGAP, *see* Islet neogenesis-associated protein
Intracellular delivery 43
Irradiation 9, 10, 190, 195, 196, 322, 323, 327,
    347, 348
Islet 43, 71, 74–75, 371–373, 375–377, 379–383,
    407, 408
Islet neogenesis-associated protein 75, 371–373,
    376, 377

K

Keratin 19, 233–234
Kinase inhibitor 8, 363
Knockdown(s) 35–37, 40–42

L

Label retaining cell 295, 311
Lentivirus 35, 41, 152, 229, 313
Lin⁻ cells 46
Liver 60, 62, 73, 74, 75, 369, 387–395, 398, 400–403,
    407–412, 417
Liver stem cells 74, 387–388, 400
Lymphocyte(s) 69, 269, 270, 276

M

Macrophages 128
Matrigel 24, 27, 28, 38–40, 289, 297, 300, 303, 304, 312–315, 364, 418
Megakaryocytes 109
Mesenchymal progenitor cells 60, 269
Mesenchymal stem cells 186
    bone marrow-derived 281–283, 285–289, 291
    ex vivo generation, human red blood
        cells 127, 131–132
    human umbilical cord perivascular
        cells 269–270
    isolation from mouse bone 259–261, 263–265, 267
    regenerative medicine 68–72
Mesoderm 71, 73, 337
Micro-contact printing 21–29, 31
Microscopy.. 43–46, 50, 134, 172, 278, 287, 288, 320, 334, 354, 356, 365
Migration assay 387, 398–400
Motor neuron 66–67, 171–174, 176–178, 181
MSC, *see* Mesenchymal stem cells
Multipotent... 66, 68–69, 71, 112, 128, 143, 145, 159–160, 186, 191, 216, 259, 281, 346
Murine embryonic stem cell 3, 333
Muscle fibre 297–298, 300, 303, 310, 319–323, 325, 327–329
Muscle regeneration 68, 310, 319, 322–323, 328–329
Mutation 4, 37, 57, 69, 299
Myelin basic protein 66
Myeloid 64, 123, 128

N

Nanog 3, 58, 61
NCSCs, *see* Neural crest stem cells
Nerve 56, 172, 177
Neural crest 66, 159, 170, 346
Neural crest stem cells 159
Neural stem cells 61, 65–66, 143, 145–149, 151, 153, 155, 185–186
Neurogenic astrocyte 185, 187, 192, 195
Neurogenic potential 144, 153, 156, 159
Neuron(s) 64, 65–67, 83, 143, 144, 153, 154, 156, 157, 159, 168, 171–174, 176–178, 181, 188, 199, 200, 205, 210, 211, 282, 358, 395
Neurosphere(s) 65, 143, 145, 147–156, 159, 164, 169, 185–190, 192, 200, 211
Niche 21, 234, 296
    HSC and their niche 93–95, 97, 99, 101, 103, 105
    mammalian stem cell, isolation 143–146, 154
    neural stem/progenitor cells,
        transplantation 186, 188, 190

    single muscle fiber, isolation 319–321
    stem cell resource, regenerative medicine 63, 67, 71
Notch signaling 73, 143, 145, 153–154
Numb 295–297, 299–300, 303, 305–306, 313, 315–316

O

OCT3/4 ....3, 25, 30, 57, 61, 418
Oocyte(s) 60, 61
Osteoblast(s) 58, 68, 69
Osteocyte(s) 281, 282
Oval cells 74, 387–389, 394–395, 398, 401

P

Pancreas 62, 73, 75, 369, 371, 372, 407, 408
Pancreatic transcription factors 75, 407–408, 417
Paracrine effect 72, 281
Parkinson's disease 65–66, 186, 199, 210
Partial hepatectomy (PH) 73–74, 387, 389–391
PCR 40, 200, 277, 278, 299, 309, 323, 327, 345, 348–350, 352–354, 357, 358, 362, 374, 375, 381, 382, 411, 412, 416, 417
Pdx1 ....75, 375, 382, 407–408, 417
Platelets 109–112, 116–121, 124
Pluripotency 3, 57, 61
Progenitor(s) 259, 269, 282, 295, 296, 363, 408
    hematopoietic 43, 53
    hepatic 387, 395
    human red blood cells 128, 134, 136
    islet 371, 372, 374–377, 379, 383
    megakaryocyte 109–110, 112–113, 116–117, 122–123
    neuron 143–145, 154, 157, 177, 185–187, 191, 200, 205, 207
    regenerative medicine 55–56, 59–60, 62–66, 68–69, 71–76
Proliferation assay 44, 289, 387, 398–399
Promoter 36, 37, 62, 72, 75, 216, 228, 348, 351, 364
Protein transduction domains 43

Q

Quantitative immunohistochemistry 21–22

R

Reconstructed skin 233–234, 253–255
Redifferentiation 371, 376–377
Regeneration 282, 289
    blood system 91
    epidermis 213, 233–235, 237, 239, 241, 243, 245, 247, 249, 251, 253, 255
    islet 371, 373, 375–377, 379, 381, 383
    muscle 310, 319, 322–323, 328

Regeneration *(continued)*
   musculoskeletal system 257
   nervous system 141
   pancreas & liver 369, 389
   stem cell resources, regenerative medicine 55–57, 62, 68, 70, 74
   vascular system 331, 334
Reprogramming...407–409
Retinoic acid 66, 172, 173, 176, 346, 358
RNA interference 35
RNAi...35–37, 39, 41–42
Real-time PCR (RT-PCR)....................200, 277, 278, 345, 348, 354, 357, 358, 362, 411, 412, 416, 417

S

Satellite cell(s)............67–68, 295–297, 300–304, 306–307, 309–315, 319–323, 328
Sca-1 .... 44, 46, 48, 53, 68, 71, 74, 95, 96, 100, 101, 260, 263, 264, 265, 267, 284, 288
Self-renewal 36, 42, 56–59, 68–69, 94, 158, 191, 215–216, 282, 295–296, 319
Short hairpin RNA (shRNA)................................4, 35–38
Simple gravity enrichment..............................387, 394–397
Single fibre graft 319
Skeletal muscle........................60, 67–68, 72, 295–296, 319
Skeletal muscle stem cells 56, 67–68, 295
Skin derived precursors................65–66, 159–160, 164–166
Skin stem cells...215
Skin substitute 233–234
SKPs, *see* Skin derived precursors
Small interfering RNA (siRNA) 35–49, 41, 56
Smooth muscle.............................58, 71, 73, 159, 168, 270, 345, 347, 362
Soft lithography 21–23, 31–32
Soluble factors..128, 281
Spinal cord 65–66, 105, 171–172, 180, 413
STAT3..57
Statistical design of experiments......................................109
Stem cells 55–57, 59–69, 71–76, 159–160, 169, 233–234, 407
   embryonic stem cells, *see* Embryonic stem cells
   epithelial stem cells, *see* Epithelial stem cells
   hematopoietic stem cells, *see* Hematopoietic stem cells
   liver stem cells, *see* Liver stem cells
   mesenchymal stem cells, *see* Mesenchymal stem cells
   neural stem cells, *see* Neural stem cells
   skeletal muscle stem cells, *see* Skeletal muscle stem cells
   skin stem cells, *see* Skin stem cells
Stem cells expansion 109
Stromal cells....................64, 127, 130–131, 133, 137–138, 259, 269, 282
Subependymal zone 185–186
Subventricular zone (SVZ)....................144, 146, 149, 156

T

TAT...43–44, 48–53
Teratoma...3, 58, 59, 362, 365, 418
Terminal differentiation 127, 319
Three step protocol...127, 131
Time-lapse imaging 312, 333–335, 340–341
Tissue engineering 71, 212, 233, 235, 237, 247, 249, 251, 253, 255, 281–282, 345, 362–364
Transcription factor(s) 56, 61, 67, 71, 75, 320, 372, 407–409, 417
Transdifferentiation 72–73, 75, 186, 407–409
Transplantation...... 109, 112, 244, 252, 321, 371, 401, 407
   bone marrow 93–94, 101–103
   ESC-derived dopaminergic neuron.199–203, 205–207, 209, 211–212
   ESC-derived motor neuron......171–172, 174–175, 177, 179–180, 182
   neural stem/progenitor cells.............................185–196
   regenerative medicine, stem cell 55, 59–60, 62–66, 68–74, 76
Tubulin 66, 157, 168, 187, 188, 418
Two step collagenase liver perfusion 407

U

Umbilical cord.............. 60, 63, 69, 132, 269–270, 273, 278

V

Vasculature................................62, 319, 334, 340, 391, 401
Vector(s)............ 12, 35, 37, 40, 41, 43, 143, 152, 281, 282, 291, 313, 378, 388, 408, 409
Videomicroscopy............................295, 300, 312–313, 315

X

Xenotransplantation.......... 66, 171, 173, 175, 177, 179, 181

Printed in the United States of America